HĀNAU KA UA

KAMEHAMEHA PUBLISHING

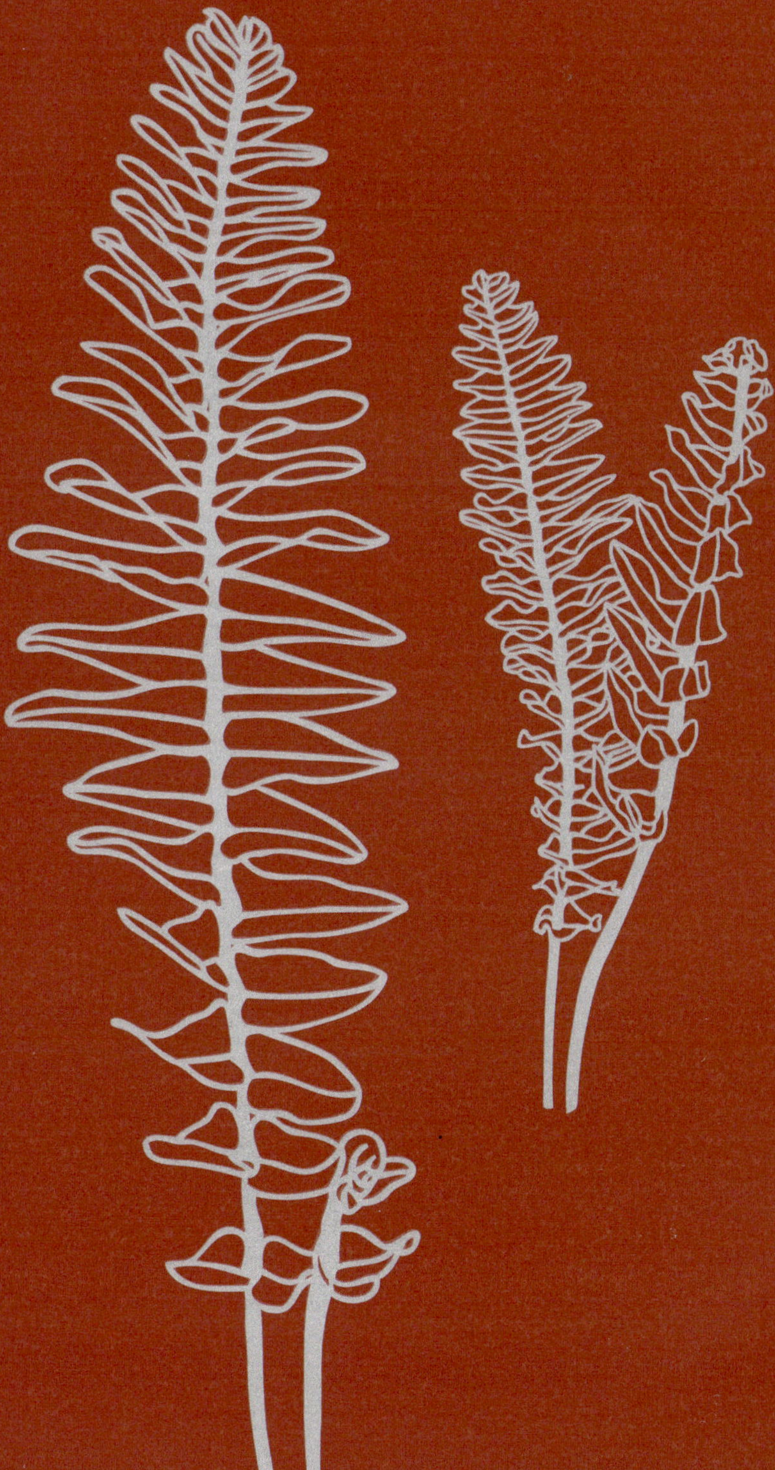

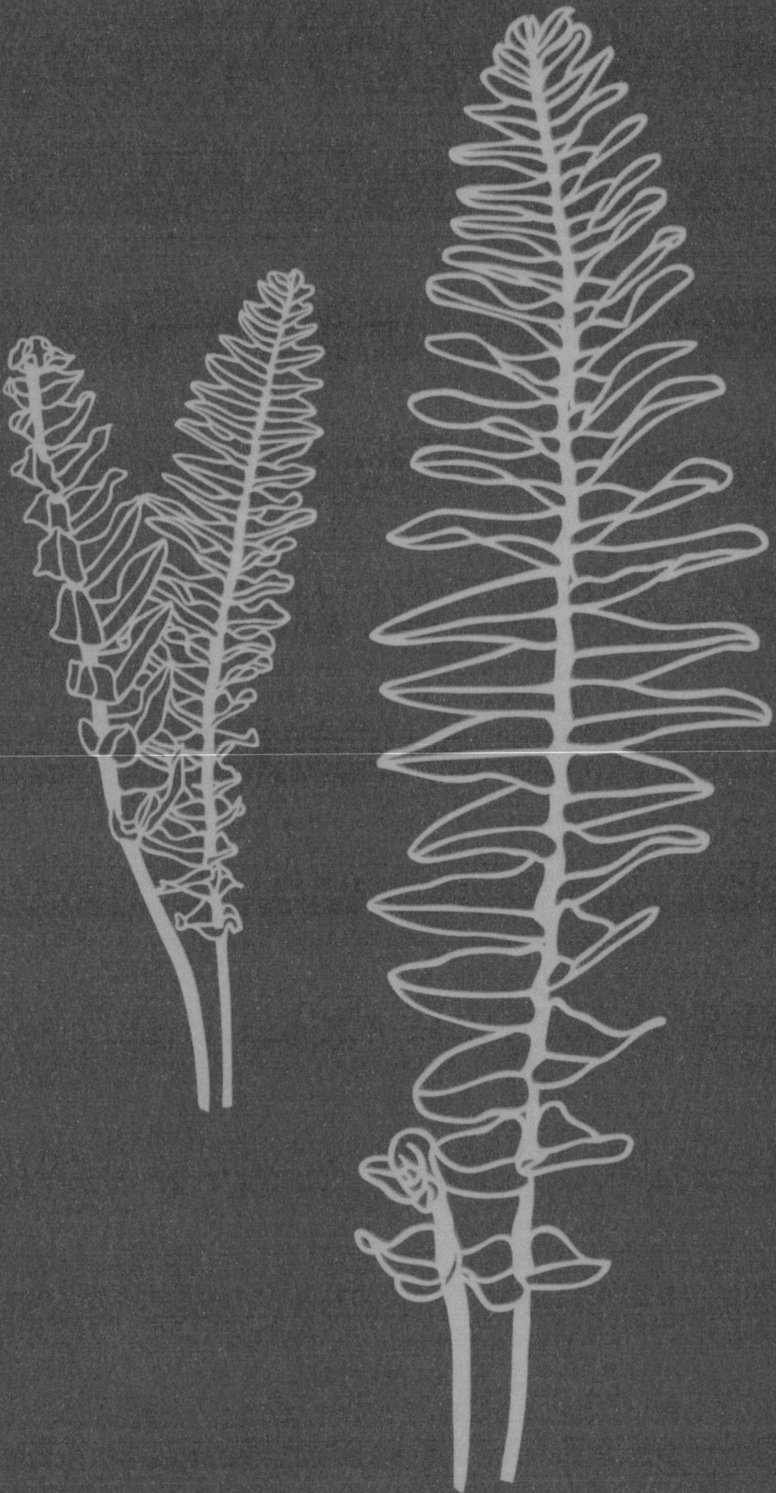

HĀNAU KA UA

HAWAIIAN RAIN NAMES

COLLETTE LEIMOMI AKANA
WITH KIELE GONZALEZ

ILLUSTRATIONS BY SIG ZANE

To my father, Clifford Mapuana Akana, and my daughter's grandfather, John Marshall Gooch, both of whom believed strongly in the value of a good education.

I also dedicate this book to my moʻopuna—may Auliawailanamalie and Alina Gonzalez be guided by the wisdom of their kūpuna, gifted with their humor, and surrounded by their love.

In the spirit of the old Hawaiian newspapers, we welcome additional information about rains in Hawaiʻi nei and hope that readers will kōkua by sharing their ʻike and adding to this collection by contacting the publisher.

PUBLISHED 2015

KAMEHAMEHA PUBLISHING
A DIVISION OF KAMEHAMEHA SCHOOLS
567 SOUTH KING STREET
HONOLULU, HAWAIʻI 96813
WWW.KAMEHAMEHAPUBLISHING.ORG

DESIGNED BY BARBARA POPE BOOK DESIGN
PRINTED IN CHINA

ISBN: 978-0-87336-246-7

21 20 19 18 17 16 2 3 4 5 6 7

CONTENTS

KAWAILANI

A SUITE OF ILLUSTRATIONS BY SIG ZANE

We honor rain! Found in song and chant, the rains of our islands speak of love, fertility, and life. Inspired by place and the beauty of each district, our designs are connected to memories and experiences. Through these illustrations, we relive those moments and the stories shared by our kūpuna.

We live on the slopes of a great mountain. The elemental rhythms of nature are clear as we watch the clouds move in from the ocean. The forests draw the clouds inland, bringing moisture to our lands. The invoking rains feed the trees, delivering fresh water to our aquifers. These blessings from our heavens nourish and enrich spirit and soul. SIG ZANE

ʻAE KŪKAHI II/III

Walking on young lava flows, I am amazed by the evolution of the land. The ʻae fern is the first to settle on the new lava flow. The unfolding of its tiny fronds is a symbol of regeneration. Imagine the contrast of fresh green in fields of black lava!

E OLA KOA! VI/VII

Native forest plants live in accord, each plant having a function in the whole. The understory is a study of harmony and balance. The koa nourishes all below its canopy. To live like a koa tree is a metaphor for caring for those around you.

MAILE KALUHEA O PANAʻEWA X/XI

Gathering maile is a pleasure beyond compare. We always search for the three-leaf maile. For me, it seems more fragrant and extraordinary. At Panaʻewa, we find the long vines, twisted amongst the ʻieʻie and ʻōhiʻa.

KALIKOHUMUʻULA XIV/XV

We often make the trek to the uplands to pick the finest for our lei. The crisp air brings forth the special budding of the ʻōhiʻa—the incredibly compact and colorful liko that we select to make our lei.

NOEʻULAKAPALAI XX/I

The palaʻā fern is lacy and delicate. Its light-green fronds make wonderful lei, and we keep secret where we gather these ferns. Hula patrons offer this fern on the kuahu because it is a kinolau of Laka, deity of hula. The name invokes "the fern in the pink-hued mist."

ULUKAU 20/21

The leaves of the breadfruit tree always stimulate creativity. "Ulu" means "to inspire, to grow, to build." "Ulu kau" means "placed inspiration." Mary Kawena Pukui wrote, "Ulukau is the unplanned acquisition of prophetic and interpretive powers without visible possession by a god."

KIAWEʻULA 24/25

When visiting mountain pools, the abundance of kī is always inspiring. Guardians of place, the simple staff of leaves honors those who came before us. Their presence is assuring and calming.

POLOLŪ 44/45

At Puʻuʻeo, in the heart of Waipiʻo Valley, the loʻi have many varieties of kalo—pololū is one. We gather this variety for our lūʻau heʻe. This design honors the generations who carry on the names of the many varieties of kalo.

KUKUIʻULAAKAMAPUAʻA 46/47

Kukui is the tree of enlightenment. The leaves dance in the wind. Distinct from the common kukui leaf form, the leaf in this design is elongated and sharply pointed. Kukui with this distinct leaf form are kinolau to Kamapuaʻa, the pig god—these trees are his likeness.

PEPEʻEʻUHOLA 132/133

The unfolding fronds are a symbol of being receptive and budding, and allude to a baby or a child. The hāpuʻu forest is a peaceful setting, a kīpuka of budding energy—an intimate place where I go for quiet and meditation.

PUA ʻĀHIHI 168/169

Famous is the pua ʻāhihi, the lehua of Nuʻuanu! The song "Pua ʻĀhihi," penned by Maddy Lam along with Mary Kawena Pukui, is a long-time favorite. Its beautiful words and hula inspired the depiction of these blossoms.

MĀLUAKIʻIWAIKEALOHA 180/181
"Māluakiʻiwai" is a hula kahiko that speaks of the māmane blossoms and reminds me of the uplands. The māmane flourishes in the arid, high-altitude areas of Maunakea and Maunaloa. When we drive through Humuʻula during the winter season, the bright-yellow blossoms of the māmane are an unforgettable sight.

PUNA PAIA ʻALA I KA HALA 212/213
Whenever we can, we visit Puna—the coastal road is always a treat. The pūhala groves of Puna, thickly forested, fragrant, and so very thorny, give the district a real personality. The birth of new land and the verdant growth stimulate creativity!

NOLU PĒ 216/217
"Nolu Pē" describes lightness—the mist on leaves and the sway of fronds in a breeze. The ʻiwaʻiwa at Wailea, North Hilo, grow along the banks of the Old Māmalahoa Highway, where the tunnel of overgrown palms and trees forms a natural greenhouse, perfect for the long stems and light green of the maidenhair ferns.

LAULOA 250/251
The dark-green, spore-less lauaʻe is a favorite. Sweet-scented lei of lauaʻe remind me of my secret patches in Keaukaha. The name chosen for this print means many things—at the time I was doing the art, the name acknowledged the long, breaking waves that stretch across an entire bay.

LEI KŪPAOA 254/255
Cherished are the fragrant lei! ʻAwapuhi is one of my favorites, always taking me back to growing up on Oʻahu and being in the back of Mānoa Valley. The flower, with its pure and simple lines, was my first hand-cut print.

HĀKEA 270/271
Spending time immersed in the ʻōhiʻa forests, we would always seek the elusive white blossoms. Typically red blossomed, it is said that the most sacred of ʻōhiʻa have white flowers. Poetically, the term "hākea" is used to describe this mysterious appearance!

Influenced by ritual, practice, and tradition, the work of Sig Zane reflects experiences. Living in Hilo has given the artist and designer a direct connection to nature. His understanding of the plants of Hawaiʻi and their cultural significance was formed through his training in hula and chant under the tutelage of Hālau O Kekuhi. The family business began in 1985 with three generations. Today, Sig, wife Nalani Kanakaole, and their son Kuhaoimaikalani Zane perpetuate kuleana through the visual arts.

Preface

I was a high school student during the 1970s, a time when more of us Native Hawaiians were beginning to embrace our cultural traditions, which, for many reasons, had been languishing for decades. I felt I had not learned very much related to Hawaiian language or culture during my previous three years at Kamehameha Schools, so as part of my senior year of studies I decided to take Hawaiian language from Sarah Quick (now Keahi), hula from Robert Cazimero, and Hawaiian music and dance from Nona Beamer. I loved learning to speak Hawaiian, and I had fun dancing hula. All of my kumu were so wonderful that I was inspired to continue my study of ʻōlelo Hawaiʻi, hula, and Hawaiian culture. That senior year in high school was truly a turning point in my life.

After graduating from Kamehameha, I attended Leeward Community College for two years, where I studied Hawaiian language under Noelani Losch. I also took every class offered there that was related to Hawaiian culture. When the time came to transfer to the University of Hawaiʻi at Mānoa, I went to see a counselor and told him I was interested in majoring in Hawaiian. His response was, "There's no future in Hawaiian." I was not deterred; I was determined to learn all I could about my Hawaiian heritage.

At UH–Mānoa I enrolled in Hawaiian-language classes taught by Haunani Bernardino, Sarah Nākoa, Pila Wilson, and Larry Kimura. I also enrolled in classes on Hawaiian culture and history taught by Rubellite Johnson, John Charlot, Pauline King, and Davianna McGregor.

My growing competency in ʻōlelo Hawaiʻi allowed me to take on interesting assignments in my other classes, several of which became springboards for long-term projects. An assignment for a class on Hawaiian religion from John Charlot had me collecting and translating articles about kāhuna from *Ka Nupepa Kuokoa*, a Hawaiian-language newspaper. Decades later, my research for this assignment was published in the *Hawaiian Journal of History* under the title "Kahuna Hoʻopunipuni Letters of 1863." While taking a class on Kamapuaʻa literature, I was asked by Professor Charlot to translate a portion of the long legend. My translation was later published by Bishop Museum Press as *The Oʻahu Exploits of Kamapuaʻa, the Hawaiian Pig-God*. Not only was I continuing my study of ʻōlelo Hawaiʻi, I was able to share what I was learning with others.

In addition to Hawaiian language, hula has been a huge part of my life. After Kamehameha, I took hula ʻauana from Maiki Aiu and hula kahiko from Hoakalei Kamauʻu. At UH–Mānoa I enrolled in hula classes with Hoʻoulu Cambra, Ed Kalāhiki, Vickie Takamine, and Noe Zuttermeister. At UH–Hilo I took a summer hula workshop with Edith Kanakaʻole. After graduating from the university, I continued taking hula from Frank Hewett, Leimomi Ho, Leimomi Maldonado, Colleen Aiu, George Holokai, Kimo Alama Keaulana, and finally from Kaleo Miyasato and Ed Kalāhiki, from whom I graduated as kumu hula. I have a deep appreciation for all these kumu who shared with me their ʻike and aloha for hula and our Hawaiian culture. When I began teaching social studies at Kamehameha Middle School in 1983, I teamed up with my fellow teacher Brad Cooper (a graduate of Robert Cazimero) to start a hula club, which is less formal than a hālau, and taught hula for thirteen years. Aunty Maiki would tell her students, "Hula is life!" and this has certainly been true for me.

Early in my hula training, I was introduced to several Hawaiian wind and rain names while learning chants and songs. Eventually, I became acquainted with the legend of Kūapākaʻa by Moses K. Nakuina, with its chants naming the winds of the islands. I thought it wondrous that my Hawaiian forebears were so attuned to their environment that they assigned individual names to the multitude of winds and rains occurring throughout the archipelago.

I remember translating passages from *Ka Leo o ka Lahui,* where Kamapuaʻa summoned the winds of Oʻahu to fill the bellies of two hungry women. The more I delved into Hawaiian literature, the more interested I became in wind names, and I started to collect them. One of my resources was the *Hawaiian Dictionary.* Another was Abraham Fornander's version of the Kūapākaʻa legend, which actually predates Nakuina's. Then I began going to the Bishop Museum to look for wind names in Hawaiian chants. Since both wind and rain names are often mentioned in chants, I started collecting rain names as well.

Eventually my research led to the Hawaiian-language newspapers, an extremely rich source of ʻike, or knowledge, from the early 1800s to the mid-1900s, when knowledge of Hawaiian culture was more plentiful and more commonly available. I started to document the wind and rain names I was collecting, little realizing the magnitude of my undertaking. The more I looked, the more names I found. Just recently I discovered a connection between the ua Waʻahila and my lifelong home, Hālawa.

So the search continues—for the names of winds and rains that are close to home and those in faraway places across Ka Pae ʻĀina. Although the work goes on, it is time to share the information I have gathered on rain names to date; the companion volume of wind names is planned to follow. This work will serve as a resource for students and others wanting to learn more about their one hānau. In addition, it is a reminder to us all that there is indeed a future in Hawaiian, and that future is now.

Mahalo i Ke Akua, mahalo i koʻu mau kūpuna a me koʻu mau mākua, mahalo i Ke Aliʻi Pauahi.

I am deeply grateful to the members of Kamehameha Publishing, directed previously by Kēhaunani Abad and now by Ron Cox, for their invaluable contributions to this book. My daughter, Kiele Gonzalez, edited my translations, researched place names, found new wind and rain names, and labored extensively in every aspect of preparing the manuscript. Matthew Corry gave guidance and editorial support throughout the process, was an amazing problem solver, and has been one of the project's greatest advocates. Keoni Kelekolio contributed his ʻōlelo Hawaiʻi expertise to the text and bravely took on the more perplexing translation queries. Doreen Jones served as a gauge for editorial consistency and helped bring the many unwieldy pieces into a unified whole.

A sincere mahalo to Kamaoli Kuwada and Kaliko Beamer-Trapp, our fresh sets of eyes, who provided insightful manaʻo and corrections to the text. Mahalo also to Kalena Silva and Nathan Napoka, who gave important feedback at an early stage.

Thanks to the Bishop Museum Library and Archives staff, including Leah Pualahaʻole Caldeira, who assisted with bibliographic citations from their collection, and to Janet "BJ" Short, who helped with the viewing of original copies of ʻōlelo Hawaiʻi newspapers.

I am very appreciative of the following individuals, who shared information or sources, or helped in various ways: Puakea Nogelmeier, Keoki Maguire, Kimo Alama Keaulana, Aaron Salā, Larry Kimura, Kainani Kahaunaele, Hiʻilei Kawelo, Keliʻi Kotubetey, Jason Kāpena Achiu, Kapali Lyon, Myoung-ja Hwa, Keola Donaghy, Kilohana Silve, and Ryan Gonzalez.

Many, many thanks to Barbara Pope, who gave excellent guidance and suggestions from the beginning, and whose vision for this book has resulted in the elegant design we now enjoy. She and her team, including Maureen Liu-Brower, also provided valuable editorial feedback. Sig Zane's classic art graces this book, and his son, Kuhao Zane, contributed his artistry to the publication. A big mahalo to my mom, Philomena Mamie Madeira Akana, who for years has had to put up with a messy living room full of books and papers.

And finally to the kūpuna whose voices, whether spoken or written, left a rich legacy from which this and countless other works have been and will be derived—you are the kahua paʻa on which we stand.

CLA

Introduction

In the minds and na'au of our Hawaiian kūpuna, every being and every thing in the universe was born. Our kūpuna respected nature because we, as kānaka, are related to all that surrounds us—to plants and creatures, to rocks and sea, to sky and earth, and to natural phenomena, including rain and wind. This worldview is evident in a birth chant for Queen Emma, which inspired the title of this book: "Hānau ke ali'i, hānau ka ua me ka makani" (The chiefess was born, the rain and wind, too, were born) (Nogelmeier 266).

Our kūpuna had an intimate relationship with the elements. They were keen observers of their environment, with all of its life-giving and life-taking forces. They had a nuanced understanding of the rains of their home. They knew that one place could have several different rains, and that each rain was distinguishable from another. They knew when a particular rain would fall, its color, duration, intensity, the path it would take, the sound it made on the trees, the scent it carried, and the effect it had on people.

Some rain terms have recognizable Polynesian counterparts and are probably ancient, connecting us to our 'ohana across the oceans and across time. "Kohu" in te reo Māori of Aotearoa is a fog, mist, haze, or smog, similar to our "'ohu" in pronunciation and meaning. "Tūāua" is a rain shower, like our "kuāua." In te reo Mā'ohi of Tahiti, "rotu" is a heavy rain that lasts for a day, similar to our "loku." "Mao" (cleared, as rain) is the same in all three languages, as is "ua" (rain). Other rain names may be more contemporary, as Thomas K. Maunupau suggests about the Hāleu'ole rain of East Maui, which is known for catching people off guard.

Some rains were named after people. The ʻĀpuakea rain of Koʻolaupoko was named after ʻĀpuakeanui, the most beautiful woman in Kailua from the moʻolelo of the goddess Hiʻiakaikapoliopele. The Waʻahila rain of Nuʻuanu was named after a chiefess who excelled in a hula that was also named after her. In the genealogy of the chiefs of Koʻolau, Oʻahu, there is mention of Hālaulani and her daughter Kauaohālaulani, whose name probably means "the rain of Hālaulani." Some rains share their name with their home, like the Mololani rain that falls in the area of Mololani in Kāneʻohe. A rain name can tell the story of its origin and the characteristics of the rain. The Loku rain of Hanalei is a pouring (loku) rain, as its name suggests. In fact, most rain names have descriptive meanings. The Nihipali rain sneaks (nihi) along the cliffs (pali). The Pōʻaihale rain encircles (pōʻai) houses (hale) as it is whisked about by the wind.

Rain names may be used for their literal or figurative meaning. Kīlauea's ʻAwa-ʻawa rain, whose name can mean "grief" or "hardship," appears in chants from the legend of Pele and Hiʻiaka as Lohiʻau is being killed by Pele's fires. The Kanilehua rain is sometimes used metaphorically to refer to gossip, perhaps because one interpretation of the name refers to the chattering of birds on lehua trees. In this way, rain names remind us of the moʻolelo of places and events.

Many rain names refer to native vegetation, mostly lehua and hala trees. Some describe the rain's interaction with these plants, like the Keʻehilehua (to strike the lehua blossoms), Kīkēhala (to break open hala fruit), and the Kīpehipehipuahala (to pelt hala fruit). There are the rains Noelehua (lehua mist) and Moanilehua (to waft the fragrance of lehua), and many more, all of which remind us of the kind of flora that is or was abundant in an area and of the importance of rain to our forests.

Rains of the same name can occur in different places. Pālolo, Oʻahu, boasts of its chilly, tiny-dropped Līlīlehua rain, and so does Kāʻanapali, Maui. Māʻeliʻeli Hill on Oʻahu gets rained upon by the Pōʻaihale rain, and so do the uplands of Māʻeliʻeli, Hawaiʻi, where a misty Pōʻaihale rain creeps between houses. Rain names, just like place names, can exist in multiple areas on multiple islands. One place can have several rains. Hilo is home to at least fifty rains. Sometimes the possessive form "o" is used to connect a rain name to the place where it resides, as in "ka ua Hōʻeha-ʻili o Waiehu" (the Hōʻehaʻili rain of Waiehu), meaning that the Hōʻehaʻili belongs to Waiehu.

Some rains may act differently at certain times or may indicate a certain season. During the month of Welehu, the Uluau and Mālua rains of Hilo break in waves, accompanied by the Māluakiʻiwai wind. The rainy month of Hilinehu sees the Nāulu rain cloud over Panaʻewa. The ʻuala farmer counts the kuāua showers to help determine when to plant. The fishermen of Hilo know by the misty Heʻenehu rain that the nehu are running.

Rains represent many things to our people. First and foremost, rains are integral to our survival. A saying on Ni'ihau is: "Kau ke ao Nāulu, ola ka 'āina" (The Nāulu rain cloud appears, the land lives). In fact, the main town of Pu'uwai was originally called Kauanāulu, meaning "the Nāulu rain." There were heiau ho'oulu ua, like Mākūkū in Nu'uanu, where people brought offerings for rain. And on Moloka'i, when there was too much rain, a woman, Pauulea, built the heiau Kaimukāluaua, where the raindrops were caught and cooked in the stones.

Rains have divine and mythical sources. Almost all the major akua are associated with rain; Lono, in particular, is associated with winter rains. The rainy season (from about October to February) is when he reigns over Hawai'i and is celebrated in the Makahiki. Heavy rain and dark rain clouds are two of his kinolau. The wet areas of the islands are associated with Kamapua'a, who is a form of Lono and who can turn into the ao pua'a rain cloud. One of the forms of the god Kāne is Kāneikaua (Kāne of the rain). Another form is Kāneikamakaoka'ōpua (Kāne of the 'ōpua cloud) (Beckwith 53). When such a cloud is seen in the morning in Kona, Hawai'i, it is a sign of rain (Pukui, *'Ōlelo* 9). Kūmauna is a rain god and one of the rain forms of Kū. In a story of Papa and Wākea, Papa gives birth to a gourd, whose parts become the sky, sun, stars, moon, and clouds. The juice Wākea pours over the clouds becomes rain (Beckwith 304–5). Kūlanihāko'i is the name of a mythical pond in the sky that sometimes overflows, falling to the earth as rain. Local news reports in the 'ōlelo Hawai'i newspapers of the nineteenth and twentieth centuries continued to poetically reference this pond.

Rains are a source of pride for one's homeland. To know one's home is to know its stories and legends, its famous ali'i, its landmarks, and its rains. An old string-figure chant says: "'O Kohala 'āina ua ha'aheo i ka ua 'Āpa'apa'a" (Kohala, a land proud of its 'Āpa'apa'a rain). The district is sometimes referred to as Kohala ua 'Āpa'apa'a. The most famous rains are synonymous with the land to which they belong. Say the name "Laniha'aha'a," and those in the know will automatically think of Hāna, Maui. Our homeland is crucial to our identity, and the rains are an integral part of our homeland, yet what do we know today about our rains?

Rain names were an important component in the traditional, largely place-based, information imparted to Hawaiian youngsters. Knowledge of local weather patterns ensured survival and proliferation, and knowledge of one's homeland was held in high esteem (Charlot 2005).

Some rains were interpreted as signs. The Koko rain could mark the birth of an important chief. Rain or rainbows might indicate the presence of a deity. Rain at a person's death or funeral was considered tears of the gods.

Rains are plentiful in Hawaiian poetry. Mist settling onto a mountain peak might express the joy and thrill of lovemaking. Some rains represented growth, life, good fortune, or beauty. Grief over the loss of a loved one might be expressed as a rainstorm, signifying tears of great sadness and pain. Rain names are therefore common in Hawaiian dirges, and many of the mele examples in this book are of that genre.

Full of imagery and metaphor, the chants and songs presented in *Hānau ka Ua* show that our kūpuna were exceptional poets. Because of space constraints, we have not included entire mele in the examples. Instead, we have selected excerpts that give context for the use of a rain name. We have also endeavored to add diacritical marks to less well-known rain names. In some entries, we offer definitions of the individual rain names, although we recognize that this is a subjective, and therefore inconclusive, process. This is especially true for examples where we have not yet found a moʻolelo, or story, as to how the ua got its name. In other cases, there may be several moʻolelo that explain different meanings of a rain name. All these are important when trying to understand the various ʻano, or characteristics, of an ua. Wherever Hawaiian is used in the examples, we have used current orthography. We also offer English translations and cite each source to facilitate further research. Whenever possible, we have noted additional sources that contain the same or similar material.

The naming of rains and the transmission of these names from one generation to the next is a process that has been happening for centuries. This ʻike can be found in various sources, including the ethereal realm, audio files, books, and articles, as well as in the minds of our treasured kūpuna who are still with us. *Hānau ka Ua* draws on an extensive body of oral tradition and literature, including approximately three hundred ʻōlelo Hawaiʻi primary sources from the nineteenth and twentieth centuries. It contains the fullest record of Hawaiian rain names and their lore to date, but it is by no means exhaustive. We hope that researchers will continue to find additional rain names in Hawaiian newspapers, mele, audio collections, and archives. Interviews of kupa o ka ʻāina (natives long attached to their land) may reveal even more rain names that are not included in this volume. In the spirit of the old Hawaiian newspapers, we welcome additional information about rains in Hawaiʻi nei and hope that readers will kōkua, sharing their ʻike and adding to this collection by contacting the publisher.

The evolution of ʻōlelo Hawaiʻi from a rich oral language to a rich written language presents a challenge for compilers. The spelling of names and terms, diacritical markings, punctuation, hyphenation, capitalization, italicization, paragraphing, variant spellings, regional differences, and word pairings are rendered differently in different source texts, making it difficult to achieve consistency and clarity in a compilation. While we believe in letting old texts be reproduced with their inherent style and idiosyncrasies, we also feel an obligation to standardize the content so today's reader can have access to information in a clear, familiar format. In an effort to strike a good balance, we observed the following guidelines:

TRANSLATIONS

In many cases a new translation is provided with a preference for literal renderings. Previously published translations included in this volume have sometimes been altered to fix typos, regularize word usage (one word versus two), alter diacritics, adjust case (upper or lower), and standardize punctuation.

ATTRIBUTION

Care has been taken to properly attribute works to their respective author(s). When authorship is not known or certain, established editorial conventions are followed.

AUTHOR NAMES AND TITLES OF WORKS

Spellings of authors' names and titles of works are reproduced as they appear in the sources, except for many ali'i and akua names that commonly take diacritics today. This has resulted in some unevenness and inconsistencies.

DIACRITICS

To the extent possible, Hawaiian diacritics are used throughout the text. This includes the insertion of diacritics into quoted materials that originally did not have the 'okina and kahakō. Exceptions may occur when pronunciations are not known, when inserting diacritics may unintentionally change the meaning, and when reproducing the names of authors, the titles of works, and the names of newspapers.

HYPHENATION

Hyphens have been removed from most terms, including those from the *Hawaiian Dictionary*. Hyphenating Hawaiian names used to be a convention to help readers understand the meaning of the term. Now that 'ōlelo Hawai'i has regained a sufficient base of fluency, hyphenations, with rare exceptions, are no longer necessary.

CAPITALIZATION

The decision to render a rain term as a proper noun is based on several criteria, including the way the rain is most commonly rendered in contemporary literature, the availability of multiple examples of the rain in a place, and the context in which the term is used. The decision to capitalize rain terms was made on a case-by-case basis, sometimes leading to inconsistencies.

PRIMARY VERSUS SECONDARY SOURCE

When information related to rain names is available in both primary and secondary sources, secondary sources often are cited. This is because primary sources are not always readily available and because contemporary readers are accustomed to modern Hawaiian orthography.

MULTIPLE CITATIONS

Multiple citations of identical or near-identical materials are included to demonstrate that Hawaiian rain names are numerous and have been used in many contexts.

REFERENCE
For word treatment and spelling, *Māmaka Kaiao* takes precedence over the *Hawaiian Dictionary.*

The handling of these practical aspects of *Hānau ka Ua* illustrates the great care that must be taken when dealing with old texts and creative works. Even so, we acknowledge that such technicalities are secondary, and we affirm that the rich linguistic tradition of ʻōlelo Hawaiʻi in many ways defies translation and remains open to multiple interpretations. The rain names and examples in this volume represent our best understanding at this point in time.

Although this is a book on Hawaiian rain names, the reader will come upon many references to wind names. During the past decade we have gathered more than six hundred wind names; *Hānau ka Makani* will be published in the near future as a companion to this volume.

As many older Hawaiians have felt a renewed sense of pride in being Hawaiian, young kamaliʻi are being raised in traditional ways, and many non-Hawaiians locally and abroad have associated deeply with our Hawaiian language, culture, and ʻāina. Today the thirst for Hawaiian ʻike has intensified, and with this ʻike comes kuleana, or responsibility, to care for the ʻāina that sustain us—and to whom the rains belong.

Rain and wind names are a precious legacy left by our kūpuna. The existence of hundreds, perhaps thousands, of rain and wind names is evidence of the value our kūpuna attached to these forces. It is hoped that this compilation will help us kānaka maoli maintain the intimate connection our kūpuna had with the Hawaiian universe. For the kumu hula and poʻe hula a oli, it is hoped that this book will contribute to an even greater understanding of the elements that are chanted and danced in mele. For each new generation of Hawaiian learners, it is hoped that this book will help inspire a commitment to nānā i ke kumu (look to the source) and to delve into the rich literature of our kūpuna. For all of our lāhui, it is hoped that this book will help us find a place for rain and wind names in our homes, offices, schools, farms, studios, in our naʻau, and beyond the tips of our tongues.

Our kūpuna believed they could influence nature: slow the setting sun, cause the waves to rise, and throw the rain to another district. In the legend of Kūapākaʻa, in order to get the winds in his wind gourd to blow, he needed first to know their names. Perhaps the same is true of the rains. Rain names are a convergence of past, present, and future, and in using them, we will summon the knowledge of our kūpuna, joining them in calling out to the lands on which the rains reside, calling on ourselves to rise to a higher consciousness, and calling to the future, where new life will continue to be born.

A

ʻĀkōlea. Rain associated with Mahiki, Hawaiʻi. Also the name of a native fern.

1. Kaka ka puʻu wai ua ʻĀkōlea
   I paheʻe Mahiki i ka wai a ka ua
   ʻAʻole e paheʻe Mahiki i ka lā
   I paheʻe nō Mahiki i ka ua

   The hill is rinsed by the water of the ʻĀkōlea rain
   Mahiki becomes slippery in the rainwater
   Mahiki would not be slippery in the sun
   Mahiki becomes slippery in the rain

   From a mele inoa, or name chant, for Keawenuiaʻumi chanted by Kūapākaʻa. Hawaiian source: Kanepuu, "He moolelo" 11/14/1867. English trans. by author. Additional sources: Nakuina, *Moolelo* 51; Nakuina, *Wind* 44–45.

ʻAla. Rain associated with Hulāʻia, Kauaʻi.

1. Mālie Hulāʻia he ani na ka Paupua
   Na ka makani hali ʻal[a], he Alaʻoli ē
   He ʻoliʻoli koʻu, ko ka mea haku
   He leo koʻu e laʻi ana ma luna
   He ua ʻAla ko Hulāʻia

Hulāʻia is tranquil, a soft blowing of the Paupua wind
Of the fragrance-bearing wind Alaʻoli
Joyfulness is mine, the composer
I have a voice that floats above
Hulāʻia has an ʻAla rain

From a makena, or lament, for the passing of Kamehameha IV. Hawaiian source: Nogelmeier 355–56. English trans. by author.

ʻAlaea. Red rain associated with Kiʻi, Hawaiʻi. "ʻAlaea" means "red."

1. *See Hāʻao (#4).*

Alanilehua. Same as *Wailehua*. Rain associated with Hilo and Puna, Hawaiʻi.

Rain of Hilo, Hawaiʻi

1. Poonahoahoa—Ua hiki mai nei ko kākou hoa o ka ʻiu o ka lā o ka wā i hala, ma Honolulu nei; a ua maikaʻi nō hoʻi kona ola, i ka nokea paha e ka ua Alanilehua, a me ka ʻai i ka momona o ka ʻāina ua nui ʻo Hilo.

   Poonahoahoa—Our friend from the height of the sun of bygone days has arrived here in Honolulu. His health is good, probably from the constant Alanilehua rain and from eating the richness of the land of much rain, Hilo.

   From an article on current events in 1862. Hawaiian source: "Na mea hou." English trans. by author. *Note: Thomas Spencer, also known as Poonahoahoa, was a Hilo-based staff member of* Ka Nupepa Kuokoa, *the newspaper in which this article appears.*

Rain of Panaʻewa and Puna, Hawaiʻi

2. Ua Alanilehua
   Ua kapa ʻia nō hoʻi kēia ua i kekahi manawa, ʻo Wailehua; ua pili nō ia i ka wai o nā pua lehua. E hoʻomaka nō kēia ua e hele mai, mai nā ʻae kai mai o Hāʻena, a mai luna aʻe o ko laila lewa lani, e hele ana ia i ke komohana, e kilihune ana i nā liko hīnano o Puna, a e loku ana hoʻi i luna o nā pōhai lehua o Panaʻewa; ʻaʻole nō ʻo ia e lihi launa mai i ke town, akā, e puka mai nō ia mau paka ua ma waho nei o ka ʻaoʻao komohana o Panaʻewa, a laila, huli aku nō a hele pōʻai i ka hema e piʻi ana i ka uka o Pāʻieʻie, a nalo aku hoʻi i loko o nā ʻohu wai o ka nahele. He lahilahi wale nō kona kūlana, a he kākaʻikahi nō hoʻi kona ʻike ʻia ʻana mai, a ma waena wale nō o ka hora 10 a me 12 o ke kakahiaka kona wā e ʻike ʻia ai.

   Alanilehua Rain
   This rain is sometimes called Wailehua. It is associated with the nectar of lehua blossoms. When this rain starts to come from the water's edge at Hāʻena and from above the upper heavens of that place, it will travel to the west, sprinkling the buds of Puna's hīnano blossoms and pouring down over the clusters of

Panaʻewa's lehua trees. It won't ever come close to town, but these raindrops will appear outside of the western border of Panaʻewa. Then it turns and circles to the south, ascending the uplands of Pāʻieʻie, and disappearing within the watery mists of the forest. It has a delicate nature, is rarely seen, and is visible only between the hours of 10 and 12 in the morning.

From a description of rains by Lilia H. Richards and company, who traveled to the uplands of Puna in 1876. Hawaiian source: Richards et al. English trans. by author.

Anulipoa. Rain associated with ʻŌmaʻolala (or Maʻolala), Hawaiʻi. "Anu lipoa" may refer to darkness and cold.

1. Aloha kuʻu hoa i ka ua Anulipoa
Huʻihuʻi, koʻekoʻe, koʻoū
He ahi ke kapa o kāua e mehana ai
E lala ai kāua i ʻŌmaʻolala
I pili wale, i haʻalele lā ē

Greetings to my companion of the Anulipoa rain
Chill, cold, damp
Fire shall be the blanket that warms us both
So we warm ourselves at ʻŌmaʻolala
To come together, then to depart

From a chant to Hiʻiakaikapoliopele and her traveling companions by Kauakahiapaoa. Hawaiian source: Hoʻoulumāhiehie, *Ka Moʻolelo* 184 and 411. English trans.: Hoʻoulumāhiehie, *Epic* 384. Additional source: Emerson, *Pele* 217–19.

ʻĀpaʻapaʻa. Rain and wind associated with Kohala, Hawaiʻi. Also the name of a hill in Kohala, Hawaiʻi. "ʻĀpaʻapaʻa" means "firm, hard."

1. Ua ʻĀpaʻapaʻa—He ua kēia i hui pū me ka makani; ma uka mai o ka ʻāina, pupuhi i ka moana, inā e holo ka waʻa i ka lā e pā ai kēia makani me ka ua, a pihō i ke kai, puhi ʻia e ka makani a make. No laila, i kapa ʻia aku ai kēia inoa, ka ua ʻĀpaʻapaʻa o Kohala.

ʻĀpaʻapaʻa rain—This is a rain that occurs with wind; from the land it blows toward the ocean. If a canoe sails out on a day when this wind and rain strike, and it starts to founder out at sea, it will be blown by the wind, and there will be destruction. And so this name has been given: the ʻĀpaʻapaʻa rain of Kohala.

From a letter to the newspaper *Kuokoa* regarding famous rains. Hawaiian source: Kalaaukumuole, "Manao wehewehe." English trans. by author. Additional source: Kalaaukumuole, "Thought explanations."

2. ʻO Kohala ʻāina ua haʻaheo
I ka ua ʻĀpaʻapaʻa

Kohala, a land that is proud of its rain
The ʻĀpaʻapaʻa rain

From a mele hei, or string-figure chant. Source: Pukui, *Nā Mele Welo* 98–99.

3. ʻO Kohala Iki, ʻo Kohala Nui
ʻO Kohala ua ʻĀpaʻapaʻa

Kohala Iki, Kohala Nui
Kohala of the ʻĀpaʻapaʻa rain

From a mele hei and mele pana, or string-figure and place-name chant, for the districts and famed aspects of Hawaiʻi Island. Hawaiian sources: "Ku hau pio"; Fornander, *Fornander* 6: 213. English trans. by author. Additional source: *Nā Leo Hawaiʻi Kahiko*, Record I, Side A, Band 12. *Note: "Ku hau pio" mentions that "Kohala Nui" and "Kohala Iki" are North and South Kohala, both of which are associated with the ʻĀpaʻapaʻa.*

4. ʻO mākou iho nō, nā hoaloha o ka ʻAhahui Aloha ʻĀina i loko o ka pōʻai o nā kāne o ka ʻāpana kaulana ʻo Kohala Nui, Kohala Iki, ʻo Kohala ua ʻĀpaʻapaʻa me nā manaʻo lōkahi, ma o ko mākou mau kōmike lā.

We are the friends of the Patriots Association from among the men of the famous district of Greater Kohala and Lesser Kohala, Kohala of the ʻĀpaʻapaʻa rain, and we are of one mind as expressed through our committee members.

From a message of condolence to the family of Joseph Nāwahīokalaniʻōpuʻu upon his passing. Hawaiian source: Sheldon, *Ka Puke* 236. English trans.: Sheldon, *Biography* 317.

apo pue kahi. Rain term related to the emptiness felt after a loved one has passed on. "Apo pue kahi" may refer to a huddling embrace with a companion.

1. Aloha aku au a pau ke aho
Uē aku hoʻi ʻo ka hoa mai ke anu
ʻO kuʻu hoa mai ka ua apo pue kahi

I will love you until I no longer have breath
Crying for my companion in the cold
My companion from the apo pue kahi rain

From a kanikau, or lament, for Keʻeaumoku by his daughter Kaʻahumanu. Hawaiian source: Kamakau, "Ka moolelo o na Kamehameha." English trans. by author. Additional sources: Kamakau, *Ke Aupuni Mōʻī* 90; Kamakau, *Ruling* 317.

ʻĀpuakea. Rain associated with Hāna, Maui, and with Koʻolaupoko, Oʻahu, and found in other areas. Also the name of a place in Kāneʻohe, Oʻahu.

Rain of Hilo, Hawaiʻi

1. Homa Hilo i ka ua ʻĀpuakea
He ua ʻĀpuakea hehi i ka moana
Hehi ihola i ke kai o Piʻihonua

Hilo is disappointed in the ʻĀpuakea rain
An ʻĀpuakea rain that treads upon the ocean
Treading upon the sea of Piʻihonua

From a kanikau, or lament, for S. P. Koko by his wife. Hawaiian source: Honolulu, "He kanikau." English trans. by author.

Rain of Hāna, Maui

2. Puīa i ke ʻala ka uka o Wākiu
Ka nae ʻaʻala o Kahalaowaka
Paoa i ke ʻala ka ua ʻĀpuakea o Hāna

The upland of Wākiu is permeated with fragrance
The sweet scent of Kahalaowaka
The ʻĀpuakea rain of Hāna is soaked with perfume

From a mele describing a sweet fragrance. Hawaiian source: "He moolelo . . . no Laukaieie." English trans. by author.

3. He ʻāina kaulana ʻo Hāna i ka wā kahiko . . . aloha nō nā aliʻi i kēia ʻāina ʻoluʻolu i hoʻopulu ʻia e ka ua ʻĀpuakea, ka ua wāwahi i luna o ka hala, ka hala mai Wākiu a Honokalani.

Hāna was a famous land in the olden times . . . the chiefs really loved this pleasant land dampened by the ʻĀpuakea rain, the rain that shatters upon the hala trees, the hala from Wākiu to Honokalani.

From the history of the Kamehameha dynasty. Hawaiian source: Kamakau, *Ke Aupuni Mōʻī* 227. English trans. by author. Additional sources: Kamakau, *Ruling* 385; Pukui, Elbert, and Mookini 273.

4. Ka wai hālana kiʻo wai a ka ua
I hoʻokiʻo ʻia e ka ua ʻĀpuakea
E ka makani Koholālele nei
ʻO ʻoe anei kahi ʻanoʻai?
Ka ua wāwahi i luna o ka hala
Ka hala mai Wākiu a Honokalani

The overflowing pool of rainwater
Gathered up by the ʻĀpuakea rain
O Koholālele wind
Are you a friend?

The rain shatters upon the hala trees
The hala from Wākiu to Honokalani

From a kanikau, or lament, for Peʻapeʻa, a son of the Maui ruler Kamehamehanui-ʻailūʻau. Hawaiian source: Fornander, *Fornander* 6: 429. English trans. by author.

Rain over the ocean off of Kaʻuiki, Maui

5. *See Lauʻawa (#2).*

Rain of Kahaluʻu, Oʻahu

6. Ua ʻĀpuakea—Kahaluʻu, Koʻolaupoko, Oʻahu; like mist; when it falls on the trees it looks like frost.

From a list of rain names and their descriptions. Source: "Names . . . rain."

Rain of Kailua, Oʻahu

7. E ka ua ʻĀpuakea
Kui ʻia mai nā ʻāhihi
Na ka Malanai e lawe mai
I wehi, i ʻohu no Kalani

O ʻĀpuakea rain
The ʻāhihi blossoms are to be strung
The Malanai wind will bring them
As a decoration, an adornment for the chief

From the song "Pela kapu o Kakae" by the Kawaihau Glee Club. Hawaiian source: Holstein 33. English trans. by author.

8. "Akā, ʻo kaʻu wahi ʻai naʻe, aia lā i ka ua ʻĀpuakea o Kailua." *"But the food I want is there in the ʻĀpuakea rain of Kailua."*

Said by Hiʻiakaikapoliopele, referring to the lūʻau leaves broiled by Kaʻanahau. Hawaiian source: Hoʻoulumāhiehie, *Ka Moʻolelo* 450. English trans.: Hoʻoulumāhiehie, *Epic* 420.

Rain of Kekele, Luluku, and Maluaka, Oʻahu

9. No kēlā inoa mai ʻo ʻĀpuakeanui i loaʻa mai ai kēlā ua kaulana o Kailua e hele mai ai a haluku iho i ka ulu hala o Kekele me Luluku, ʻo ia hoʻi ka ua ʻĀpuakea, i holo ma ko ke mele, penei:

. . . .

Hele haʻaheo ka ua ʻĀpuakea
Holo ʻaui i ke kai o Maluaka ē, i laila
Kaʻa ʻōlelo ka ua i luna o ka hala
Ke poʻo o ka hala o ʻĀhulimanu

From that name, ʻĀpuakeanui, came the name of the famous rain of Kailua that pummels the hala groves of Kekele and Luluku, namely the ʻĀpuakea, which goes like this in song:

. . . .
The ʻĀpuakea rain moves proudly along
Slipping off into the sea of Maluaka, ah, there
Words are spoken by the rain on the hala
The uppermost hala of ʻĀhulimanu

From the legend of Hiʻiakaikapoliopele. Hawaiian source: Hoʻoulumāhiehie, *Ka Moʻolelo* 146. English trans.: Hoʻoulumāhiehie, *Epic* 137–38. *Note: Hoʻoulumāhiehie says that "ʻĀpuakeanui" is the name of a woman who was considered the most beautiful in all of Kailua, Oʻahu.*

Rain of Koʻolau, Oʻahu

10. E hoʻi e ka uʻi o Koʻolau
ʻOiai ua malu nā pali
ʻO ka neʻe a ka ua ʻĀpuakea
Kāhiko i ke oho o ka palai

Let the youth of Koʻolau return home
For the cliffs are shaded
The creeping of the ʻĀpuakea rain
That adorns the fronds of the palai ferns

From the song "Pali Koolau." Hawaiian source: Holstein 74. English trans. by author.

11. Aloha wale ka leo ua makani
Ka leo heahea o ka ua ʻĀpuakea
E hea ana i ke ao makani kualau

So beloved is the windy, rainy voice
The calling voice of the ʻĀpuakea rain
Calling to the windy kualau rain cloud

From an affectionate greeting by Kahelekūlani to her child. Hawaiian source: Kaualilinoe, "Ka moolelo" 11/12/1870. English trans. by author. Additional source: Kaualilinoe, "Legend" 89.

Rain of Malaukaʻa, Oʻahu

12. Hele haʻaheo ka ua ʻĀpuakea
Holo nani i ke kai o Malaukaʻa ē—i laila

The ʻĀpuakea rain goes proudly
Traveling gracefully to the sea of Malaukaʻa—there

From a mele lei, or song presenting a lei; originally composed for Lunalilo and inherited by Kalākaua. This portion of the mele was composed by Kapauakanoa. Hawaiian source: *Na Mele Aimoku* 170. English trans. by author.

Rain of Malaukaʻa and Mololani, Koʻolaupoko, Oʻahu

13. Hoʻoipo i Mololani ka ua ʻĀpuakea, ʻeā
He ua pūlumi kai no Malaukaʻa, ʻeā

Romancing at Mololani is the ʻĀpuakea rain
A sea-sweeping rain of Malaukaʻa

From a mele mākaʻikaʻi, or travel chant, for ʻEmalani Kaleleonālani by Moehonua. Source: Nogelmeier 153.

14. Ke ʻike aku i ke koʻa māʻokiʻoki o Heʻeia
Me ka ua ʻĀpuakea o Mololani
Ke hehi maila lā i Malaukaʻa ē

Seeing the broken coral beds of Heʻeia
And the ʻĀpuakea rain of Mololani
Trampling upon Malaukaʻa

From a kanikau, or lament, for Kamehameha IV by one of his sisters. Hawaiian source: "He kanikau no ka Moi Iolani Kamehameha IV." English trans. by author.

Rain of Mololani, Kāneʻohe, Oʻahu

15. Kila kuʻi i ka ua ʻĀpuakea i Mololani
E hehi ana i ka ʻōnohi o ke koa

Majestically the ʻĀpuakea rain reaches up to Mololani
Treading on the leaf buds of the koa trees

From a mele mākaʻikaʻi, or travel chant, for ʻEmalani Kaleleonālani by her aunt Kamaʻikuʻi Young. Source: Nogelmeier 99.

16. Auē kuʻu makua kāne ē
Mai ka ua ʻĀpuakea o Mololani ē
Mai ka ua Hōliʻo i ka moana lā
Ke hehi maila i ka ʻili o ke kai ē

Pity for my dear father
From the ʻĀpuakea rain of Mololani
From the Hōliʻo rain over the ocean
Treading upon the surface of the sea

From a kanikau, or lament, for Ioseba Nawai. Hawaiian source: Kamoakeawe. English trans. by author.

17. Me ka ua ʻĀpuakea
Ka laʻi aʻo Malulani
Me ke anu o ke Koʻolau

With the ʻĀpuakea rain
The calm of Malulani
And the coolness of the Koʻolau

From the song "Kāneʻohe" by Abbie Kong and Johnny Noble. Source: Wilcox et al. 110. Additional source: Elbert and Mahoe 61. *Note: "Malulani" is probably "Mololani."*

18. He kanikau aloha kēia nou, e B. Mahune

. . . .

Kū ka ʻuhane, nānā aku ʻo ka ua ʻĀpuakea o Mololani lā
Ke hehi maila i ka maka o ke kai a luhi

This is an affectionate chant of mourning for you, B. Mahune

. . . .

The spirit stands and observes the ʻĀpuakea rain of Mololani
Trampling upon the spray of the sea until weary

From a kanikau, or lament, for B. Mahune. Hawaiian source: Wailea. English trans. by author.

19. *For another example of ʻĀpuakea of Mololani, Kāneʻohe, see Kuahine (#9).*

Rain of Mololani, Nuʻuanu, Oʻahu

20. Ka ua ʻĀpuakea o Mololani. *The ʻĀpuakea rain of Mololani.*

A traditional saying. Source: Pukui, *ʻŌlelo* 167. *Note: Pukui says, "ʻĀpuakea was once a beautiful maiden who was changed by Hiʻiaka into the rain that bears her name. Mololani is in Nuʻuanu."*

Rain of Nuʻuanu, Oʻahu

21. Aia lā ʻo ka ua ʻĀpuakea
Ke hehi ala i ke kāmakahala
Ke wehe ala i ka pua o ka ʻāhihi
ʻĀkoakoa ka ua i Nuʻuanu
E hoʻolulu ana i Kawānanakoa
E ʻaha ʻōlelo ana i Kahuailana

There indeed is the ʻĀpuakea rain
Traipsing over the kāmakahala flowers
Opening up the blooms [of] the ʻāhihi lehua
The rain gathers at Nuʻuanu
Waiting calmly at Kawānanakoa
Assembling there at Kahuailana

From a mele mākaʻikaʻi, or travel chant, for ʻEmalani Kaleleonālani credited to P. Nāhaolelua. Source: Nogelmeier 138.

Other examples

22. Ke uē maila ka ʻōhiʻa kūmakua i ke anu
Heleleʻi nā waimaka a ka ua ʻĀpuakea
Ke holoi lā i ke oho o ka mahiki

The ʻōhiʻa bush cries in the cold
The tears of the ʻĀpuakea rain fall
Washing the blades of the mahiki grass

From a greeting chant to Kamaakamahiʻai by his daughter Kahelekūlani. From the legend of Kamaakamahiʻai. Hawaiian source: Kaualilinoe, "Ka moolelo" 10/8/1870. English trans. by author. Additional source: Kaualilinoe, "Legend" 69.

23. He wā ʻōlelo na ka ua ʻĀpuakea
Ua noho like lāua me ka ua Kiaweʻula

Loud talking by the ʻĀpuakea rain
That dwelt with the Kiaweʻula rain

From a mele inoa, or name chant, for Heneri W. Auld. Hawaiian source: Hoopai. English trans. by author.

ʻĀpuʻupuʻu. Same as *Kīpū*, *Kīpuʻu*, and *Kīpuʻupuʻu*. Rain and wind associated with Waimea, Hawaiʻi. "ʻĀpuʻupuʻu" refers to the bumpiness of the skin in the cold, wind-driven rain. It may also refer to the hilly land of Waimea.

1. E oʻu hoa o ka ua ʻĀpuʻupuʻu o Waimea, ka mea mākau o ka ʻaihue, ua manaʻo anei ʻoe ʻo ka ʻaihue ka mea e waiwai ai? ʻAʻole! ʻO kou mea ia e ʻilihune loa ai.

My dear friend of the ʻĀpuʻupuʻu rain of Waimea, the one who is clever in stealing, did you think that stealing would make you rich? Not so! This is what will make you very poor.

From a newspaper article reporting a store theft. Hawaiian source: "He aihue." English trans. by author.

2. Kuʻu hoa o ke kai hāwanawana o Kawaihae
Kai pili aloha a māua me kuʻu kāne
Mai ka ua ʻĀpuʻupuʻu o Waimea

My dear companion of the whispering sea of Kawaihae
Sea of tenderness for my husband and me
From the ʻĀpuʻupuʻu rain of Waimea

From a kanikau, or lament, for J. W. Keawehunahala by his wife. Hawaiian source: Kauiomanoa. English trans. by author.

3. Aloha ke kāne, ka hoa hele o ke alanui ē
Kuʻu hoa pili i ka ua ʻĀpuʻupuʻu anu o Waimea ē

Love to the husband, the traveling companion on the road
My intimate friend in the cold ʻĀpuʻupuʻu rain of Waimea

From a kānaenae, or chant of eulogy, for Makia by his wife. Hawaiian source: Kalehua. English trans. by author.

ʻAwa. Same as *ʻAwaʻawa.* Bitter, cold, dark rain, mist, fog, vog, or smoke of the mountains; acidic rain; associated with Kīlauea, Hawaiʻi, and found in other areas. Also the name of a wind. "ʻAwa" means "bitter" and can refer to grief, tragedy, or misfortune. It is both the name of a specific rain and a generally descriptive term; its various usages are determined by the context.

Rain of Kaʻauea, Hawaiʻi

1. He mau aku nō ka ihona a hiki i Kaʻauea
   Hoʻomaha aku i ka hale kama lau kī
   E hoʻopumehana ana i ke ahi a Kūkaʻōhiʻalaka
   I ka hoʻopulu ʻana mai a ka ua ʻAwa a me ka hau anu o ka mauna
   Aloha wale ia mauna anu mehameha a kāua e hele ai

   The descent continues down to Kaʻauea
   To rest at a hut made of tied kī leaves
   Warming at the fire of Kūkaʻōhiʻalaka
   After being soaked by the ʻAwa rain and the cold hau dew of the mountain
   So beloved is that isolated, cold mountain we travel

   From a kanikau, or lament, for Mrs. Maleka Ii. Hawaiian source: Kaapa. English trans. by author.

Rain of Kahoʻokū, Hawaiʻi

2. *See Nahunahu (#5).*

Rain of Kaʻū, Hawaiʻi

3. A i loko o kēia kaua ʻana, eia ke iho nei ka ua ʻAwa, ka ua ʻaʻaki anu o nā mauna. Hui pū maila ka pouli poʻipū a kēia ua, pouli nā wahi a pau, a i loko o kēlā pouli i neʻe hope aku ai nā koa o Keōua i nā koa o Kamehameha.

   Ma muli o ke kamaʻāina loa o ko Kaʻū poʻe i nā waihona ʻāina a pau o ko lākou wahi—kamaʻāina i nā ʻālualua, nā owāwa, nā meki, nā pōhāhā nunui a ka pele i hoʻonakaka ai—ua palekana loa kā lākou ʻauheʻe ʻana.

   A ma muli nō hoʻi o ke kamaʻāina ʻole o nā koa o Kamehameha—e alakaʻi ʻia ana e Kaʻiana—i kēia mau ʻālualua a me nā hāoʻeoʻe pōhaku o kēia ʻāina, i loko o kēia wā e uhi paʻa ana kēia ua ʻAwa, ua hiki ʻole iā lākou ke hoʻohikiwawe māmā aku i kā lākou uhai ʻana i nā koa ʻauheʻe o Keōua.

ʻO kēia ua, wahi a ke kūʻauhau S. L. Peleiōhōlani, he ua anu huʻi; he ʻohu kilihune liʻiliʻi, e kolili lelehune ana i ka lau o ka lihilihi. A pēlā nō hoʻi ka ʻona o ka ʻapu ʻawa ke inu. I ka wā e noe mai ai ka ʻona a ka ʻawa, pōhina maila ka ʻike a ka maka me he ua noenoe liʻiliʻi pōhina ala. A pēia i kapa aku ai nā kūpuna o Hawaiʻi i kēia ua, he ua ʻAwa.

Ma muli o kēia poʻipū ʻana a kēia ua, ua kuemi hope nā koa o Kaʻiana no ko lākou mau oʻioʻina, a mai laila aku ua neʻe kekahi poʻe o lākou a loaʻa ʻo Kamehameha i Laupāhoehoe, a ʻo kekahi poʻe o lākou, ua hoʻi no nā Kona no ke kiaʻi a me ka mālama ʻana i ia wahi.

And during this battle, here was the ʻAwa rain pouring down, the cold, biting rain of the mountains. A smothering darkness accompanied this rain; darkness was everywhere. Within that gloom, the warriors of Keōua retreated from the warriors of Kamehameha.

Since the Kaʻū people were very familiar with the entire topography of their place—the irregularities, the gulches, the pits, the huge volcanic ejecta that the eruption cracked open—they were able to safely escape.

And because Kamehameha's fighters, led by Kaʻiana, were unfamiliar with these pitfalls and uneven terrain of the land, and at this time, the ʻAwa rain was creating a solid cover, they could not make haste and speed up their pursuit of Keōua's fleeing warriors.

This rain, says the genealogist S. L. Peleiōhōlani, is a chilly, cold rain, a fine, light mist fluttering gently onto the eyelash tips. And that is the same as the intoxication one feels when drinking from an ʻawa cup. When intoxication from the ʻawa makes one feel foggy, the eye's vision becomes dim like the tiny, gray drops of a foggy noenoe rain. And this is why our Hawaiian ancestors named this rain the ʻAwa rain.

Because this rain was overwhelming, Kaʻiana's troops withdrew to their resting places, and from there some moved on until they reached Kamehameha at Laupāhoehoe, while others returned to the Kona districts to guard and protect those places.

From the story of Kamehameha I. Hawaiian source: Hoʻoulumāhiehie, "He moolelo" 3/3/1906. English trans. by author. Additional sources: Desha, *He Moʻolelo* 1: 262; Desha, *Kamehameha* 182.

Rain of Kīlauea, Hawaiʻi

4. Hulihia Kīlauea, pō i ka uahi
   Nalowale i ke ʻAwa ka uka o ka lua

   Kīlauea is overturned, obscured by smoke
   Disappearing in the ʻAwa is the upper area of the crater

From a chant about Pele by her sister Hiʻiakaikapoliopele as their sisters toss cinders on Lohiʻau. Hawaiian sources: Hoʻoulumāhiehie, *Ka Moʻolelo* 368; Emerson, *Pele* 197. English trans. by author. Additional source: Hoʻoulumāhiehie, *Epic* 343. *Note: Emerson suggests that the ʻAwa rain refers to bad weather.*

Rain of Kīlauea and ʻŌlaʻa, Hawaiʻi

5. Ka ua ʻAwa o Kīlauea. *The ʻAwa rain of Kīlauea.*

A traditional saying. Source: Pukui, *ʻŌlelo* 167. *Note: Pukui says, "The ʻAwa is a bitterly cold rain of ʻŌlaʻa and Kīlauea, Hawaiʻi."*

Rain of Kiloi, Hawaiʻi

6. Ke hānai aʻela ka ua i ka lani
Makaʻu au i ka ua ʻAwa i ka uka o Kiloi

The rain is feeding the heavens
I fear the ʻAwa rain of the uplands of Kiloi

From a retort by Hiʻiakaikapoliopele to her sister Pele, who impatiently awaits Hiʻiaka's departure to fetch Lohiʻau. Hawaiian source: Emerson, *Pele* 17. English trans. by author. *Note: In another version of this mele (see Kapihenui 1/9/1862), "awa" refers to the plant and not the rain.*

Rain of Leleiwi, Hawaiʻi

7. ʻAwa—Leleiwi, Hawaiʻi; a cold mountain rain.

From a glossary of terms. Source: Nakuina, *Wind* 138.

Rain of ʻŌmaʻolala (or Maʻolala), Hawaiʻi

8. *See Nahunahu (#4).*

Rain of ʻŌʻōkuauli, Hawaiʻi

9. He ua Noe, he ua ʻAwa no ka mauna
Uli ka nahele o ʻŌʻōkuauli
Uli i ke nahu hoʻomau a ka makani
Makani ʻami kiʻelei a leʻa ke one
Leʻa i nā kāuna ʻami a ka ua
I nā kuʻi lua a ka ua he ʻAwaʻawa

The Noe rain, the ʻAwa rain from the mountains
Dark is the forest of ʻŌʻōkuauli
Dark from the constant biting of the wind
Wind that dances a squatting ʻami, revolving to the delight of the sand
Delighted by the rain spinning in quadruple sets of ʻami
By the repeated stomping of the rain, an ʻAwaʻawa

From a kanikau, or lament, for Kamehameha IV. Hawaiian source: Moehonua and Lilikalani. English trans. by author.

Rain of Wahinekapu, Hawai‘i

10. *See Noe (#5).*

Rain of East Maui

11. Ua ‘Awa—He ua po‘ipū, ua li‘ili‘i a pō ka lā ma ka mauna.

    ‘Awa rain—It is a rain that covers everything, raining a little at a time until day turns to night on the mountains.

    From an account of East Maui rains and their descriptions, as told to Thomas K. Maunupau by Josua Ahulii of Kaupō, Maui. Hawaiian source: Maunupau, *Huakai* 27. English trans. by author.

Rain of Nu‘uanu, O‘ahu

12. *See Kāmakahala (#3).*

Bitter rain

13. Ka ua ‘awa. *The bitter rain (of grief, tragedy, hardship).*

    Source: Pukui and Elbert, 1986, 509.

14. ‘Aka‘aka ka malihini, he mea hewa ia
    Puni ho‘i i ka ua ‘Awa a ka wahine
    I wahine au i kāne ‘oe

    The newcomer laughs, a wrongful thing to do
    Surrounded by the bitter ‘Awa rain of the woman
    I should be the wife and you the husband

    From a mele by Kamalālāwalu upon seeing her former partner Halemano being playfully bitten by a new lover, Kīkēka‘alā. Hawaiian source: Fornander, *Fornander* 5: 257. English trans. by author.

Mountain rain

15. Ua ‘Awa—He ua kēia ma ka mauna, he ua ‘awa, he ua li‘ili‘i, ‘o ke ehu wale nō; ‘a‘ole i loku ‘o ka ua, ‘o ke po‘o ke kuakea me he kanaka po‘o hina lā, he ua po‘o nui o ke kuahiwi, ‘o ke anu na‘e ka mea nui, no laila mai kēia inoa ua ‘Awa, he ua noe.

    ‘Awa rain—This is a rain in the mountains, a bitter rain, a fine rain; it is only mist; it doesn’t pour down; the head is white like a gray-haired man, a troublesome “po‘o nui” rain in the mountains, though the cold is the main problem, and therefore this name, ua ‘Awa, a misty rain.

    From a letter to the newspaper *Kuokoa* regarding famous rains. Hawaiian source: Kalaaukumuole, “Manao wehewehe.” English trans. by author. Additional sources: Kalaaukumuole, “Thought explanations”; Pukui, *‘Ōlelo* 173. *Note: In this rendering, “‘a‘ole i loku ‘o ka ua” is used instead of “aole iloko o ka ua,” which was printed in the Hawaiian source. In Pukui’s reference, we learn that the ‘Awa “falls in fine, icy cold drops that make one’s head appear white.”*

16. He ua ʻawa, ma ka mauna ia ua. *An ʻawa rain, this rain falls on the mountains.*

From a description of different rains. Hawaiian source: Malo, *Ka Moolelo* 20. English trans. by author. Additional source: Malo, *Hawaiian* 14.

17. ʻO ka uahi noe lehua ē
ʻO ke ʻAwa nui i ka mauna

The vapor that mists the lehua blossoms
The thick ʻAwa fog of the mountain

From a mele by Lohiʻau as the fires of his former lover Pele approach him. Hawaiian source: Emerson, *Pele* 205. English trans. by author. Additional sources: Pukui and Elbert, 1986, 33–34; Emerson, *Pele* 206.

18. Cold mountain rain, fog, mist; to rain or mist. *Fig.*, tragic misfortune or ordeal. When ʻawa is preceded by *ke*, it may refer to a volcanic eruption: Uē au, puni ʻā i ke ʻawa (*I weep, surrounded by lava in the downpour*).

Source: Pukui and Elbert, 1986, 33.

Rain associated with divinity
19. *See ʻUlalena (#14).*

Rain associated with great beauty
20. *See Koko (#6).*

Rain associated with high station
21. He lupe lele a pulu i ka ua ʻawa. *A kite that flies till it is dampened by icy cold [ʻawa] raindrops.*

A traditional saying. Source: Pukui, *ʻŌlelo* 85. *Note: Pukui says that this phrase describes someone "whose station has risen very high."*

Rain associated with kapu
22. He lani, he kapu ka lā i ka ua ʻAwa o Keohokapu
A kapa nō he keiki, i hānau i loko o Kū Kahi

The sun is chiefly, sacred in the ʻAwa rain of Keohokapu
And is called a child, born on Kū Kahi

From a pule, or prayer, attributed to Pele and shared by Malule of Hanapēpē, Kauaʻi. Hawaiian source: Pukui, *Nā Mele Welo* 12. English trans. by author. *Note: Kū Kahi is the third day of the lunar month.*

Vog associated with Pele
23. Hilda Hoohila Kawelo: Hele mai ʻo Pele mai Kahiki. Kona manawa i hele mai ai, hele mai nō ka ua. Hele mai ka ua Noe me ka ua ʻAwa. He aha ia mea he ua ʻAwa? Maopopo anei iā ʻoe?

Clinton Kanahele: ʻAʻole.

Kawelo: ʻAe, ka ua ʻAwa, ʻo ia nō ka uahi.

Kanahele: Oh.

Kawelo: Ka uahi o kēlā ahi, o kēlā, no ka mea, he ahi hoʻi ʻo ia i hele maila. Kēlā uahi e pōhina nei ka ʻāina āna i hele mai nei, e ʻike ʻia aku nei i ka maka o nā keiki a kānaka, mea, inā ʻoe e honi i kēlā, ʻano, mea, pōniuniu nō ʻoe.

Kanahele: Ua ʻAwa kēlā?

Kawelo: Ka ua ʻAwa hoʻi ia. Kēlā ua, ʻaʻole ʻoe pulu. ʻAʻole ʻoe pulu i kēlā ʻAwa.

Kanahele: He hanu wale nō ʻoe i kēlā ua.

Kawelo: ʻAe, hanu wale nō. "Hele mai ʻo Pele mai Kahiki, hele mai ka ua ʻAwa me ka ua Noe. Uhi aʻela, paʻa nā pali paʻa o Kahiki, ʻakahi ka hikina ʻana mai o kō aloha iaʻu."

Kanahele: He hīmeni kēlā?

Kawelo: ʻAe, he olioli hoʻi paha.

Kanahele: He olioli?

Kawelo: ʻAe, he olioli kēlā.

. . . .

Kawelo [e olioli ana]: Hele mai ʻo Pele mai Kahiki
Hele mai ka ua ʻAwa me ka ua Noe
Uhi ʻia aʻela, paʻa nā pali paʻa o Kahiki
ʻAkahi ka hikina ʻana mai o kou aloha i oʻu nei
E Kalani mā ē
E iho koke mai i lalo nei
I maha aʻe au
E Kāne mā ē
ʻOki ʻia ka lili, ka inaina o lākou nei lā ē
ʻAe

Hilda Hoohila Kawelo: Pele came from Kahiki. When she arrived, so too came the rain. The Noe rain and the ʻAwa rain came. What is this thing known as an ʻAwa rain? Do you know?

Clinton Kanahele: No.

Kawelo: The ʻAwa rain, it is smoke.

Kanahele: Oh.

Kawelo: The smoke of that fire, of that, because she came as fire. That smoke that creates a haze over the land she came to, that is seen and felt by the eyes of the children of men, if you were to inhale that, you would get dizzy.

Kanahele: That's the ʻAwa rain?

Kawelo: The ʻAwa rain, indeed. That rain, you don't get wet. You don't get wet from that ʻAwa.

Kanahele: You just inhale that rain.

Kawelo: Yes, it is just inhaled. "Pele came from Kahiki; the ʻAwa rain and the Noe rain came; completely smothering the cliffs of Kahiki; your love has finally reached me."

Kanahele: Is that a song?

Kawelo: Yes, an olioli chant.

Kanahele: An olioli?

Kawelo: Yes, it is an olioli.

. . . .

Kawelo [chanting]: Pele comes from Kahiki
The ʻAwa rain and the Noe rain come
Completely smothering the cliffs of Kahiki
Your love has finally reached me
O Heavenly One and company
Come quickly down here
That I may be at ease
O Kāne and company
End their jealousy and rage
Yes

From an interview of Hilda Hoohila Kawelo by Clinton Kanahele. Hawaiian source: Hilda Hoohila Kawelo. Transcription and English trans. by author.

ʻAwaʻawa. Same as *ʻAwa*. Bitter, cold, dark rain, vog, or smoke of the mountains; acidic rain. Associated with Puna, Hawaiʻi, and found on other parts of Hawaiʻi and on other islands. Also the name of a wind. "ʻAwaʻawa" means "bitter," and it can refer to grief, tragedy, or misfortune. It is both the name of a specific rain and a generally descriptive term; its various usages are determined by the context.

Rain of Hainoa, Hawaiʻi

1. Mikihilina Hainoa i ka ʻAwaʻawa
Ka pakī mau i ka luna o Hualālai

Hainoa is dressed in the finery of the ʻAwaʻawa rain
Constantly splashing on the upper reaches of Hualālai

From a mele given to the Bishop Museum by Thomas E. Cooke of Hilo, Hawaiʻi, who received it from Judge Simon H. Haʻaheo of Kalapana, Hawaiʻi. Hawaiian source: "Mikihilina." English trans. by author. Additional source: Moehonua and Lilikalani.

Rain of Hilo, Hawaiʻi

2. Kū aku i ka pahu, kū a ka ʻawaʻawa
Hananeʻe ke kīkala o ko Hilo kini
Hoʻi luʻuluʻu i ke one o Hanakahi

Hit by the thrusts, hit by the bitter cold [ʻawaʻawa]
The hips of Hilo's throngs sag
As they return burdened to the sands of Hanakahi

From the song "Hole Waimea," which comes from a mele inoa, or name chant, for Kamehameha I. Source: Wilcox et al. 68. Additional sources: "Hoe Puna"; *Nā Leo Hawaiʻi Kahiko,* Record II, Side A, Band 5; Elbert and Mahoe 52.

Rain of Kahakulani, Hawaiʻi

3. I Kahakulani aku nei
Kaha ʻo Kalani, noho i ka ua ʻawaʻawa
Noho nō, kui i nā lehua pua limu
He pāpahi lei no ka uka i Kaʻūmana

At Kahakulani recently
The Royal One moved along, dwelling in the bitter [ʻawaʻawa] rain
Resting, stringing the lehua that grow among the mosses
A lei of honor for the highlands of Kaʻūmana

From an uē, or wailing call of love and grief, probably for Col. Kapaakea, credited to ʻEmalani Kaleleonālani. Source: Nogelmeier 334. *Note: In this rendering, "Kaʻūmana" is used instead of "Kaūmana," which was printed in the source.*

Rain of Kapōhākau, Hawaiʻi

4. Keiki ʻalo i ka ua o Kapōhākau
Kama iki noho i ka ua a ka ʻAwaʻawa
Hoʻolono leo leʻa o ka pololei
Paha leo leʻa i Māʻuluaʻe
ʻO ʻoe ana, ʻo wau ana kekahi ē
Momoe aku kāua i ka ua a ka ʻAwaʻawa
Ke anu hoʻi ē

Boy who endures the rain of Kapōhākau
Child who dwells in the bitter ʻAwaʻawa rains
Listening to the happy trill of land shells
To the joyful chanting at Māʻuluaʻe
You will be there, I will be there too
Let us push on in the bitter ʻAwaʻawa rains
And the cold, as well

From a mele by Lohiʻau as Pele's sisters toss embers on him. Hawaiian source: Hoʻoulumāhiehie, *Ka Moʻolelo* 378. English trans.: Hoʻoulumāhiehie, *Epic* 353.

Rain of ʻŌʻōkuauli, Hawaiʻi

5. *See ʻAwa (#9).*

Rain of Puna, Hawaiʻi

6. Pō Puna i ka ua a ka ʻAwaʻawa
Pākuʻi i ka uahi a ka lua
Hauna i ka uahi a Pele lā ē
Aloha ka wahine o ka lua

Puna is obscured by the rain of the ʻAwaʻawa
Foul-smelling from the smoke of the pit
Reeking from the smoke of Pele
Aloha to the woman of the pit

From a mele to Pele by Kamapuaʻa upon seeing Pele's brother Lonomakua light a fire to consume him. Hawaiian source: Fornander, *Fornander* 5: 341. English trans. by author. Additional sources: Kahiolo, *He Moolelo* 66–67; Emerson, *Pele* 163; Pukui and Elbert, 1986, 34; Elbert 224–25.

7. Pō Puna i ka ua ʻAwaʻawa
Pōhina Puna i ka ua Noenoe

Puna is obscured by the ʻAwaʻawa rain
Puna is hazy in the Noenoe rain

From a mele by Hiʻiakaikapoliopele when she sees her friend Hōpoe and the lehua forest of Puna destroyed by Pele. Hawaiian source: Emerson, *Pele* 163. English trans. by author.

8. ʻO ʻoe ia, e nani kauona Puna i ka hala
Hiehie ʻokoʻa nō i ka ʻAwaʻawa

It is you, O beautiful one, like the fragrant hala of Puna
Standing majestically in the cold [ʻAwaʻawa] rain

From a mele for Victoria Kamāmalu. Hawaiian source: Ii, "Ke ola" 6/16/1866. English trans.: Ii, "Death" 2951.

Rain of Kulaokahuʻa, Oʻahu

9. *See Makaʻūpili (#3).*

Mountain rain

10. Nani Wahinekapu i ka Puʻulena
I ka ua ʻAwaʻawa o ka mauna

Wahinekapu is beautified by the Puʻulena breeze
By the icy cold [ʻAwaʻawa] mountain rain

From a mele for Victoria Kamāmalu. Hawaiian source: Ii, "Victoria." English trans.: Ii, "Death" 2963. *Note: "Wahinekapu" refers to the bluff near Kīlauea, Hawaiʻi.*

11. Ka ua ʻawaʻawa e hoʻonū lā i uka. *The bitterly cold [ʻawaʻawa] rain pattering in the uplands.*

Source: Pukui and Elbert, 1986, 271.

E

ʻEʻelekoa. Related to *Mālana*, *Mālanalana*, and *Mālanaʻeʻelekoa*. Rain associated with Waimea, Hawaiʻi. Also the name of a wind. "ʻEʻelekoa" means "stormy; hardship."

Rain of Uli (or ʻŌuli), Hawaiʻi

1. A ka hoʻāʻi i ka pua o ka māmane
Ke pili lā i ka ua ʻEʻelekoa

Polishing the attractive flower of the māmane tree
Clinging to the ʻEʻelekoa rain

From a mele inoa, or name chant, for Leleiōhoku. This portion of the mele was composed by Kamoonohu. Hawaiian source: *Na Mele Aimoku* 306. English trans. by author.

Rain of Waimea, Hawaiʻi

2. Ua ʻEʻelekoa—Waimea, Hawaiʻi.

From a list of rain names and their descriptions. Source: "Names . . . rain."

3. *For more examples of ʻEʻelekoa of Waimea, see Kīpuʻupuʻu (#25, #32).*

Rain associated with the Puaneiea rain

4. Paʻa pono nā ʻopi wai a ka ua ʻEʻelekoa
E hoʻokō mai ʻoe, e ka ua Puaneiea

The embankments of the 'E'elekoa rain are perfectly solid
Come fill them, Puaneiea rain

From a mele inoa, or name chant, for Heneri W. Auld. Hawaiian source: Hoopai. English trans. by author.

'ehu. Or ehu. Similar to *'ehuehu*. Spray, foam; mist that is lighter than the uhiwai, noe, and 'ohu, but heavier than the 'ehuehu.

1. Me he manu lā nō ka ua e ka'akepa nei
E wili nei i luna o ka lā'au
Luhe ke oho o ke kukui mā'e'ele i ka ua
Naku ka ua ho'okawewe i ka moana
Olokē, oloolokē, pi'oloke ka leo o ka palila
Kūkūkū ka 'ehu o ka ua i ka nahele
E kāpapa ana i ke alo o Maunakea
Palalē ka leo o ka ua i luna o ka lā'au
E hiohio haole ana i ke alo o nā pali
Ka ua ho'okina, ho'okinakina, ho'okikina

Like a bird, the rain swoops diagonally
Swirling above the trees
The leaves of the kukui droop, numb from the rain
The pouring rain roils the ocean
Ruffled, flustered, frightened is the voice of the palila bird
The 'ehu mist of the rain rises to the forest
Extending to the face of Maunakea
The voice of the rain on the trees is indiscernable
Speaking gibberish near the face of the cliffs
The constant, unceasing, persistent rain

From a mele inoa, or name chant, for Albert Ka Haku O Hawai'i. Hawaiian source: Paauilo. English trans. by author.

'ehuehu. Similar to *'ehu*. Spray, foam; mist that is lighter than the uhiwai, noe, 'ohu, and 'ehu.

1. Ku'u hoa i ka 'ehuehu a ka noe. *My companion in the ['ehuehu] spray of mist.*

Source: Pukui and Elbert, 1986, 38.

'Elehe'i. Rain associated with Liliko'i and Makawao, Maui.

Rain of Liliko'i, Maui

1. Ka ua 'Elehe'i i Liliko'i. *The 'Elehe'i rain at Liliko'i.*

From a wind chant by Kūapākaʻa. Hawaiian source: Nakuina, *Moolelo* 68. English trans. by author. Additional sources: Kuapuu, "He wahi moolelo" 5/15/1861; Kanepuu, "Kaahele"; Nakuina, *Wind* 63.

Rain of Makawao, Maui

2. Kauaʻeleheʻiomakawao. *The ʻEleheʻi rain of Makawao.*

From a list of names of lua fighting techniques. Hawaiian source: Kaawa 4/27/1865. English trans. by author. Additional source: Paglinawan et al. 112.

ʻElemalino. Rain associated with Hāna, Maui.

Rain of Hīnaʻi, Maui

1. Ka lau likoliko ʻōhiʻa o Hīnaʻi
Ke oho pālālā lena uli o ke kī
I hoa ʻia e ka ua ʻElemalino ē, i laila

The glistening ʻōhiʻa leaf buds of Hīnaʻi
The broad, dark gold leaf of the ti plant
Held fast by the ʻElemalino rain, there

From a mele lei, or song presenting a lei, originally composed for Lunalilo and inherited by Kalākaua. This portion of the mele was composed by M. Kekauonohi. Hawaiian source: *Na Mele Aimoku* 160. English trans.: Reichel, *Keʻalaokamaile*. *Note: "Hīnaʻi" is the name of a hill in Hāna.*

Rain of Kaʻuiki, Maui

2. Niniu Kaʻuiki, ke poʻo lanakila
Hāliuliu i ka ua ʻElemalino ē

Kaʻuiki, triumphant summit, reels
Leaning into the ʻElemalino rain

From a mele recalled by Hoʻoulumāhiehie. Hawaiian source: Hoʻoulumāhiehie, *Ka Moʻolelo* 417–18. English trans.: Hoʻoulumāhiehie, *Epic* 389–90.

H

Hā'ao. Rain that falls in successive showers; associated with Ka'ū, Hawai'i, and with Nu'uanu, O'ahu. Also found on other parts of O'ahu and on Kaua'i. Also the name of an area and a spring in Wai'ōhinu, Hawai'i, and the name of a kapa pattern (uahā'ao or naouahā'ao). "Hā'ao" refers to the retinue of a chief following in procession, as the showers of this rain follow one another like a chief's retinue.

Rain of 'Au'aulele, Hawai'i

1. A'u lei o ka ua Hā'ao
   E lele a'ela ma uka o 'Au'aulele

   My lei of the Hā'ao rain
   Flying above 'Au'aulele

   From an uē helu, or wailing call of love, by Ululani to Kamehameha I upon his arrival at her home in Hilo, Hawai'i. Hawaiian source: Ho'oulumāhiehie, "He moolelo" 12/27/1905. English trans. by author. Additional sources: Desha, *He Mo'olelo* 1: 114; Desha, *Kamehameha* 77; Pukui and Korn 10–11; *I Mua e nā Pōki'i.*

2. Ku'u haku i ka ua Hā'ao ē
   Ke lele a'ela ka ua
   Ma uka o 'Au'aulele
   Lele ka ua, lele pū nō me ka makani
   E lelepo'o ana ka wai o Kahā

Kuʻu haku mai ka wai
Hāʻule poʻo ē

My lord in the Hāʻao rain
The rain flies up
Above ʻAuʻaulele
The rain flies, flies with the wind
The water of Kahā dives headfirst
My lord from the water
That falls on the head

From a mele aloha, or song of love, by the warrior Kaihekīʻoi on seeing Keōuakū-ʻahuʻula being taken to Puʻukoholā Heiau to be sacrificed. Hawaiian source: Kamakau, *Ke Kumu Aupuni* 112–13. English trans. by author. Additional sources: Kamakau, "Ka moolelo o Kamehameha I" 5/11/1867; Kamakau, *Ruling* 158; Fornander, *Fornander* 5: 692–93; Hoʻoulumāhiehie, "He moolelo" 7/10/1906; Desha, *Kamehameha* 338. *Note: Kamakau,* Ruling *158, says, "Ua Hāʻao (rain of Hāʻao) is the name of a rain that comes down at Waiʻōhinu in Kaʻū and keeps that district green. Hāʻao is the name of the spring second in size of the five springs that water Waiʻōhinu. The chant is still chanted by the old people of Kaʻū who retain their love of Keōua and hatred for Kamehameha." Similar versions of this mele are by Wailama (see Fornander) and Kuakāhela (see Hoʻoulumāhiehie and Desha). The last three lines recall the saying, "Ka wai nāʻuke poʻo o Kahā. The water of Kahā that removed head lice." Pukui,* ʻŌlelo *179, explains, "The water of Kahā is in Waiʻōhinu, Kaʻū. The chief Keōuakūʻahuʻula once discovered that he had lice on his head. Not wanting others to know, he went to Kahā where he washed his head and had the pests removed."*

Rain of Hāʻao, Hawaiʻi

3. Ka ua o Hāʻao—This rain name was used in poetic expressions and made famous by the chant of Kuakāhela announcing the killing of the Kaʻū [and Puna] chief, Keōuakūʻahuʻula, by Keʻeaumoku: Kuʻu haku i ka ua o Hāʻao (*My lord in the Hāʻao rain*). Kuakāhela was Keōua's kahuna.

   Source: Pukui, Elbert, and Mookini 91. *Note: See versions of Kuakāhela's chant, a kanikau, or lament, in Hoʻoulumāhiehie, "He moolelo" 7/10/1906; Desha,* Kamehameha *338. The lines of the chant are very similar to those of the example above.*

4. ʻO ka maka i ka ua o Hāʻao
   A ka heʻe i ka ua ʻAlaea o Kiʻi

   Starting in the Hāʻao rain
   And retreating in the ʻAlaea rain of Kiʻi

   From a chant for Kūaliʻi by his guardians, Kumahukia and Kaiwiokaekaha. Hawaiian source: Kamakau, "He mele no Kualii." English trans. by author. Additional source: Fornander, *Fornander* 4: 398–99. *Note: Fornander says that the Hāʻao is "a wind driving rain, said to be peculiar to ʻAuʻaulele."*

Rain of Kaʻū, Hawaiʻi

5. ʻO kekahi mea kupaianaha nō i hoʻomaopopo ʻia, ʻo nā ʻaʻahu hulu manu o ua aliʻi nei, a me kekahi mau pono ʻē aʻe o kona kūlana aliʻi kiʻekiʻe, ua hoʻokau ʻia akula i luna o ka waʻa o Keaweaheulu mā i holo aʻe ai i Kaʻū e kiʻi iā ia. ʻO kekahi hana ʻano ʻē nō ia a ua aliʻi nei o ka ua Hāʻao, a e hōʻike mai ana nō paha i kona hāliʻaliʻa ʻē nō i kona make.

Another unusual thing that was seen: the bird-feather capes of this aliʻi and some other belongings related to his high rank were placed on board the canoe Keaweaheulu and company had sailed to Kaʻū to fetch him. This was a very odd thing for this aliʻi of the Hāʻao rain to do, and perhaps it showed his premonition of death.

From a reference to the high chief Keōuakūʻahuʻula of Kaʻū, Hawaiʻi. Hawaiian source: Desha, *He Moʻolelo* 2: 87. English trans. by author. Additional source: Desha, *Kamehameha* 329. *Note: For another example of the Hāʻao as a metaphor that stood in for the place Kaʻū, see Desha,* He Moʻolelo *2: 97; Desha,* Kamehameha *336.*

Rain of Waiʻōhinu, Hawaiʻi

6. Nani Kaʻū, hanohano i ka makani
I hanohano nō i pā ʻia e ka Moaʻe
I ka nihi kokolo a ka ʻohu i nā pali
Ke kīpū maila i luna o Lauhue
Kahua heihei ʻAuʻaulele na ka ʻino
He hale kelepona na ka ua i ka nahele
He wā ʻōlelo na ka mamo i ka pua lehua
Hoʻihoʻi mai nō e pili me ka ua Hāʻao
I holoholo nō a nui me ka ua Nihipali

Beautiful Kaʻū, magnificent in the wind
Majestic in the blowing of the Moaʻe wind
In the quiet creeping of the ʻohu mist along the cliffs
The mist settles atop Lauhue
ʻAuʻaulele is like a racetrack for the storm
A telephone exchange for the rain in the forest
Noisy chatter by the mamo bird to the lehua flower
Brought back to be with the Hāʻao rain
That traveled often with the Nihipali rain

From a chant by Monika Keawe for the passing of Hehena Puoanui. Hawaiian source: "Nani Ka-u." English trans. by author. Additional source: Keawe.

7. He malihini ka ua ma kēia ʻāina ʻo Kaʻū nei, a ʻo kahi noho mau a ka ua, aia wale iho nō ia ma uka o Waiʻōhinu, a ua kapa ʻia ka inoa o ia ua, he "ua Hāʻao." A inā hoʻi e nui loa mai ka ua ʻana, a kīpalalē aʻe i nā lae ʻōhiʻa o ka

uka, a laila, e kapa aku nō ko ʻoneʻi mau keiki kupa, he "ua Puni," akā, ʻaʻole naʻe ia he ua mau loa.

Rain is foreign to this land, Kaʻū, and the only place where rain always exists is in the upland of Waiʻōhinu, and the name of this rain is called Hāʻao. And if it rains heavily, spreading haphazardly upon the promontories of ʻōhiʻa trees in the uplands, then those native to this land call it a Puni rain; however, it is not a rain that lasts very long.

From an article sent to a newspaper from Paiahaʻa, Hawaiʻi. Hawaiian source: Pahee, "No ka ua." English trans. by author. Additional source: Pahee, "Of the rain."

8. Hanini ke poʻi wai o Kūlanihākoʻi i ka poli o Waiʻōhinu ē
Ke hoʻopē hala ʻole ʻia ala e nā kulu waimaka o ka ua Hāʻao

The heavenly pool of Kūlanihākoʻi overflows onto the bosom of Waiʻōhinu
Ever soaked by the tears of the Hāʻao rain

From a poem of grief for Joseph Nāwahīokalaniʻōpuʻu. Hawaiian source: Sheldon, *Ka Puke* xx. English trans.: Sheldon, *Biography* 013.

9. Ka ua Hāʻao o Waiʻōhinu. *The Hāʻao rain of Waiʻōhinu.*

A traditional saying. Source: Pukui, *ʻŌlelo* 167. *Note: Pukui says that this phrase is a poetic reference to Waiʻōhinu. "The Hāʻao rain comes down from the mountain in columns to Waiʻōhinu. It is mentioned in songs and chants of Kaʻū."*

10. Kuʻu kaikuahine mai ka ua Hāʻao o Waiʻōhinu
Ke lele aʻela ma uka o ʻAuʻaulele

My dear sister from the Hāʻao rain of Waiʻōhinu
Flying above ʻAuʻaulele

From a kanikau, or lament, for Mrs. Maleka Ii. Hawaiian source: Kaapa. English trans. by author.

11. Hāʻao—spring, land section, and church in Waiʻōhinu . . . named for a Kaʻū rain and a supernatural girl.

Source: Pukui, Elbert, and Mookini 34.

Rain of Kaʻanaokāhinahina, Oʻahu

12. E nānā iho ana i Waipuʻilani
E noho iho ana i Kaʻanaokāhinahina
Eia au i ka ua a ka Hāʻao
I walea ai i ke kui pua ʻāhihi
He lei no Lea, wahine i ke kuahiwi

Gazing down on Waipuʻilani
Residing there at Kaʻanaokāhinahina
Here am I in the Hāʻao rain

Delightedly stringing lehua ʻāhihi blossoms
As a lei for Lea, woman of the mountain

From a mele by Kapauakanoa. Hawaiian source: Hoʻoulumāhiehie, *Ka Moʻolelo* 146. English trans.: Hoʻoulumāhiehie, *Epic* 138. *Note: "Lea" is both the name of a goddess of canoe makers and the name of a fish god who stands on Wailea ("Water of Lea") Point between Kaʻōhao in Kailua and Waimānalo, Oʻahu (Kailua Historical Society 259).*

Rain of Kahuailanawai, Oʻahu

13. Pau ka hihia o ke oho iā Kahuailana
Ke kahikū ʻia ala ke oho ʻawapuhi
Ua hoʻopewapewa ʻia e ka Hāʻao

The tangles of the locks are smoothed out by Kahuailana
The ginger-like strands have been separated out
Brought together in clusters by the Hāʻao rain

From a mele hōʻihi, or honorific chant, praising the hair of ʻEmalani Kaleleonālani. Source: Nogelmeier 53.

Rain of Leleaʻanae (or Leleʻanae), Oʻahu

14. Aia lā i Kahuailanawai
Ke hoʻi aʻela ma uka o Leleaʻanae
E kipa o pulu ʻoe i ka ua Hāʻao ē
Auē kuʻu kāne ē

There at Kahuailanawai
Going back upland to Leleaʻanae
Come and visit, lest you be soaked by the Hāʻao rains, oh
Alas! My beloved husband, oh

From an uē, or wailing call of love and grief, for Kamehameha IV by his wife, ʻEmalani Kaleleonālani. Source: Nogelmeier 330.

15. E ʻāhaʻi ana ma ke kua, o pulu i ka ua Hāʻao ē
Ke lele aʻela ma luna o Leleaʻanae
Lele ka ua, lele pū nō me ka makani

Carrying off on the back to avoid getting drenched by the Hāʻao rain
Flying above Leleaʻanae
The rain flies, flies with the wind

From a kanikau, or lament, for Kamehameha IV by his wife, ʻEmalani Kaleleonālani. Hawaiian source: Nogelmeier 336. English trans. by author.

Rain of Malailua, Oʻahu

16. Welu ka ʻāhihi o Malailua
I ka heheia e ka ua Hāʻao
I kīpalaheʻa i ka ua Kāmakahala

The ʻāhihi bushes of Malailua are tattered
Tangled up by the Hāʻao rain
Stained by the Kāmakahala rain

From a mele for Kauilanuimakehaikalani Kamaialii. Hawaiian source: Kamakau, "He mau mele koihonua." English trans. by author.

Rain of Nuʻuanu, Oʻahu

17. He aha hoʻi kāu, e ka Hāʻao
E ka ua lei i ka mākāhala
Ua makua i luna o ke kī
Ulu ka lau, kaha ka pua
Māewa i ka ua o Nuʻuanu

What are you up to, O Hāʻao
Rain that wears the mākāhala flower
Like a parent above the kī plants
The leaves grow, the flowers develop
Swaying in the rain of Nuʻuanu

From a mele mākaʻikaʻi, or travel chant, for ʻEmalani Kaleleonālani. Hawaiian source: Nogelmeier 67. English trans. by author. Additional source: Fornander, *Fornander* 6: 543.

18. Kuleana i Nuʻuanu ka ua Hāʻao
He ālai ka Malailua no Lanihuli

The Hāʻao rain belongs in Nuʻuanu
The Malailua wind is a screen for Lanihuli

From a mele kālaiʻāina, or political chant, for ʻEmalani Kaleleonālani credited to Mrs. Nāmāhana. Hawaiian source: Nogelmeier 207. English trans. by author.

19. Hana mākāhā Nuʻuanu na ka makani
He pāuli hiwa na ka ua Hāʻao
He loko pā pōhaku no ke Kiʻoao
Na kuʻu ʻanae no Leleʻanae

Nuʻuanu, a sluice gate for the wind
A gloomy blackness from the Hāʻao rain
A stone-enclosed pond for the Kiʻoao
By my beloved ʻanae fish of Leleʻanae

From the song "He mele aloha." Hawaiian source: Fornander, *Fornander* 6: 544. English trans. by author. *Note: Fornander says that "Hāʻao" is the "name of a kind of rain at Koʻolau accompanied by black darkness," and that "Kiʻoao" is the "name of the rain at Nuʻuanu; poetically speaking, Nuʻuanu in the great rain is swimming with water."*

20. Hoʻoipo i Nuʻuanu ka ua Hāʻao, ʻeā
He ua haʻa anu no Luakaha, ʻeā

Romancing at Nuʻuanu is the Hāʻao rain
A rain of Luakaha that dances coldly

From a mele mākaʻikaʻi, or travel chant, for ʻEmalani Kaleleonālani by Kamaʻikuʻi. Source: Nogelmeier 152.

21. *For another example of Hāʻao of Nuʻuanu, see Waʻahila (#4).*

Rain of Waipuhia, Oʻahu

22. Ua komo ka leʻaleʻa iā Waipuhia
Ke kani nei ka ʻaka a ka ua Hāʻao

Waipuhia has become joyous
The Hāʻao rain roars with laughter

From a mele for Kanaʻina. Hawaiian source: Fornander, *Fornander* 6: 540. English trans. by author.

Rain of Kauaʻi

23. Aloha ke kapa huki palai
I ka lupea e ka ua Hāʻao

Greetings to the shores that conceal and confuse
Made attractive by the Hāʻao rain

From the song "Kawaikini." Hawaiian source: Holstein 121. English trans. by author.

Rain of Hīhīmanu, Kauaʻi

24. Hui nā ʻuhane i ka ua o Koʻolau
Halakau ana i Hīhīmanu
ʻIkea maila e ka ua Hāʻao

The spirits meet in the rain of Koʻolau
Settling at Hīhīmanu
Seen by the Hāʻao rain

From a kanikau, or lament, for Kaʻahumanu. Hawaiian source: Niauehoe. English trans. by author. Additional source: Fornander, *Fornander* 6: 454.

Other examples

25. He mai!
He mai, e kuʻu pua lehua o ka wao
I pōhai ʻia e nā manu o uka
Kuʻu lehua i mohala i ka ua o Hāʻao
Ua ao ka hale nei, ua hiki maila ʻoe
Mai! Mai! Eia nō mākou nei

Come!
You are welcome, O lehua blossom of mine from the upland forest

A blossom around whom the birds gather
My lehua that bloomed in the Hā'ao rain
Light comes to our house, for you are here
Come! Come! We are here

From a mele kāhea, or chant of welcome. Source: Handy and Pukui 172.

26. Unuhi 'o Kalani, 'onou a ka'awale
Waiho iho ka hua no ka Hā'ao

The Heavenly One drew it out, pushed it away to separate from it
He left his seeds to the Hā'ao rains

From a mele kū'auhau, or genealogical chant, for 'Emalani Kaleleonālani. Source: Nogelmeier 286–87.

27. 'Auhea wale ana 'oe lā
E ka nani o nā pua lā?
I partner 'oe no ku'u kino lā
I hoa no ka ua Hā'ao

Where are you
O beauty of the flowers?
Be a companion for my person
And a companion for the Hā'ao rain

From the song "Ko hanu ka'u e li'a nei." Source: Lili'uokalani 213. Additional source: Holstein 74.

Hālaulani. Rain associated with Hilo and Pana'ewa, Hawai'i. "Hālau lani" means "heavenly structure."

Rain of Haili, Hawai'i

1. Anoano aloha ia'u ka ua Hālaulani
'O ia ua ha'aheo i ka mu'o o ka hala
Hala pua ho'okahi a ka manu i Haili
I kui a ho'olawa i nā lehua o Mōkaulele
I lei kāhiko 'ula i luna o ka lā'au

The Hālaulani rain is adored and held sacred by me
It is the rain that falls proudly on the budding hala
The very same hala fruit that the birds of Haili
String with the lehua of Mōkaulele until satisfied
A red lei to adorn the tops of the trees

From a mele aloha, or song of love. Hawaiian source: Wahinealoha. English trans. by author.

Rain of Hilo, Hawaiʻi

2. I ke ahiahi, ʻo ia ka lā 2 o Iulai, o ia makahiki nō, kau lākou i luna o Keōua moku, a holo akula, pau ka ʻike ʻana iā Hilo ʻāina ua Hālaulani; ʻekolu pō, ʻelua ao i ka ʻōpū o ka moana.

   In the evening of July 2 of that year, they boarded the ship Keōua *and sailed off. Hilo, land of the Hālaulani rain, could no longer be seen. They were in the belly of the ocean for three nights and two days.*

   From an obituary for E.J.W. Keawehunahala. Hawaiian source: Kaaipaka. English trans. by author.

3. Aloha ka wailele ʻo Waiānuenue, wai ʻau a ka ʻuhane, wai heʻe o ka puʻe no ke kino ē
   I honi au i ka hala moani ke ʻala, ʻaweʻawe me he ua Hālaulani lā ē

   Beloved is the waterfall Waiānuenue, bathing water of the spirit, surfing water for the body to ride to the mouth of the stream
   I smell the wind-blown perfume of the hala, spreading like fingers of the Hālaulani rain

   From a mele ʻuhane, or lament, for Kealohapauole, presumably by Kaikamahineole. Hawaiian source: "He uhane." English trans. by author.

Rain of Panaʻewa, Hawaiʻi

4. Ua Hālaulani
   No kona hāʻule mai mai luna mai ma ke ʻano kaupoku hale, a ʻoiai, e loku ʻia iho ana nā kama hele e nēia mau paka ua, i ia wā na kamaʻāina e pane aku ai: "He ua Hālaulani kēia, he ua kaupoku hale lehua no Panaʻewa." ʻO nā pua lehua nō naʻe ke ʻaho a nēia ua, i laila nō ia e noke ai a lawa kona makemake; he kākaʻikahi loa kona paneʻe mai ma waho nei o nā ululāʻau. E hoʻomaka kona iho ʻana iho mai nā ao mai, ma luna pono aʻe o ka ʻaoʻao ʻākau hikina o ka moku lehua nui o Panaʻewa, a e pōʻai hele iho nō ʻo ia ma loko o Panaʻewa a pau ke aho, hoʻi ana i luna.

   Hālaulani Rain
   Because it originates above and is like the roof of a house, and since travelers are drenched by these raindrops, the locals say, "This is a Hālaulani rain, a rain like the roof of a lehua house of Panaʻewa." The lehua blossoms are like the purlins of this rain; that is where the rain persists until its desires are satisfied. Rarely does it move beyond the forests. It starts to descend from the clouds right above the northeastern side of the large lehua forest of Panaʻewa, and it goes around in circles within Panaʻewa until it has fully discharged, then it returns above.

   From a description of rains by Lilia H. Richards and company, who traveled to the uplands of Puna in 1876. Hawaiian source: Richards et al. English trans. by author.

Hāleuʻole. Rain associated with Mokuola, Hawaiʻi, and with East Maui. "Hāleu ʻole" means "without toilet paper" and refers to not having enough time to wipe because the rain is so sudden.

Rain of Mokuola, Hawaiʻi

1. Ka ua Hāleuʻole is a swift rain that comes over the ocean toward Mokuola.

   From a list of rain names and their descriptions. Source: Kelsey, "Rains."

Rain of East Maui

2. Ua Hāleuʻole—He ua hana kolohe, ʻo ia hoʻi i kou wā e makemake ai e hoʻopau pilikia, ʻo ia kona wā e ʻiliki iho ai, a mao koke aʻe nō, a ua hou mai nō, a pēlā i kapa ʻia ai kēia inoa. ʻO ka manaʻo o ka mea kākau, he inoa kēia no kēia au hou aʻe nei nō, a no ka loaʻa ʻole o ka inoa kūpono, no laila kapa ʻia kēia inoa ʻeʻepa.

   Hāleuʻole rain—This is a naughty rain because when you want to use the bathroom, that is when it pours down suddenly and clears up quickly, and then falls again, thus giving it this name. The writer thinks that it is a modern name, and because the correct name is not available, it was therefore given this peculiar name.

   From an account of East Maui rains and their descriptions, as told to Thomas K. Maunupau by Josua Ahulii of Kaupō, Maui. Hawaiian source: Maunupau, *Huakai* 27. English trans. by author.

Sudden rain

3. Ua Hāleuʻole. *A rain without toilet paper (so sudden that there is not time to use toilet paper).*

   Source: Pukui and Elbert, 1986, 54.

4. *For another example of Hāleuʻole as a sudden rain, see Peʻepāpōhaku (#1).*

Hāliʻipili. Rain associated with Lele (or Lāhainā), Maui. Also found on Hawaiʻi and Oʻahu. "Hāliʻi pili" means "to spread over pili grass." It is both the name of a specific rain and a generally descriptive term; its various usages are determined by the context.

Rain of Umauma, Hawaiʻi

1. Auē kuʻu keiki
   ʻAʻohe keiki—He hoa pili nō ā
   Kuʻu keiki mai ka ua Hāliʻipili o Umauma
   Mai ka hala ʻāwili pū nō me ka hīnano
   Aloha ia hala a kāua e lei ai

   Pity for my dear child
   He wasn't just a child—But a very close companion

My beloved child from the Hāli'ipili rain of Umauma
From the hala fruit entwined together with the hīnano flower
Love for this hala that we wear as a lei

From a mele 'uhane, or lament, for Keliikuewa. Hawaiian source: Hoakimoa. English trans. by author.

Rain of Lele (or Lāhainā), Maui

2. To spread over a region, as a shower, like the spreading of a mat. Hāli'ipili i ke kula o Lele (*The shower extends over the plain of Lāhainā*).

   Source: Andrews 145.

3. Ku'u kaikamahine mai nā kai 'ewalu
   Mai ka ua he'e koko 'ula i ka moana
   Mai ka ua Hāli'ipili o Lele

   My beloved daughter from the eight seas
   From the fleeting rainbow rain over the ocean
   From the Hāli'ipili rain of Lele

   From a mele 'uhane, or lament, for Moenahele. Hawaiian source: Kaleo. English trans. by author.

Rain of 'Ualaka'a, O'ahu

4. *See Kuahine (#17).*

Shower

5. Hāli'i pili. *To spread over pili grass, said poetically of a rain shower.*

   Source: Pukui and Elbert, 1986, 54. Additional source: Andrews 145.

Hāmākualoa. Rain associated with East Maui. "Hāmākua loa" means "long Hāmākua."

1. *See Kiu (#1).*

hau. Related to *Kēhau.* Dew, frost, snow, ice; a cool breeze; to blow, of a cool breeze.

Rain of Kēkē, O'ahu

1. 'O ka ua kili noe 'ana i ka hau anu o Kēkē
   I uka ho'i ka hala me ka lehua
   I kupu i uka o Kaliuwa'a
   I 'ane'i kou inoa, e ō mai

   The misting in the chilly hau frost of Kēkē
   Upland are the hala and lehua trees
   Sprouting above Kaliuwa'a
   Here is your name chant, respond

From a mele to Kamapua'a by his older brother, Kaikihonuakele. Hawaiian source: Kahiolo, "He moolelo." English trans. by author. Additional sources: Kahiolo, *He Moolelo* 176; Pukui and Elbert, 1986, 151; Elbert 248–49; Fornander, *Fornander* 6: 516; Kent 378.

Cold dew

2. *See 'Awa (#1), Lehua (#1), and Ma'akua (#3).*

Mountain snow

3. *See Po'onui (#1).*

Hauapēpē. Rain associated with Kalihi, O'ahu. Possibly related to the Haupe'e-pe'e wind, also associated with Kalihi.

1. Aloha 'ino nō ku'u makuahine
Mai ka ua Hauapēpē o Kalihi
Mai ka lā wela o Kaiwi'ula

What great pity for my dear mother
From the Hauapēpē rain of Kalihi
From the hot sun of Kaiwi'ula

From a kanikau, or lament, for Luakauwahine. Hawaiian source: Pawai. English trans. by author.

he'e koko 'ula. Or he'e koko. Rain over the ocean; also descriptive of rains over the plains. "He'e koko 'ula" may refer to a fleeting rainbow.

Rain over the ocean

1. *See Hāli'ipili (#3).*

He'enehu. Same as *Hukihe'enehu*. Rain associated with Hilo, Hawai'i. "He'e nehu" refers to the time when nehu fish run.

1. Name of a misty rain observed off the coast of Hilo when nehu fish are running.

Source: Pukui and Elbert, 1986, 63.

2. Ka ua He'enehu o Hilo. *The He'enehu rain of Hilo.*

A traditional saying. Hawaiian source: Pukui, *'Ōlelo* 167. English trans. by author. *Note: Pukui says, "The people knew the season when the schools of nehu fish followed the rain."*

3. Ka ua He'enehu—a Hilo rain. It comes early in the morning when the nehu nets are drawn.

From a list of rain names and their descriptions. Source: Kelsey, "Rains."

Hehilauʻulu. Same as *Hehiʻulu.* Rain associated with Piʻihonua, Hawaiʻi. "Hehi lau ʻulu" means "to tread upon ʻulu leaves."

1. Auē kuʻu hoa pili ʻo ke kāne ē!
   Kuʻu hoa o ka hale wai anu o Hilo
   No Hilo hoʻi au, no ka ipu a Kulukuluʻā
   No ke one holu i Waiolama
   No ka ua Hehilauʻulu o Piʻihonua

   Oh, my intimate companion, my husband!
   My companion of the cold, watery house of Hilo
   I am from Hilo, from the gourd of Chief Kulukuluʻā
   From the shifting sands at Waiolama
   From the Hehilauʻulu rain of Piʻihonua

   From a mele by Kamalālāwalu as her former partner, Halemano, departed with a new lover, Kīkēkaʻalā. Hawaiian source: Fornander, *Fornander* 5: 257. English trans. by author.

Hehipuahala. Rain associated with Poʻokū, Kauaʻi. "Hehi pua hala" means "to tread upon hala fruit."

1. Ua Hehipuahala o Poʻokū—rain that treads on the pandanus flowers of Poʻokū, Hanalei, Kauaʻi.

   Source: Kent 378.

Hehipuapili. Rain associated with Waʻahila, Oʻahu. "Hehi pua pili" means "to tread upon pili flowers."

1. He ipo i ka hale hālau loa
   Mai ka ua Hehipuapili Waʻahila

   A sweetheart in the long house
   From the Hehipuapili rain of Waʻahila

   From a mele aloha, or song of love. Hawaiian source: Fornander, *Fornander* 6: 541. English trans. by author.

Hehiʻulu. Same as *Hehilauʻulu.* Rain associated with Piʻihonua, Hawaiʻi. "Hehi ʻulu" means "to tread upon ʻulu."

1. Ka ua Hehiʻulu o Piʻihonua. *The [Hehiʻulu] rain that treads on the breadfruit leaves of Piʻihonua.*

   A traditional saying. Source: Pukui, *ʻŌlelo* 167.

Hōʻehaʻili. Rain associated with Waiehu, Maui. Also found on Kauaʻi. "Hōʻeha ʻili" means "to hurt the skin." It is both the name of a specific rain and a generally descriptive term; its various usages are determined by the context.

Rain of Waiehu, Maui

1. Ka ua Hōʻehaʻili o Waiehu. *The skin-hurting [Hōʻehaʻili] rain of Waiehu.*

   A traditional saying. Source: Pukui, *ʻŌlelo* 167. *Note: Pukui describes the Hōʻehaʻili as "a chilly, pelting rain."*

2. He aloha, he lihaliha, he kūmākena
   He ʻū iā ʻoe
   E Hon. Iosepa Kahoʻoluhi Nāwahīokalaniʻōpuʻu
   A haʻo ē!
   I uē ʻia mai nei ʻoe e Nā Wai ʻEhā
   E ka makani Kiliʻoʻopu o Waiheʻe
   Ka ua Hōʻehaʻili o Waiehu

   Loving, heartsick, grief-stricken
   Mourning for you
   O Hon. Joseph Kahoʻoluhi Nāwahīokalaniʻōpuʻu
   We shall truly miss you!
   You have been mourned by the lands of the four waters
   By the Kiliʻoʻopu wind of Waiheʻe
   And the Hōʻehaʻili rain of Waiehu

   From a message of condolence for the passing of Joseph Nāwahīokalaniʻōpuʻu from people of Nā Wai ʻEhā, Maui. Hawaiian source: Sheldon, *Ka Puke* 208. English trans.: Sheldon, *Biography* 288–89.

Rain of Hanalei, Kauaʻi

3. Hanohano wale ʻo Hanalei
   I ka ua nui hōʻeha ʻili
   I ka wai ʻuʻinakolo
   I ka poli o Nāmolokama

   Hanalei is simply magnificent
   In the heavy rain that hurts the skin
   And the roaring water
   In the bosom of Nāmolokama

   From the song "Maikaʻi Kauaʻi." Hawaiian source: Wilcox et al. 160. English trans. by author.

Hōliʻo. Rain associated with Hawaiʻi, Oʻahu, and Kauaʻi. Also the name of a wind.

Rain of Hawaiʻi and Oʻahu

1. Rain name associated with Hawaiʻi and Oʻahu.

   Source: Pukui and Elbert, 1986, 77.

Rain of Oʻahu

2. Ua hānupa wale i ke aloha i ka ʻehu kai
I ka uaina maila, kaʻawale ka Malanai
Kaʻawale ka pili aloha me ka makua
He makua, he keiki na ka ua Hōliʻo
I hoʻopumehana i ka poli o ka mauʻu ē
Aloha kāua ē

Swelling with love for the sea spray
Pushed aside, the Malanai wind is separated
Separated from the tender closeness with the parent
A parent, a child of the Hōliʻo rain
Warming in the bosom of the grass
Love to us both

From a greeting chant to Kamaakamahiʻai by his daughter Kahelekūlani. From the legend of Kamaakamahiʻai. Hawaiian source: Kaualilinoe, "Ka moolelo" 10/8/1870. English trans. by author. Additional source: Kaualilinoe, "Legend" 69.

Rain of ʻĀhulimanu, Oʻahu

3. Ke poʻo o ka hala o ʻĀhulimanu
E nā ana i ka ʻūhā puluhi
He kāmaʻa puki na ka ua Hōliʻo

The uppermost hala of ʻĀhulimanu
Respite for thighs that are wearied and drenched
A protective boot to wear in the Hōliʻo rain

From a mele for the goddess Lea by Kapauakanoa. Hawaiian source: Hoʻoulumāhiehie, *Ka Moʻolelo* 146. English trans.: Hoʻoulumāhiehie, *Epic* 138.

Rain of Kānehoalani, Oʻahu

4. He kanikau aloha kēia nou, e B. Mahune
. . . .
Noho nani ka ʻuhane i luna o Kānehoalani
He nani ʻoe, he milimili na ka ua Hōliʻo

This is an affectionate chant of mourning for you, B. Mahune
. . . .
The spirit sits pretty atop Kānehoalani
You are beautiful, a darling of the Hōliʻo rain

Hawaiian source: Wailea. English trans. by author.

Rain of Kauoha, Oʻahu

5. Aia ma Wailupe, he wahi e kapa ʻia nei ka inoa ʻo Kauoha . . . ua kūkulu ʻia kekahi mau hale nui no kekahi mau aliʻi no Kapueo a me Kepoʻonui, ua

makemake loa kēia mau aliʻi i ia wahi, no ka ʻoluʻolu maikaʻi o ka makani he Mālualua . . . kūkulu kalaʻihi a ka ua Hōliʻo i ua mau wahi ʻelemākule nei.

At Wailupe is a place called Kauoha . . . several large houses were built for chiefs, for Kapueo and Kepoʻonui. These chiefs really liked this place because of the perfect coolness of the wind, a Mālualua . . . these old men were oppressed by the Hōliʻo rain.

From an article about the places on Oʻahu where chiefs liked to stay in times of old. Hawaiian source: Kalanikuihonoinamoku. English trans. by author.

Rain of Mololani, Kāneʻohe, Oʻahu

6. *See ʻĀpuakea (#16).*

Rain of Nuʻuanu, Oʻahu

7. E hoʻi ka nani i Nuʻuanu
Ua anu i ka ua Hōliʻo
Ke hahi maila i ka moana

May the beauty return to Nuʻuanu
Chilled by the Hōliʻo rain
Treading upon the ocean

From a mele inoa, or name chant, for ʻEmalani Kaleleonālani. Source: Nogelmeier 93.

Rain of Waolani, Oʻahu

8. E hoʻi ka nani i Nuʻuanu
Ua anu i ka ua Hōliʻo
Ke nihi aʻela i Waolani

May the beauty return to Nuʻuanu
Chilled by the Hōliʻo rain
Moving quietly along to Waolani

From a mele mākaʻikaʻi, or travel chant, for ʻEmalani Kaleleonālani. Source: Nogelmeier 96.

Rain of Nuʻalolo, Kauaʻi

9. Nū ka leo o ke kai i ka haka lewa o Nuʻalolo
Kū ka ʻehu o ka huna o ke kai i nā pali
Hū ka ʻōmaka wai a ka ua i ka makani
Makani halihali i ka ua Hōliʻo

The voice of the sea roars upon the floating platform of Nuʻalolo
The mist of the sea ascends the cliffs
The source of the rain gushes in the wind
Wind that carries the Hōliʻo rain

From a mele mākaʻikaʻi, or travel chant, for ʻEmalani Kaleleonālani and her travels on Kauaʻi. Hawaiian source: Nogelmeier 72. English trans. by author.

Rain that causes trees to huddle

10. *See Makaʻūpili (#3).*

Hoʻokamumuhala. Roaring rain associated with Hanalei, Kauaʻi. "Hoʻokamumu hala" refers to the crunching sound of hala.

1. E nānā mai i ou mau pōkiʻi
I nā hoa ukali o ke ala
O ke ala nui, ala iki
O ka ua hāʻawe kua
Me he keiki lā
O ka ua Hoʻokamumuhala
Hoʻokamumuhala o Hanalei ē

Look upon your little sisters
Friends who have followed you on your journey
Over the wide way, the narrow way
In the rain, heavy on the back
Like a child
In the [Hoʻokamumuhala] rain that roars in the hala trees
Roaring in the hala trees of Hanalei ē

From a mele in the legend of Lāʻieikawai, recited by Mailekaluhea to her brother ʻAiwohikupua as he was leaving on a canoe after forsaking her and her sisters. Hawaiian source: Haleole, *Ke Kaao* 70. English trans.: Haleole, *Lāʻieikawai* 29.

Hoʻolua. Heavy rain associated with a strong north wind of the same name; associated with Hilo, Hawaiʻi. Also found on Maui, Kauaʻi, and on other islands. "Hoʻolua" means "to talk loudly and pointlessly."

Rain of Hilo, Hawaiʻi

1. Kau i ka lani ka liko ua o Hilo
I hana ʻia a maikaʻi ka wai
Nani Hilo i ka Hoʻolua i ka Mālua
Huli mai ka ʻamaʻu, ke poʻo o ka lehua
Luli mai i ka hana ʻia e ka wai nui

Placed in the heavens are the rain buds of Hilo
Fashioned to perfection is the water
Beautiful is Hilo in the Hoʻolua and the Mālua
The ʻamaʻu fern and the crown of the lehua blossom bend
Swaying in reaction to the great water

From a mele shared by Z. P. Kalokuokamaile. Hawaiian source: Kaheiheimalie. English trans. by author. Additional source: *Na Mele Aimoku* 150. *Note: The Hoʻolua rain is sometimes associated with the Mālua rain, especially in Hilo.*

2. Kau i ka lani ka holoaʻa ua o Hilo
I hoʻowaʻa, ʻeli, a moku kahawai
Komo Hilo i ka Hoʻolua i ka Mālua ē
Huli ke ʻamaʻu, ke poʻo o ka lehua
I huli nō i h[e]hia e ka wai nui

Hung in the heavens is the rain trough of Hilo
That gouges, digs, and cuts the ravine
Hilo enters into the Hoʻolua and the Mālua
The ʻamaʻu fern and the crown of the lehua bend
Bending, crushed by the great water

From a message of condolence for the passing of Joseph Nāwahīokalaniʻōpuʻu from the Women's Patriotic League of Hilo. Hawaiian source: Sheldon, *Ka Puke* 268–69. English trans. by author. Additional source: Sheldon, *Biography* 355.

3. Nae iki Hilo i ka pehia e ka Hoʻolua
I ka pehia mau ʻia e ka ua Polopuahīnano
I kuʻi a lima ʻia e ka ua Kanilehua
Kuʻi lua, kuʻi kolu ka leo o ke kai o Hāʻena

Hilo is a bit out of breath, being pelted by the Hoʻolua
Continuously pelted by the Polopuahīnano rain
Punched by the Kanilehua rain
Two times, three times stronger is the voice of the sea of Hāʻena

From the legend of Nāmakaokapāoʻo. Hawaiian source: Ilihia 10/13/1894. English trans. by author.

Rain of Kaʻuiki, Maui
4. *See Lauʻawa (#2).*

Rain of Nāpili, Maui
5. He ʻōpalapala ʻulu na ka ua Hoʻolua
Ke hoʻopiha maila i nā kahawai
E mau auaneʻi i kai o Nāpili ē—i laila

ʻUlu tree rubbish made by the Hoʻolua rain
Fills the streams
That will be clogged below Nāpili—there

From a mele by C. Kanaʻina for Lunalilo. Hawaiian source: *Na Mele Aimoku* 157. English trans. by author. Additional source: Kanaʻina.

Rain of Malelewaʻa, Kauaʻi

6. Ke oho o ke kī lipo o Malelewaʻa
Ke oho o ka lāʻau limua i ka ua
Ua hoʻopewa ʻia e ka Hoʻolua

The leaves of the dark kī plant of Malelewaʻa
The leaves of the plants covered with moss from the rain
Clumped together by the Hoʻolua

From a mele hōʻihi, or honorific chant, praising the hair of ʻEmalani Kaleleonālani. Hawaiian source: Nogelmeier 59. English trans. by author.

Heavy rain

7. He ua Hoʻolua, he ua nui nō ia. *A Hoʻolua rain, it is a very heavy rain.*

From a chapter on the Hawaiian names of spatial areas above and below. Hawaiian source: Malo, *Ka Moolelo* 20. English trans. by author. Additional sources: Malo, *Hawaiian* 14; Andrews 194.

8. Ua Hoʻolua—a rainstorm from the north; generally with wind.

From a list of rain names and their descriptions. Source: "Names . . . rain."

Rain brought by the Hoʻolua wind

9. Hoʻolua ka makani mai ka ʻākau mai, holo aku ma ka hema o nā mokupuni, he ua kahi Hoʻolua, a ua ʻole kahi Hoʻolua.

Hoʻolua, the wind from the north that blows along the southern parts of the islands. Some Hoʻolua winds bring rain, some do not.

From a chapter on the Hawaiian names of spatial areas above and below. Hawaiian source: Malo, *Ka Moolelo* 20. English trans. by author. Additional sources: Malo, *Hawaiian* 14; Andrews 194.

Hoʻonuʻupua. Rain associated with Hāna, Maui. "Hoʻonuʻu pua" means "to heap up the flowers."

1. Noho iki i Hāna ka ua Hoʻonuʻupua
Hoʻowehiwehi i ka lāʻau ka hunahuna wai
Ka hunahuna ao hekili
He ua kākua pua lau hīnalo
I pē ʻia i ke kapa lehua na ka Moani

Dwelling for a bit in Hāna is the flower-heaping [Hoʻonuʻupua] rain
The fine droplets of water adorn the trees
The minute bits from the thundercloud
A rain that drapes the bracts of the hīnano blossom
Pressed into the blanket of lehua by the [Moani] breeze

From a mele kūʻauhau, or genealogical chant, for ʻEmalani Kaleleonālani. Source: Nogelmeier 258.

hoʻopala ʻōhiʻa. Rain associated with ʻōhiʻa ʻai trees. "Hoʻopala ʻōhiʻa" means "to ripen the ʻōhiʻa."

1. Ka ua hoʻopala ʻōhiʻa. *The rain that ripens mountain apples.*

   A traditional saying. Source: Pukui, *ʻŌlelo* 167. *Note: Pukui says that the hoʻopala ʻōhiʻa is "the rain that comes just as the mountain apple is beginning to ripen."*

Hoʻopuluhīnano. Rain associated with Naue, Kauaʻi. "Hoʻopulu hīnano" means "to moisten the hīnano flower."

1. Ua ʻike aʻela ʻo Kauakahiapaoa i ka nani o Halemaʻumaʻu, ua inu i ka wai ʻono hoʻomalule o ka pūnāwai Koʻolihilihi o Puna, ua wela ke kīkala o ua keiki nei o ka ua Hoʻopuluhīnano o Naue . . .

   Kauakahiapaoa saw the beauty of Halemaʻumaʻu, drank the delicious waters of Koʻolihilihi Spring of Puna that leave one weak; the loins of this son of the Hoʻopuluhīnano rain of Naue were on fire . . .

   From the legend of Hiʻiakaikapoliopele as Pele and Kauakahiapaoa made love. Hawaiian source: Hoʻoulumāhiehie, "Ka moolelo o Hiiaka-i-ka-poli-o-pele." English trans. by author. Additional sources: Hoʻoulumāhiehie, *Ka Moʻolelo* 418; Hoʻoulumāhiehie, *Epic* 390.

Hukiheʻenehu. Same as *Heʻenehu*. Rain associated with Hilo, Hawaiʻi. "Hukiheʻe nehu" refers to a fine-meshed net for catching nehu fish.

Rain of Huia, Hawaiʻi

1. *See Kanilehua (#19).*

Rain of Paikākā, Hawaiʻi

2. Kuʻu makuahine hoʻi

   . . . .

   Mai ka papa kāhulihuli o Wailuku
   Mai ka ua Hukiheʻenehu a ka lawaiʻa
   ʻUhane ʻauʻau kai o Paikākā ā

   My dear mother, indeed!

   . . . .

   From the bridge of Wailuku
   From the Hukiheʻenehu rain of the fisherman
   Spirit who swims the sea of Paikākā, yes

   From a kānaenae, or chant of eulogy, for H. Kamahiai by her son. Hawaiian source: Makuaole. English trans. by author.

Rain of Punahoa, Hawaiʻi

3. E ala, e ala, ua ao, ua mālamalama
Ke hoʻolale mai nei ke keiki hukiheʻe nehu
He ua Hukiheʻenehu no kai o Punahoa

Arise, awaken, it is day, it is bright
The boy who pulls in the nehu fish net is making an appeal
A Hukiheʻenehu rain from lower Punahoa

From a call to the chiefs and people of Hilo by Kūapākaʻa, who was trying to awaken them to set sail. Hawaiian source: Nakuina, *Moolelo* 91. English trans. by author. Additional source: Nakuina, *Wind* 87–88.

I

ʻIwilani. Rain associated with the god Kāne.

1. ʻO ka hiʻa i pākela ka loa
ʻO Kāne i ka ua ʻIwilani

The weapon that excelled in length
Was Kāne of the ʻIwilani rain

From the mele "Kanaʻina." Source: Fornander, *Fornander* 6: 539. *Note: "I ka ua ʻIwilani" may be part of the name of a Kāne god form, as in Kāneikauaʻiwilani, and may not be a specific rain name.*

K

Ka‘au. A fine rain in Kohala, Hawai‘i.

Source: Pukui and Elbert, 1986, 108.

Kā‘eleloli. Also known as Kā‘ekeloi and Kā‘eke‘ekeloi. Also the name of a wind. Kā‘eleloli, Kā‘ekeloi, and Kā‘eke‘ekeloi sound similar to the words "kā‘eleloi" and "kā‘eke‘eke," which refer to the rolling or ruffling sound of a drum or of kā‘eke‘eke bamboo pipes.

1. Name of a rain at Makiki, O‘ahu.

Source: Pukui and Elbert, 1986, 109.

Kāhikohala. Rain associated with Kea‘au, Hawai‘i, and with Kekele, O‘ahu. "Kāhiko hala" means "to adorn the hala trees."

Rain of Kea‘au, Hawai‘i

1. He ua Kāhikohala ko Kea‘au
   Ke kāhiko maila i ka papa o Mali‘o

   Kea‘au has the Kāhikohala rain
   Adorning the plain of Mali‘o

   From a chant for Kawaikini. Hawaiian source: Lohiau, "He mele no Kawaikini." English trans. by author.

Rain of Kekele, Oʻahu

2. Kuʻu wahine mai ka nuku o Nuʻuanu
   Mai ka ua Kāhikohala o Kekele

   My darling wife from the Nuʻuanu Gap
   From the Kāhikohala rain of Kekele

   From a kanikau, or lament, for a woman whose name is rendered "ke Aupuni" in the newspaper. Hawaiian source: Konaaihele. English trans. by author.

Kālepa. Rain associated with Heʻeia, Oʻahu.

1. Ke nihi aʻela ka ua Kālepa
   Ka ua ʻokiʻoki koʻa o Heʻeia
   Ke hehi maila i ka maka o ke koʻa
   A luhi ē—Luhi i ka ua ke koʻa
   O Heʻeia Kea ē

   The Kālepa rain moves quietly
   The coral-breaking rain of Heʻeia
   Trampling on the bud of the coral
   Until weary—Weary from the rain is the coral
   Of Heʻeia Kea

   From a letter describing Queen Kapiʻolani's journey from Honolulu to Koʻolau, Oʻahu. Hawaiian source: "I ka Poalua iho nei." English trans. by author. Additional sources: Kapihenui 2/13/1862; Hoʻoulumāhiehie, *Ka Moʻolelo* 157; Hoʻoulumāhiehie, *Epic* 148.

Kāmakahala. Rain associated with Nuʻuanu, Oʻahu. Also the name of a native forest tree.

Rain of Malailua, Oʻahu

1. *See Hāʻao (#16).*

Rain of Nuʻuanu, Oʻahu

2. Lei Nuʻuanu i ka ua Kāmakahala
   Kāhiko nō i ke alo o Malailua

   Nuʻuanu is crowned with the Kāmakahala rain
   Adorning the front of Malailua

   From a mele lei, or song presenting a lei, originally composed for Lunalilo and inherited by Kalākaua. This portion of the mele was composed by Kapauakanoa. Hawaiian source: *Na Mele Aimoku* 170. English trans. by author. *Note: In this rendering, "Malailua," a place name associated with Nuʻuanu, Oʻahu, is used instead of "malaila," which was printed in the Hawaiian source.*

3. Auē kuʻu hānai
   Kuʻu kaikūnane hoʻi
   Mai ka ua Kāmakahala o Nuʻuanu
   E ʻāhaʻi ana ma ke kua ka malu
   A pulu kākou i ka ua ʻawa
   Ke hehi maila i ka pua o ka ʻāhihi
   Kuʻu hoa o ka wai o Kahuailanawai ē

   Oh, my beloved hānai
   My dear brother indeed
   From the Kāmakahala rain of Nuʻuanu
   Carrying off on the back, where there is shelter
   Till we get soaked by the ʻawa rain
   Treading over the blossoms of the ʻāhihi bush
   My companion of the waters of Kahuailanawai

   From a kanikau, or lament, for Kamehameha IV by one of his sisters. Hawaiian source: "He kanikau no ka Moi Iolani Kamehameha IV." English trans. by author.

Kanikaniʻāʻula. Rain associated with Hilo, Hawaiʻi. "Kanikaniʻāʻula" means "insomnia," "ringing in the ears," or "to mourn in chanting."

1. *See Kanilehua (#18).*

Kanikanilehua. Same as *Kanilehua.* Rain associated with Kumukahi and Mākaulele (or Mōkaulele), Hawaiʻi, and with Lāhainā, Maui. "Kanikani lehua" can refer to the chattering of birds on ʻōhiʻa lehua trees, the rustling of lehua flowers, or the drinking of the rain by lehua flowers.

Rain of Kumukahi and Mākaulele (or Mōkaulele), Hawaiʻi

1. Kuʻu kāne o ka ua Kanikanilehua o Mākaulele
   Ke haki manuā maila i ka lae o Kumukahi

   My dear husband of the Kanikanilehua rain of Mākaulele
   Breaking at the cape of Kumukahi

   From a mele by Kamalālāwalu upon seeing her former partner Halemano being playfully bitten by a new lover, Kīkēkaʻalā. Hawaiian source: Fornander, *Fornander* 5: 257. English trans. by author.

Rain of Kuʻia and Mākila, Maui

2. Ka holo o Kuʻia a me Mākila
   Kuʻu makua o ka ua Kanikanilehua
   Mākila ka makua o ka ʻohu noenoe o Kuʻia

And the trail of Kuʻia and Mākila
My mother of the pattering [Kanikanilehua] rain
Mākila was the mother of the [ʻohu] fog of Kuʻia

From a mele aloha, or song of love, for Lāhaināluna Seminary. Source: Fornander, *Fornander* 6: 529.

Kanikaukū. Rain associated with the sea at Manolau, Kauaʻi. "Kanikau kū" means "abrupt lamentation" or "to lament freely."

1. *See Kukupaʻu (#2).*

Kanikoʻo. Rain associated with Heʻeia, Oʻahu, and with the Koʻolau wind. "Kani koʻo" refers to the tapping sound of a cane.

Rain of Heʻeia, Oʻahu

1. A kuʻu hoa e noho lā i ka laʻi
I walea i ka ua Kanikoʻo o Heʻeia
I ka ua Pōʻailauʻawa o Moelana

And my companion who resides in the calm
Relaxing in the Kanikoʻo rain of Heʻeia
In the Pōʻailauʻawa rain of Moelana

From a mele by Halemano as he played kilu and recalled the time when he and his wife, Kamalālāwalu, lived in the forests of Moelana in Kekele, Oʻahu. Hawaiian source: Fornander, *Fornander* 5: 255. English trans. by author.

2. Ka ua Kanikoʻo o Heʻeia. *The Kanikoʻo rain of Heʻeia.*

A traditional saying. Hawaiian source: Pukui, *ʻŌlelo* 168. English trans. by author. *Note: Pukui describes the Kanikoʻo as "the rain of Heʻeia that sounds like the tapping of walking canes" and says that the rain of Hilo is similarly described.*

Rain of Koʻolau, Oʻahu

3. Ka ua Kanikoʻo, ua o ke Koʻolau
Ua kani koʻo i ke kai a mālie

The Kanikoʻo rain, rain of the Koʻolau
Rain tapping on the sea like a cane until it is calm

From a mele for Kaumualiʻi. Hawaiian source: Fornander, *Fornander* 6: 474. English trans. by author. Additional source: Pukui and Elbert, 1986, 129. *Note: Fornander says that the Kanikoʻo produces "the sound of heavy raindrops on dry leaves, or dry thatching of the pandanus leaf," and that this rain accompanies "the Koʻolau wind, which calms the troubled waters."*

Kanilehua. Same as *Kanikanilehua.* Rain associated with Hilo, Hawai'i. Also found on other parts of Hawai'i and on O'ahu and Kaua'i. "Kani lehua" can refer to the chattering of birds on 'ōhi'a lehua trees, the rustling of lehua flowers, or the drinking of the rain by lehua flowers.

Rain of Hilo, Hawai'i

1. Ua Kanilehua

   I ka wā e la'i mālie ai nā ululā'au, a e la'i mālie ana ka wai o ka lehua i loko o kēlā a me kēia 'ōmaka lehua, a laila, ua piha hau'oli a'e nā manu, a ua inu a kena, a lawa ka 'i'ini (pēlā ka ho'omakākiu mau a nā po'e kia manu), akā, i ka wā e loku mai ai nēia mau paka ua, a e luaiele 'ia ana ho'i e nā ahe makani Pu'ulena a Malanai paha, a laila, ua pīhoihoi nā manu, a ho'omaka lākou e uē, me he mea lā i loko o ka pilikia nui; i ia wā e lohe aku ai 'oe, e kakani ana nā manu li'ili'i ma 'ō a ma 'ane'i o ka ululā'au, e uē ana no ka nele i ka wai o ka lehua. A 'o ia kona kapa mua 'ia 'ana, "ua Kanilehua."

   Ua ho'omaka pinepine nō kēia ua ma waena o ka hora 8 a me 11 o ke kakahiaka, a ma waena o ka 1 a me 3 o ka 'auinalā. He pānanaiki wale nō kona kūlana ke hele mai, a he lokuloku a kilihune aku ma kona pau 'ana. E ho'omaka kona 'ike 'ia e hele mai ana, ma ka wēlau 'ākau o ka moku lehua o Pana'ewa, a e hele mai i ke komohana ma ka pili lā'au mai, a ma luna mai o nā ulu hala, a i kekahi wā, e komo a'e nō ka mokupuni 'o Mokuola i loko o ka wēlau 'ākau o kēia ua, a laila, ne'e hele a'e 'o ia i ke komohana, a kilihune mai nō i luna o ka hapa hikina o ke town o Hilo, a huli ho'i akula ma luna o nā lehua o Mōkaulele, kahi nona nā lei lehua e lei mau 'ia nei e ko Hilo kini. A ma laila aku, pi'i a nalowale i loko o nā 'ohu wai lehua o Haili. He wahi ululā'au ia, 'elua mile me 'ekolu hapahā paha ma ka hema o ke town o Hilo. 'A'ole e 'oi aku ka manawa e loku ai kēia ua ma mua o ho'okahi hora mai kona 'ikea 'ia 'ana, akā, e pau nō nā moku lehua i ka ho'opulu 'ia e ia i loko o ia manawa.

   Ua komo pinepine kēia inoa i loko o ka hapa nui o nā mele o kēia 'āina, a ua lilo nō ho'i i inoa ho'omāhie no ke town o Hilo. Aia a 'ōlelo 'ia mai, "Kāua i ka ua Kanilehua," a laila, 'o ka mana'o nō ma ia pane 'ana mai, "Kāua i Hilo," a 'o ia aku nō ho'i, "Kāua i Hanakahi."

   Kanilehua Rain

   When the forest is quiet and still, and the nectar within each and every lehua blossom is also quiet and still, then the birds are full of happiness for they have drunk their fill and satisfied their desires (that's what is always observed by the bird catchers). However, when these raindrops pour in a torrent, and the gently wafting Pu'ulena or Malanai causes the trees to sway, then the birds get alarmed and begin to cry as if they were in the midst of some calamity. That is when you will hear the little birds make a lot of noise throughout the forest, crying over the lack of lehua nectar. And this is why it was first called "Kanilehua rain."

This rain frequently begins between the hours of 8 and 11 in the morning, and between 1 and 3 in the afternoon. It is very compact in the way it moves along, pouring down heavily and finally showering lightly. It becomes visible as it starts to move over the northern extremity of the lehua-laden forest of Panaʻewa. Then it travels westward to the edge of the forest and above the hala groves. Sometimes the island Mokuola will be part of the northernmost extremity of this rain's reach, and then the rain moves to the west and sprinkles over the eastern half of Hilo town. It turns and alights on the lehua of Mōkaulele, the source of the lei lehua always worn by the multitudes of Hilo. And from there it climbs and disappears within the lehua nectar mists of Haili. It is a small forest perhaps two and three-quarters of a mile south of Hilo town. The time it takes for this rain to spill out does not exceed one hour from when it is sighted; however, the lehua forests are completely saturated within that period of time.

This name is used in most of the songs of this land and it has even become a charming name for Hilo town. When it is said, "We are in the Kanilehua rain," the meaning of that saying is, "We are in Hilo" and "We are in Hanakahi."

From a description of rains by Lilia H. Richards and company, who traveled to the uplands of Puna in 1876. Hawaiian source: Richards et al. English trans. by author.

2. Ka ua Kanilehua o Hilo
Kui lei lehua e Panaʻewa

The Kanilehua rain of Hilo
Strung into a lei of lehua by Panaʻewa

From a kanikau, or lament, for Kamehameha II on his departure for England. Hawaiian source: Fornander, *Fornander* 6: 437. English trans. by author.

3. He kuleana ko ka ua Kanilehua
He maʻū makamaka ia na ka Moani
He pohu lā ia na ka Puʻulena
E pē ana i ke kai o Hanakahi
Ua honi ke kini o ka Hilo Pali Kū

The Kanilehua rain has a prerogative
It is a wetness that accompanies the Moani breeze
A calmness for the Puʻulena wind
Carrying fragrance over the sea of Hanakahi
The multitudes of Hilo Pali Kū have smelled it

From a mele kālaiʻāina, or political chant, for ʻEmalani Kaleleonālani by O. Lumaheihei. Hawaiian source: Nogelmeier 210. English trans. by author. *Note: In mele, the Kanilehua rain sometimes is associated with the Moani breeze.*

4. Naue au me ka ua Kanilehua
Me ka huakaʻi ʻala a ka Moani
Ke kuahaua maila i ka lehua
Hiki ka Puʻulena ʻo ke onaona ka nui
Lālani ʻia e hele nei i Hilo

I move with the Kanilehua rain
With the fragrant journey of the Moani breeze
Announcing to the lehua blossoms
That the Puʻulena wind has arrived, consisting of a strong, sweet aroma
Formed in lines going to Hilo

From a mele inoa, or name chant, for Keahi. Hawaiian source: Lohiau, "He inoa." English trans. by author.

5. Mahalo piha, ʻai wahi moi nō ka poʻe palaualelo, i nā keiki o hanaʻiʻo ka ua iā Kohala, me ke keiki o ka ua kaulana Kanilehua o Hilo, inā paha i hāwanawana iki mai i ke keiki o [ka] ua kaulana Paʻūpili, inā hoʻi ua ʻikea nā ʻai a Wahinemanuʻa.

Much thanks—the idle people were able to eat some moi fish—to the children of "hanaʻiʻo ka ua iā Kohala" (Kohala, where the rain pours hard) and the child of the famous Kanilehua rain of Hilo. Had someone but whispered to the child of the famous Paʻūpili rain, perhaps a wager for Wahinemanuʻa's pile (of fish) would have been made (to see who is the better fisherman).

From a reference by Daniel Kahāʻulelio of Lāhainā, Maui, to Rev. E. S. Timoteo and Rev. S. Kapu of Hawaiʻi Island, who showed their skill in catching moi at Launiupoko, Maui. Hawaiian source: Kahāʻulelio 312. English trans. by author.

6. ʻAʻole nō i liʻuliʻu loa ma hope iho o kēia manawa, ua huli hoʻi akula ʻo Iosepa me kona kōkoʻolua no nā kaiaulu o ka ʻāina kulaiwi a hanu hou akula i nā kipona ea waianuhea o Hilo i ka ua Kanilehua.

Not long after this time, Joseph and his bride returned to the area of his homeland and breathed again the cool, sweet air of Hilo, of the Kanilehua rain.

From a biography of Joseph Nāwahīokalaniʻōpuʻu. Hawaiian source: Sheldon, *Ka Puke* 11. English trans.: Sheldon, *Biography* 037.

7. He ukana luʻuluʻu nāu, e ka ua Kanilehua, e hiʻipoi aku ai a hūnā iho i nā iwi o ke kanaka āu i kukuni paʻa ai i kona aloha ma kou puʻuwai. Eia ke hoʻi luʻuluʻu akula i ke one o Hanakahi.

A sad burden for you, O Kanilehua rain, to cherish and to secrete the bones of the man whose love is branded in your heart, here returning sadly to the sands of Hanakahi.

From an account of Joseph Nāwahīokalaniʻōpuʻu's funeral at Hilo Harbor. Hawaiian source: Sheldon, *Ka Puke* 179. English trans.: Sheldon, *Biography* 257–58.

8. Luʻuluʻu wale ko Hilo kini
Me ka ua Kanilehua

Hilo's multitudes are overwhelmed by sadness
Along with the Kanilehua rain

From a letter of sympathy to Mrs. Emma ʻAʻima Nāwahīokalaniʻōpuʻu from the residents of Baldwin Home of Kalawao, Molokaʻi, where people with Hansen's disease were sent to live. Hawaiian source: Sheldon, *Ka Puke* 258. English trans.: Sheldon, *Biography* 344.

9. No ka lā hiki ʻoe ma Kumukahi
No ka ua Kanilehua o Hilo
Hanohano ke kanaka o ka ʻāina

You are from the sunrise at Kumukahi
From the Kanilehua rain of Hilo
Noble indeed is the man of the land

From a mele for Joseph Nāwahīokalaniʻōpuʻu. Hawaiian source: Sheldon, *Ka Puke* 362. English trans.: Sheldon, *Biography* 460.

10. Nani wale nō Hilo
I ka ua Kanilehua

Hilo is beautiful indeed
In the Kanilehua rain

From the song "Hilo march" by Joseph K. Aeʻa. Hawaiian source: Elbert and Mahoe 51. English trans. by author.

11. Kaulana mai nei ʻo Hilo ʻeā
Ka ua Kanilehua ʻeā
Ka ua hoʻopulu ʻili ʻeā
Ka ʻili o ka malihini ʻeā

. . . .

Haʻina mai ka puana ʻeā
No ka ua Kanilehua ʻeā

Renowned is Hilo
For the [Kanilehua] rain that sings upon the lehua
The rain that soaks the skin
The skin of the visitors

. . . .

The story is told
Of the [Kanilehua] rain that sings upon the lehua

From the song "Hilo hula" by Joe Kalima. Source: Wilcox et al. 66.

12. No Hilo e ka ua Kanilehua
Popohe lehua a i Hanakahi
Ho'okahi a'u mea nui aia 'oe
'O kou aloha kai hiki mai

Hilo the home of Kanilehua rain
Lehua bloom to perfection at Hanakahi's land
There's only one thing I treasure, it is you
For your love has come here to me

From the song "Manu 'ō'ō." Source: Wilcox et al. 167. Additional source: Elbert and Mahoe 77. *Note: Wilcox et al. say, "The Kanilehua rain is a symbol of chatter and gossip."*

13. Hilo i ka ua Kanilehua. *Hilo of the Kanilehua rain.*

A traditional saying. Source: Pukui, *'Ōlelo* 107. *Note: Pukui says, "The Kanilehua rain, or the rain that patters in the lehua forest, is frequently referred to in the chants and songs of Hilo."*

14. Ka ua Kanilehua o Hilo. *The Kanilehua rain of Hilo.*

A traditional saying. Source: Pukui, *'Ōlelo* 168. *Note: Pukui refers to "Hilo, where the rain moistens the lehua blossoms."*

15. Pō Hilo i ka ua Kanilehua. *Hilo is darkened by the Kanilehua rain.*

A traditional saying. Source: Pukui, *'Ōlelo* 293. *Note: Pukui says that this phrase is "said of one who is weighted by sorrow and grief."*

16. Ua like nō a like
Me ka ua Kanilehua
Me he ala e 'ī mai ana
Aia i laila ke aloha

So like, so very like
The Kanilehua rain
That seems to say
There is the one to love

From the song "Ua like nō a like" by Alice Everett. Source: Wilcox et al. 258. Additional source: Emerson, *Unwritten* 168. *Note: Wilcox et al. say that the composer "compares love to the cherished attributes of Hilo, its rain, birds, and flowers."*

17. He makana na ka ua Kanilehua
Iā 'oe, e ka ua Kūkalahale
Lawe 'ia, lei 'ia, pūlama 'ia
I pa'a ke aloha pili me 'oe
I aloha nāu no nā kau a kau

A gift of the rain Kanilehua (Hilo)
To you O rain Kūkalahale (Honolulu)

Take it, wear it, cherish it
A love sealed with you
A love for you, forever

From the song "He makana" for Helen Henderson, presented on her wedding day by Helen Desha Beamer. Source: Kaaihue 59.

18. E Pāpā ē, e hoʻi kāua i ka ʻāina lupe, i Kuaihelani. Alia, e hoʻi kāua a keʻehi iā Hawaiʻi o Keawe, i ke poʻo o ka moku, i luna o Mailehahei, kū iho kāua i laila, nānā aku i ka ua Kanilehua o Hilo, ua Mailani, ua Kanikaniʻāʻula, ua Lūlehuahala o Panaʻewa, ua Leilehua noho i uka o Mākaulele, i kahi a kō makuahine [e] noho ai.

 O Father, let us return to the land of kites, Kuaihelani. Wait, before we return, let us tread upon Hawaiʻi of Keawe, to the summit of the island, above Mailehahei. We should stop there to observe the Kanilehua rain of Hilo, the Mailani rain, the Kanikaniʻāʻula rain, the Lūlehuahala rain of Panaʻewa, the Leilehua rain that dwells upland of Mākaulele, where your mother lives.

 From a kanikau, or lament, for Kamehameha IV. Hawaiian source: Moehonua and Lilikalani. English trans. by author.

19. ʻO kuʻu ʻāina ua Kanilehua
Ke hehi maila i ka liko ʻōhiʻa
Ma uka ia ua, ma kai ka ua Hukiheʻenehu
Ke huhuhui aʻela i ke kai ʻo Huia ē

 My beloved land of the Kanilehua rain
 Treading upon the ʻōhiʻa leaf buds
 In the uplands is that rain, in the lowlands is the Hukiheʻenehu rain
 Gathering at the sea of Huia

 From a mele by a chief of Hilo in the legend of Kaʻehuikimanōopuʻuloa. Hawaiian source: Uaua. English trans. by author.

20. Ka ua Kanilehua may be heard some distance away. It starts at Leleiwi Point, goes ma uka along the forest to the lava flow at Mōkaulele (one of the places where the noted lehua of Hilo were obtained), then turns toward Hilo town. It comes in big drops beating on the lehua and is of short duration. Its coming indicates stormy weather on this side of Hawaiʻi.

 From a list of rain names and their descriptions. Source: Kelsey, "Rains."

21. Aloha ʻoe! Aloha ka lehulehu e haiamū ana ma kou mau kahua momona, aloha ʻo Kahuā ʻāina makani, aloha ke kai hāwanawana o Kawaihae; ʻo ʻoe hoʻi, e ka nalu o Huia, aia mākou i laila e moe nei; ke hoʻomanaʻo aʻela mākou i nā hoa maikaʻi o ka ʻāina ua nui ʻo Hilo, nani wale ko lākou noho ʻana, e hehi ʻia ai e ke ehu o ka ua Kanilehua, a me ka ua Moaniʻala.

Love to you! Love to the many people crowding on your rich grounds, love to the windy land of Kahuā, love to the whispering sea of Kawaihae, and to you, O surf of Huia; we are there resting, recalling our good friends of Hilo, the land of plentiful rain; their lives are glorious, trodden upon by the mist of the Kanilehua rain and the Moaniʻala rain.

From an editorial in the newspaper *Kuokoa* written prior to a change in editorial staff. Hawaiian source: "Honolulu, Dekemaba." English trans. by author.

22. *For more examples of Kanilehua of Hilo, see Hoʻolua (#3) and Polohīnano (#2).*

Rain of Hilo and Panaʻewa, Hawaiʻi

23. I ka ua Kanilehua o Hilo
I ka nihi hele aʻe ma Panaʻewa
E nihi aʻe ana nō e pili
Me ka liko lehua i Mōkaulele

To the Kanilehua rain of Hilo
Which moves quietly along at Panaʻewa
Softly stealing along to unite
With the buds of the lehua of Mōkaulele

Written in memory of Joseph Nāwahīokalaniʻōpuʻu by Rev. Iosepa B. Hanaʻike. Hawaiian source: Sheldon, *Ka Puke* 329. English trans.: Sheldon, *Biography* 422.

Rain of Hilo Hanakahi, Hawaiʻi

24. ʻAnoʻai kuʻu wehi lā
I ka ua Kanilehua
Pua nani ka lehua lā
No Hilo Hanakahi
. . . .
Haʻina kuʻu wehi lā
I ka ua Kanilehua

Greetings to my adornment
In the Kanilehua rain
The lehua is a beautiful flower
From Hilo Hanakahi
. . . .
Tell of my adornment
In the Kanilehua rain

From the song "ʻAnoʻai." Hawaiian source: "ʻAnoʻai." English trans. by author.

25. A e ka ua Kanilehua o Hilo Hanakahi, ua pau kou hoʻomaʻū ʻana i kona mau pāpālina. *And as for you, O Kanilehua rain of Hanakahi's Hilo, your wetting of his cheeks is thus ended.*

From a composition written by the members of the Hawaiian National Band after hearing news of the passing of Joseph Nāwahīokalaniʻōpuʻu. Hawaiian source: Sheldon, *Ka Puke* 311. English trans.: Sheldon, *Biography* 400.

26. Luʻuluʻu Hanakahi i ka ua nui
I nā kīhene waimaka o ka ua Kanilehua
He waimaka luʻuluʻu, he ʻanoʻai aloha ē!

Hanakahi is overladen with the heavy rain
With the bundles of tears of the Kanilehua rain
Tears of sorrow, a tender greeting!

From an expression of sympathy written by the patriots of Nā Wai ʻEhā after the passing of Joseph Nāwahīokalaniʻōpuʻu. Hawaiian source: Sheldon, *Ka Puke* 207–8. English trans. by author. Additional source: Sheldon, *Biography* 288.

27. Aloha ʻino nō kuʻu kaikamahine, kuʻu lei ʻāʻī, kuʻu hoa hele o ke alanui ē

. . . .

Aloha ka ua Kanilehua, ua kāhiko ia i ke one o Hanakahi ē

Pity for my daughter, the lei that adorns my neck, my traveling companion on the road

. . . .

Love to the Kanilehua rain, it has adorned the sands of Hanakahi

From a mele ʻuhane, or lament, for Kealohapauole, presumably by Kaikamahineole. Hawaiian source: "He uhane." English trans. by author.

28. Aia i ka wēkiu lā
Kuʻu lei lehua puakea
Māholehole lua i ka laʻi lā
I ka luna kapu o Hanakahi
Uluhua, uluhua wale lā
I ka ua Kanilehua
I ka loku hala ʻole i ka pua lā
Ke pua maila i ka ʻiu

High above
Is my darling lei of white lehua blossoms
Scraped, bruised in the calm
On the sacred heights of Hanakahi
Annoyed, I am completely annoyed
With the Kanilehua rain
With the intense, unceasing downpour upon the flowers
As they bloom in the heights

From the song "Uluhua." Hawaiian source: Holstein 30. English trans. by author. *Note: In this rendering, "pua" is used in the last line instead of "pu-a," which was printed in the Hawaiian source.*

29. Hilo Hanakahi, i ka ua Kanilehua

. . . .

Ha'ina ka puana, i ka ua Kanilehua

Hilo of Chief Hanakahi, in the Kanilehua rain

. . . .

Tell the story in the refrain, of the Kanilehua rain

From the song "Hilo Hanakahi" by Keola Nālimu. Source: Wilcox et al. 65. Additional sources: Pukui and Elbert, 1986, 129; Elbert and Mahoe 50.

Rain of Hilo Hanakahi and Mōkaulele, Hawai'i

30. 'O Hilo Hanakahi, one 'ōiwi
'Āina kaulana me ke kilakila
Ho'opē mau 'ia e ke Kanilehua
Pulu 'elo nā lehua i Mōkaulele

Hilo Hanakahi, the native sands
A famous and majestic land
Continuously drenched by the Kanilehua
The lehua at Mōkaulele are soaked

From a memorial chant for Joseph Nāwahīokalani'ōpu'u. Hawaiian source: Sheldon, *Ka Puke* 371–72. English trans. by author. Additional source: Sheldon, *Biography* 469.

Rain of Hilo One, Hawai'i

31. Eia Hilo One ka pua o Kina lā
I loku a'e ka ua Kanilehua

Here at Hilo One is a Chinese bloom
That the Kanilehua rain soaks

From a mele inoa, or name chant, for Mrs. Emma 'A'ima Nāwahīokalani'ōpu'u. Hawaiian source: *Nā Leo Hawai'i Kahiko*, Record I, Side B, Band 16. English trans. by author. Additional sources: "Aia i Hilo One"; Pukui, *Nā Mele Welo* 224–25.

Rain of Ka'ena, Hawai'i

32. Kū ho'i Ka'ena 'o ka ua Kanilehua, 'eā
He lili puka ia na ka ua Kua'o'e, 'eā
No ka Ulumano i Hilo Hanakahi, 'eā

Ka'ena is struck by the Kanilehua rain
Jealousy issues forth from the Kua'o'e rain
For the Ulumano of Hilo Hanakahi

From a mele mākaʻikaʻi, or travel chant, for ʻEmalani Kaleleonālani by Kapaʻakea. Hawaiian source: Nogelmeier 159. English trans. by author.

Rain of Kapaʻa, Hawaiʻi

33. ʻIke wale i kuʻu keiki hala i Kapaʻa
Ua ua ʻia e ka ua kakahiaka
Ua hoʻopulu ʻia e ka ua Kanilehua

See my boy, headed off to Kapaʻa
Rained upon by the morning rain
Drenched by the Kanilehua rain

From a mele in the memoirs of John Papa Ii. Hawaiian source: Ii, "Ke ola" 6/30/1866. English trans. by author. Additional source: Ii, "Death" 2960.

Rain of Kūkiʻi, Hawaiʻi

34. Kuʻu kāne mai ka lā o Kūkiʻi ē
Uē ana i ke aloha o ka wahine
He wahine, he hoa no ka pō
Kuʻu kāne mai ka ua Kanilehua

My dear kāne from the sun of Kūkiʻi
I weep with a wife's love
A wife, a companion for the nighttime
My husband of the Kanilehua rains

From a kanikau, or lament, to Lohiʻau by Hiʻiakaikapoliopele. Hawaiian source: Hoʻoulumāhiehie, *Ka Moʻolelo* 383. English trans.: Hoʻoulumāhiehie, *Epic* 358.

Rain of Leleiwi, Hawaiʻi

35. Kuʻu kaikuahine o ka ʻāina hānau
O ka ʻāina ua nui ʻo Hilo
E hehi haʻaheo mai ana i ka lau o ka ʻulu
Ke lulumi aʻela i uka, piha nā kahawai
Kuʻu kaikuahine mai ka ua Kanilehua
Mai ka ua nihi mai ma Leleiwi
Ke hoʻonuʻu maila i luna o ka ʻōhiʻa

My beloved sister of our birth land
The land of Hilo with its great rain
Treading proudly over the breadfruit leaves
Crushing its way upland, the streams are full
My darling sister from the Kanilehua rain
From the creeping rains at Leleiwi
Causing torrents over the ʻōhiʻa trees

From a kanikau, or lament, for Mrs. Maleka Ii. Hawaiian source: Kaapa. English trans. by author.

Rain of Leleiwi and Mokuola, Hawai'i

36. Nani wale ka nui i Mokuola
Ka hehia e ka ua Kanilehua
Ke nihi a'ela i Leleiwi

So many at Mokuola
Trampled upon by the Kanilehua rain
Softly creeping to Leleiwi

From the song "Laieikawai" by Mrs. R. P. Waipa. Hawaiian source: Holstein 142. English trans. by author.

Rain of Leleiwi and Pana'ewa, Hawai'i

37. No ka ua Kanilehua—Ua puka mai kēia ua mai loko mai o nā ua kaulana 'elua o Hilo nei; 'o ka ua Kualau me ka ua Ulumano o Pana'ewa. Ua pā mai kēia ua ma ka hikina aku o Hilo, a hehi maila i ka 'āina ma ka lae kaulana 'o Leleiwi, a nihi a'ela i ka uka wehiwehi o Pana'ewa, a lū helele'i ihola i luna o ka lihilihi kukū o 'Aipō. 'A'ohe nui loa 'o kēia ua, 'o ka lelehuna wale iho nō i luna o ka lā'au, a e kapa 'ia ma nā inoa 'elua, 'o ia ho'i, ua Kanilehua a ua Lūlehua o Pana'ewa nui moku lehua.

Regarding the Kanilehua rain—This rain results from the two famous rains of Hilo: the Kualau and the Ulumano of Pana'ewa. This rain falls to the east of Hilo and treads the land at the famous cape of Leleiwi, and tiptoes to the verdant uplands of Pana'ewa, and scatters atop the protruding eyelashes of 'Aipō. This rain does not get heavy, it only falls delicately atop the trees, and it is called by two names, the Kanilehua rain and the Lūlehua rain of great Pana'ewa with its lehua groves.

From an article on the famous places of the Kanilehua rain. Hawaiian source: Laweliilii, "Na wahi." English trans. by author. Additional source: Laweliilii, "Relating." *Note: Laweliilii, "Relating," says that the Kualau are heavy showers, the Ulumano are the violent rains, and the Kanilehua is mist-like.*

Rain of Mōkaulele, Hawai'i

38. E ka ua Kanilehua a me nā lehua o Mōkaulele, 'a'ole 'o ia e pāpahi hou ana iā 'oe. *O Kanilehua rain and the lehua of Mōkaulele, he will never again be adorned by you.*

From a kānaenae, or chant of eulogy, for Joseph Nāwahīokalani'ōpu'u by Moke Manu. Hawaiian source: Sheldon, *Ka Puke* 303. English trans. by author. Additional source: Sheldon, *Biography* 392.

39. Ke uē aku nei mākou iā 'oe, e Hanohano Hilo i ka ua Kanilehua
Ke pehi a'ela ma uka o Mōkaulele
Ua helele'i i ke onaona wai a ka manu
Nā kipona wai lehua o Haili ē

We weep for you, O Pride of Hilo in the Kanilehua rain
Pelting away at the uplands of Mōkaulele
It drizzles down upon the fragrance, nectar of the birds
The sweet lehua honey of Haili

Written in praise of the late Joseph Nāwahīokalaniʻōpuʻu by members of the Women's Patriotic League. Hawaiian source: Sheldon, *Ka Puke* 279. English trans.: Sheldon, *Biography* 365.

Rain of ʻŌlaʻa, Hawaiʻi

40. Na wai hoʻi e ʻole ka makeʻe, he iwi hilo no ke kuamoʻo
He wohi i pehia a pulu ʻelo me nā ʻōmaka ua Kanilehua
I pāwehi hoʻoipo ʻia a pulu ʻelo me ka lehua o ʻŌlaʻa

Who indeed could deny the affection for him, his nature drawn from ancient custom
A wohi chief, pelted and soaked in the first drops of the Kanilehua rain
Lovingly adorned and drenched with the lehua of ʻŌlaʻa

From a kaniʻuhū, or expression of grief, dedicated to Joseph Nāwahīokalaniʻōpuʻu. Hawaiian source: Sheldon, *Ka Puke* xix. English trans.: Sheldon, *Biography* 011.

Rain of Panaʻewa, Hawaiʻi

41. Ua Kanilehua, he ua kēia mai ka moana mai, a kiola aʻela i uka o ka ʻāina, nihi aʻela ma nā lae lāʻau, hehi kū maila i ka liko o ka ʻōhiʻa, me ka pua o ka ʻōhiʻa, he ua hehi lehua o Panaʻewa, ma laila i kapa ʻia aku ai e ka poʻe kahiko, he ua Kanilehua.

Kanilehua rain is a rain from the ocean, thrown above on the land, reaching the forest belt, treading the bud of the ʻōhiʻa, and its blossoms, a pelting lehua rain of Panaʻewa, whereby is given this name by the ancients, a Kanilehua rain.

From a letter to the newspaper *Kuokoa* regarding famous rains. Hawaiian source: Kalaaukumuole, "Manao wehewehe." English trans.: Kalaaukumuole, "Thought explanations."

42. He hoa ʻoe noʻu i ka pua o ke ʻaʻaliʻi
He aliʻi, he hiwahiwa na ka pūnohu ʻula
He punahele hoʻi au na Panaʻewa
He lani no nā ua Kanilehua

You are a companion of mine with the blossom of the ʻaʻaliʻi
A chief, a darling of the red rising mist
I am indeed a favorite of Panaʻewa
A royal one of the Kanilehua rain

From a mele aloha, or song of love, chanted by Kuluwaimaka. Hawaiian source: "Kuu aloha." English trans. by author.

43. [ʻUla] ʻā ka ʻōhiʻa o Panaʻewa
Ka ʻuo i ka ua Kanilehua
Pūloku Hilo, hemolele i ka Moani

Fiery red is the ʻōhiʻa of Panaʻewa
Like small bunches of feathers in the Kanilehua rain
Hilo sparkles, flawless in the Moani breeze

From condolences to Mrs. Emma ʻAʻima Nāwahīokalaniʻōpuʻu and her children by the patriots of Wānanalua and Nāhiku after the passing of Joseph Nāwahīokalaniʻōpuʻu. Hawaiian source: Sheldon, *Ka Puke* 283. English trans. by author. Additional source: Sheldon, *Biography* 370.

44. E ka ua Kanilehua
Kui ʻia nā lehua o Panaʻewa
Na ka Puʻulena e lawe mai
I wehi, i ʻohu no Kalani

O Kanilehua rain
The lehua blossoms of Panaʻewa are to be strung
The Puʻulena wind will bring them
As decorations, adornments for the chief

From the song "Pela kapu o Kakae" by the Hawaiian Glee Club. Hawaiian source: Holstein 33. English trans. by author.

45. Onaona ʻo Panaʻewa
He sweet darling ʻoe, he lei na ke aloha
No ka ua Kanilehua

Panaʻewa is softly fragrant
You are a sweet darling, a lei made by love
For the Kanilehua rain

From the song "Pua magnolia." Hawaiian source: Holstein 96. English trans. by author.

46. Kuʻu makuahine hoʻi
Kuʻu makuahine mai ka ua Kanilehua o Panaʻewa

My beloved mother indeed
My dear mother from the Kanilehua rain of Panaʻewa

From a kānaenae, or chant of eulogy, for H. Kamahiai. Hawaiian source: Makuaole. English trans. by author.

Rain of Waiākea, Hawaiʻi

47. Sweet onaona o Hanakahi
E wehi nei lā i Waiākea
Pulu pē i ka ua Kanilehua
A Panaʻewa lā e hiʻi mai nei

The sweet redolent one of Hanakahi's land
Adorned with verdure at Waiākea
Drenched in the Kanilehua rains
Which Pana'ewa so cherishes

From a kānaenae, or chant of eulogy, for Joseph Nāwahīokalani'ōpu'u. Hawaiian source: Sheldon, *Ka Puke* 369. English trans.: Sheldon, *Biography* 467.

Rain of Lē'ahi, O'ahu

48. Hanohano Lē'ahi i ka'u 'ike
E pa'ihi i ka ua Kanilehua
Ua [ani] uka i ke ahe a ke Kiu

Grand is Lē'ahi in my sight
Finely dressed in the Kanilehua rain
The uplands beckon in the gently wafting Kiu wind

From the song "Uwila i ka luna o Manoa" by Kanalu. Hawaiian source: *Buke Mele* 80. English trans. by author.

Rain of Hanalei, Kaua'i

49. Mauna pu'u mamau a ka ua Kanilehua
I ka ho'owali kī'ililī 'ia e ke Pahelehala
Ki'i papani hone i ke kula o Hanalei
I pauma wai 'ia komo i Lumaha'i

The rough, hilly mountainside of the Kanilehua rain
Lazily mixed by the Pahelehala wind
Playfully attempting to block the plains of Hanalei
Water is pumped to enter Lumaha'i

From a mele inoa, or name chant, for Keliihelemauna. Hawaiian source: Holaniku. English trans. by author. Additional source: "Pā ka leo" 11.

Rain of Kaupe'a, Kaua'i

50. Nā wāhine kaha pu'ewai o Ko'olau
Kui lei lehua
'Ō'ū Kanilehua ākea
'Oki hala lei o Kaupe'a lā

The women who surf the river channels of Ko'olau
Who string garlands of lehua
And like 'ō'ū birds in the expansive Kanilehua rain
Cutting the hala of Kaupe'a for lei

From a mele by Hi'iakaikapoliopele to bring about a fatal illness to the Maui chief 'Olepau. Hawaiian source: Ho'oulumāhiehie, *Ka Mo'olelo* 333. English trans.: Ho'oulumāhiehie, *Epic* 309.

Drenching rain

51. Pulu ʻelo i ka ua Kanilehua. *Drenched in the Kanilehua rain.*

A traditional saying. Source: Pukui, *ʻŌlelo* 300. Additional source: Judd 7. *Note: Pukui says that this phrase means "drenched by the rain or thoroughly drunk."*

Fragrant rain

52. Ua ʻike maka aʻe nei hoʻi au
I ka ua Kanilehua
Ke neʻe maila i ka uka
A hoʻopē ʻaʻala i ka nahele

I definitely saw with my own eyes
The Kanilehua rain
Marching to the uplands
And drenching the forest in a sweet perfume

From the song "He ala nei he ala" by Iwa. Hawaiian source: Holstein 48. English trans. by author.

Misty rain

53. Name of a mist-like rain famous at Hilo.

Source: Pukui and Elbert, 1986, 129.

54. Ua lohe i ka noe a ka ua Kanilehua
I nā ʻaka iki a ka ua koʻiawe i nā pali

Hearing the mist of the Kanilehua rain
The chuckling of the rain showering lightly on the cliffs

From a mele inoa, or name chant, for Heneri W. Auld. Hawaiian source: Hoopai. English trans. by author.

Other examples

55. Ua Kanilehua i pāwehi hoʻoipo ʻia . . . me ka lehua. *Kanilehua rain lovingly adorned . . . with lehua.*

Source: Pukui and Elbert, 1986, 322.

56. ʻO ka poʻe Haole eʻe pākeke a mākou i holo pū ai, komo maila lākou lā i nā kapa mehana—ʻelua a ʻekolu kuka o ka mea hoʻokahi. Akā, ʻo wau, he ʻoluʻolu loa ihola au i ia ʻano huʻihuʻi, no ka mea, ua ʻano like nō ka huʻihuʻi me ko nā kakahiaka huʻihuʻi loa o ka ua Kanilehua, a ua like nō me ko Kīlauea ma Hawaiʻi.

The foreign passengers with whom we sailed wore warm clothing, two or three coats per person. But as for me, I was very content in this sort of cold because it was similar to the frigid mornings of the Kanilehua rain, and also similar to that of Kīlauea on Hawaiʻi.

From an account written by Joseph Nāwahīokalaniʻōpuʻu on his way to San Francisco. Hawaiian source: Sheldon, *Ka Puke* 50. English trans. by author. Additional source: Sheldon, *Biography* 102.

57. Kuʻu pua malihini, my darling
I koʻiʻi i ka ua Kanilehua
Kuko loko a hoʻokō
Just you and I
Just you and I

My rare new blossom, my darling
Freshened in the rain Kanilehua
A strong desire within awaits fulfillment
Just you and I
Just you and I

From the song "Pua malihini" for Charles William Dahlberg by Helen Desha Beamer. Source: Kaaihue 68.

58. *For another example of Kanilehua, see Līlīlehua (#30).*

Kapaʻahoa. Same as *Kapakapaʻahoa*. Rain associated with Kauaʻi. Also the name of a wind.

Rain of Kahana, Kauaʻi

1. Aia ka ua i Olokele
He ua Kapaʻahoa ko Kahana
He ua hanahanai ʻoʻopu ʻai lehua
He iʻa kia ahi manu no Mokihana

There is the rain at Olokele
A Kapaʻahoa rain belongs to Kahana
A rain that nurtures the lehua-eating ʻoʻopu fish
A fish that attracts birds of Mokihana

From a makena, or lament, for ʻEmalani Kaleleonālani. Source: Nogelmeier 356.

2. Kuʻu hoa i ka ua Kapaʻahoa o Kahana
Kuʻu hoa pili i ke anu o ka mauna

My companion in the Kapaʻahoa rain of Kahana
My close companion in the cold of the mountain

From a lament for Lono-opio by Pelekaia. Hawaiian source: Fornander, *Fornander* 6: 425. English trans. by author.

Rain of Līhau, Kauaʻi

3. I aloha i ka ua a Kapaʻahoa
I ka nihi ʻana aʻe ma Līhau

Love for the rain of Kapaʻahoa
That creeps softly at Līhau

From a mele for Miss Ane Bell. Hawaiian source: Kaanaana, "He wehi." English trans. by author.

Rain of Pōhākomo, Kauaʻi

4. He nani hemolele ka pali ʻo Lehua
Ka paheʻe i ka wai o Pōhākomo
He wai honi ʻili no ka ua Kapaʻahoa
E hehi ana i ka lau o ka lāʻau

A perfect beauty is the cliff Lehua
Slipping in the waters of Pōhākomo
Water from the Kapaʻahoa rain that kisses the skin
Treading upon the leaves of the trees

From an account by William Hoapili Kaauwai of a trip around the world. Hawaiian source: Kaauwai 9/19/1868. English trans. by author.

Rain of Waimea, Kauaʻi

5. ʻO Luʻanuʻu a Laka, ʻo Luʻanuʻu ke keiki a Laka, ʻo Hīkāwaelena ka makuahine, he aliʻi wahine ʻo ia no ka ua Kapaʻahoa no Waimea i Kauaʻi.

Luʻanuʻu of Laka, Luʻanuʻu is the son of Laka; Hīkāwaelena is his mother; she is a chiefess of the Kapaʻahoa rain of Waimea in Kauaʻi.

From the legend of Luʻanuʻu. Hawaiian source: Kamakau, "Ka moolelo Hawaii" 10/28/1869. English trans. by author. Additional source: Kamakau, *Tales* 147.

6. Kuʻu kāne, e kuʻu kāne hoʻi
Kuʻu kāne mai ka wai ʻula ʻiliahi o Waimea
Wai nono ʻula a ka ua Kapaʻahoa

My beloved husband, oh, my dear husband indeed
My dear husband of the red sandalwood waters of Waimea
Red-glowing water of the Kapaʻahoa rain

From a kanikau, or lament, for Kamehameha IV by his wife, ʻEmalani Kaleleonālani. Source: Nogelmeier 339. *Note: Pukui,* ʻŌlelo *179, says that "ka wai ʻula ʻiliahi o Waimea" refers to Waimea Stream, which runs red following a storm. "Where it meets Makaweli Stream to form Waimea River, the water is sometimes red on one side and clear on the other. The red side is called ʻwai ʻula ʻiliahi.'"*

7. Kau ke Kiuwaiʻahulu o Waimea
Wai nono ʻula a ka ua Kapaʻahoa
I hoʻoluʻu a kohu i ka pili
A ʻula mai heʻa ka uka o Kahana

The Kiuwai'ahulu wind of Waimea settles
Blushing water of the Kapa'ahoa rain
Dyed and stained by the closeness
Becoming red, stained red are the uplands of Kahana

From a chant originally composed for Lunalilo and inherited by Kalākaua. This portion of the mele was composed by Ka'ahumanu. Hawaiian source: *Na Mele Aimoku* 147–48. English trans. by author.

8. *For another example of Kapa'ahoa of Waimea, see Noe (#12).*

Kapaeloaihiki. Mist associated with Kawaikini, Kaua'i.

1. Me he polope'a ala Kawaikini i ka mālie
I ka lalawe a ka Wai'ōpua makani
I hea luna 'ia e Kapaeloaihiki

Kawaikini in the calm is like the thick stem of the pandanus
Against the workings of the Wai'ōpua wind
Covered by the mist, Kapaeloaihiki

From a makena, or lament, for 'Emalani Kaleleonālani. Source: Nogelmeier 350.

Kapakapa'ahoa. Same as *Kapa'ahoa*. Rain associated with Kahana, Kaua'i.

1. Ku'u hoa o ka wai 'ula me ka wai 'iliahi
Kapakapa'ahoa o Kahana
Ua 'Ūkiukiu o Hikilei
Maka'ūpili o Pe'ape'a, 'oia ['a]ne'i

My dear friend of the red water and the 'iliahi-colored water
Kapakapa'ahoa rain of Kahana
'Ūkiukiu rain of Hikilei
Maka'ūpili rain of Pe'ape'a, so it is

From a kanikau, or lament, for Ka'ahumanu. Hawaiian source: Niauehoe. English trans. by author. Additional source: Fornander, *Fornander* 6: 454. *Note: Fornander says that the 'Ūkiukiu is "a teasing rain, rain with driving wind."*

Kapakea. Rain associated with Hawai'i and O'ahu. "Kapa kea" means "white kapa."

Rain of Lanakila, Hawai'i

1. Hala 'oe, ku'u hoa ho'i ē
He hoa i ka ua Kapakea o Lanakila
Mai ke kapa kīhae a mehana nō i ka 'ili
Ho'omehana aku i ka poli o ke hoa, ku'u kāne

You have passed away, my companion
A friend in the Kapakea rain of Lanakila
From the strips of kapa that warm the skin
Warming the bosom of my companion, my dear husband

From the legend of Mokulehua. Hawaiian source: "He wahi kaao." English trans. by author.

Rain of Mololani, Nuʻuanu, Oʻahu

2. Ka ua Kapakea o Mololani. *The Kapakea rain of Mololani.*

A traditional saying. Hawaiian source: Pukui, *ʻŌlelo* 168. English trans. by author. *Note: Pukui says that "the rain and mist at Mololani, Nuʻuanu, resemble a white sheet."*

Kapuaʻikanaka. Rain associated with Pālāwai, Oʻahu. "Kapuaʻi kanaka" means "footsteps."

1. Ka ua Kapuaʻikanaka o Pālāwai. *The Kapuaʻikanaka rain of Pālāwai.*

A traditional saying. Hawaiian source: Pukui, *ʻŌlelo* 169. English trans. by author. *Note: Pukui describes the Kapuaʻikanaka as a rain that sounds like human footsteps.*

2. I ia wā ʻo ia i ʻike aku ai i ka hele kawewe ʻana aʻe a ka ua i Pālāwai.... I kēia wā i paeaea aʻe ai ʻo ia i kēia kau e pili ana i ke kāne, iā Kaʻanahau, a iā Pele nō hoʻi.

Kuʻu kāne i ke ala pili o Mahinui
Mai ka ua Kapuaʻikanaka i Pālāwai
Ka ua o Kailua i kai ē

At that point, she recognized the thrumming rain of Pālāwai.... At this time, she presented the following chant about Kaʻanahau, which also pertained to Pele.

My man of the clinging path of Mahinui
From the [Kapuaʻikanaka] rain of Pālāwai that follows like footsteps
The rain of Kailua by the sea

From the legend of Hiʻiakaikapoliopele. Kaʻanahau of Kailua, Oʻahu, was Hiʻiaka's lover. Hawaiian source: Hoʻoulumāhiehie, *Ka Moʻolelo* 154. English trans.: Hoʻoulumāhiehie, *Epic* 145.

Kauaolelea'anae. Rain associated with Leleaʻanae (or Leleʻanae), Oʻahu. "Ka ua o Leleaʻanae" means "the rain of Leleaʻanae."

1. Kauaoleleaʻanae—a rain that forms above the cliff of Leleaʻanae and travels toward Nuʻuanu Pali from the direction of Makapuʻu Point.

From a list of rain names and their descriptions. Source: Kelsey, "Rains."

Kauaʻula. Same as *ʻUla.* Rain associated with Lāhainā, Maui, and found on other parts of Maui. Also the name of a wind and of an ahupuaʻa in Lāhainā. "Ka ua ʻula" means "the red rain."

Rain of ʻAuwaiawao, Maui

1. *See Nahua (#10).*

Rain of Kaimiau, Maui

2. Ke ʻaʻe Kāhili, pakele i Kaimiau
   Mōlaʻe ke alo o ka ua Kauaʻula

   When Kāhili transgresses, escape to Kaimiau
   Clear and unobstructed is the face of the Kauaʻula rain

   From a mele hoʻoilina, or legacy song, for ʻEmalani Kaleleonālani. Source: Nogelmeier 102. *Note: Sterling 25 says that "Kāhili" is the name of a pali, or cliff, at the head of Kauaʻula Valley in Lāhainā, Maui.*

Rain of Kanahā, Maui

3. Kauaʻula ka ua noho i uka
   Noho i uka o Kanahā

   Kauaʻula is the rain that dwells inland
   Occupying the uplands of Kanahā

   From a mele for Kaumualiʻi. Hawaiian source: Fornander, *Fornander* 6: 476. English trans. by author. *Note: Fornander refers to "Kauaʻula, name of a wind and rain in one of the valleys back of Lāhainā, hence the name of that ridge, Kanahā, which lies above."*

Rain of Kānehoa, Maui

4. Aloha ka lehua ʻaki popohe i ke kula
   I walea ka noho ʻana i ka hoʻopiha ʻia e Kauaʻula
   Ke nū maila i Kānehoa
   He hoa ʻoe i ka lā ʻīnea o Lāhaināluna

   Love for the round and shapely lehua high on the plain
   Enjoying life made complete by the Kauaʻula rain
   Which patters down at Kānehoa
   You are a friend on the day of hardship at Lāhaināluna

   From a kanikau, or lament, for Lāhaināluna Seminary by its students upon hearing that the building was decaying. Hawaiian source: Fornander, *Fornander* 6: 528. English trans. by author.

Rain of Luaʻehu, Maui

5. Aloha ʻino nō kuʻu kaikamahine, kuʻu lei ʻāʻī, kuʻu hoa hele o ke alanui ē

   . . . .

   Hoʻohihi ka ʻuhane i ka ua Kauaʻula, i ka nani o Luaʻehu, kini kohu ke noho ʻoe ē

   Pity for my daughter, the lei that adorns my neck, my traveling companion on the road

   . . . .

The spirit is enchanted with the Kauaʻula rain in the beauty of Luaʻehu, well-suited for your dwelling

From a mele ʻuhane, or lament, for Kealohapauole, presumably by Kaikamahineole. Hawaiian source: "He uhane." English trans. by author.

Kaulāʻau. Rain associated with Pāhala, Hawaiʻi. "Kau lāʻau" means "to rest on trees."

1. Ka ua Kaulāʻau o Pāhala. *The tree-resting [Kaulāʻau] rain of Pāhala.*

   A traditional saying. Source: Pukui, *ʻŌlelo* 169. *Note: Pukui says, "The rain of Pāhala, in Kaʻū, Hawaiʻi, seems to rest on the treetops."*

Kea. Misty rain associated with Hāna, Maui, and with the *Mālualua* rain. Also found at Koʻolau, Maui, and at Puna, Kauaʻi. "Kea" means "white."

Rain of Hāna, Maui

1. Ka ua Kea o Hāna. *The white [Kea] rain of Hāna.*

   A traditional saying. Source: Pukui, *ʻŌlelo* 169. *Note: Pukui says that this saying "refers to the misty rain of Hāna, Maui, that comes in from the sea."*

2. No ka ua Kea o Hāna
   ʻO kēia hoʻi ka ua kaulana o kēia ʻāina ʻo Hāna nei a hiki i kēia lā. He ua kupanaha loa kēia, ʻelua nō manawa e hiki mai ai ʻo kēia ua—ʻo ke kakahiaka nui a ʻo ke awakea; he ua kūpuni kapa paʻūpaʻū kēia a ko ʻoneʻi mau keiki kamaʻāina e heʻe puʻewai ai i Lāʻie i Waiohonu a me ʻUmiahu i Kahawaiokapia.

   Relating to the Kea rain of Hāna
   It is indeed the most famous rain of this land of Hāna, even till today. This very unusual rain comes twice, in the early morning and at midday. This is a rain that causes even the local children who surf the mouth of the stream at Lāʻie at Waiohonu and ʻUmiahu at Kahawaiokapia to clutch tightly their heavy, cold-weather kapa.

   From an article on the news of Hāna, Maui. Hawaiian source: Paahao. English trans. by author. *Note: The references to stream water and clutching kapa in this example suggest that the Kea is a very cold rain.*

3. E paʻa mai ana ka noe i uka a me ka ua liʻiliʻi, a ʻo ia nō paha ka mea i ʻōlelo ʻia ai, "Noenoe ua Kea o Hāna, hoa pili o ka Mālualua."

   . . . Noenoe ua Kea o Hāna—he ua, noenoe (noe) ua kea (keʻokeʻo) kēia i kakahiaka wale nō a ʻano awakea aʻe nō, pau. ʻO kēia ka ua i mele mau ʻia e ka poʻe hīmeni, a ʻoi loa aku ʻo ka poʻe i kāmau kīʻaha a ihoiho ka noe a ʻike pōwehiwehi i kou poʻe hoa, ʻo ia ka wā naʻauao loa o ke kanaka a me ka leʻa i ka hīmeni ʻana a penei e kani ai ke mele o kēia ua:

Noenoe ua Kea o Hāna
Hoa pili o ka Mālualua

The mist stays inland, along with the fine rain, and that may be what is talked about in the saying, "Misty Kea rain of Hāna, close companion of the Mālualua."

... Misty Kea rain of Hāna—this is a rain that is misty and white that appears only in the morning; just before midday, it's gone. This is the rain that singers always sing about, especially the people who tilt their glasses and the mist descends and your friends look blurry. That is when people sing the most skillfully and merrily, and this is how the song for this rain goes:

Misty Kea rain of Hāna
Close companion of the Mālualua

From an account by Thomas K. Maunupau of an expedition through Kaupō, Maui. Hawaiian source: Maunupau, *Huakai* 13, 27. English trans. by author. Additional source: Holstein 63.

4. ʻO ka noe a ka ua Kea
Kaulana nei aʻo Hāna
Me ka ua Mālualua
E uhai ana i ka noe

The misting of the white [Kea] rain
For which Hāna is famous
Then the Mālualua rain
Seems to follow the mist

From the song "Waikaloa" by John Piʻilani Watkins. Source: Wilcox et al. 274. *Note: Wilcox et al. say that the Kea rain is "the famous misty white rain of the Hāna district."*

5. ʻO ka ua Kea hoʻonani pua
Puaʻi aʻela i ke kai

The Kea rain adorns the blossoms
Then flows into the sea

From a mele lei, or song presenting a lei, originally composed for Lunalilo and inherited by Kalākaua. This portion of the mele speaks of Hāna, Maui, and was composed by M. Kekauonohi. Hawaiian source: *Na Mele Aimoku* 160. English trans. by author. Additional source: Reichel, *Keʻalaokamaile.*

6. "Ke aloha ʻia aku nei ʻoukou a pau loa e Maui Hikina i ka ua Kea hoʻonani pua o Hāna, a i hoʻoipo ʻia e ka ua kilihune o ka makani Kāʻilialoha o Kīpahulu," aloha ʻoukou a nui loa.

"Greetings are being expressed to all of you by East Maui, in the Kea rain of Hāna that adorns flowers, wooed by the light kilihune rain of the Kāʻilialoha wind of Kīpahulu"; much love to all of you.

From a message of condolence for the passing of Joseph Nāwahīokalaniʻōpuʻu by the Hui Aloha ʻĀina o nā Wāhine. Hawaiian source: Sheldon, *Ka Puke* 278. English trans. by author. Additional source: Sheldon, *Biography* 364.

7. Kuʻu makua kāne mai ka ua Kea o Hāna ē
He ua kūpuni kapa ia no ia ʻāina
Aloha wale ia ʻāina iʻa iki a kāua e noho ai

My beloved father from the Kea rain of Hāna
A rain of this land that causes people to clutch their kapa
So beloved is this land of little fish where we live

From a kānaenae, or chant of eulogy, for Paulo Ku by Mrs. Hakaleleponi Paulo. Hawaiian source: Paulo. English trans. by author. *Note: The phrase "kūpuni kapa," meaning "to clutch kapa," suggests a very cold rain. Pukui,* ʻŌlelo 54, *says, "Hāna was known in ancient times as the land where fish were scarce. Believing slanderous tales about Kūʻula and his wife, Hinahele, the ruling chief of Hāna ordered them destroyed. Having mana over the fish of the sea, the two caused a scarcity until their son ʻAiʻai brought them back to life. Kūʻula and Hinahele were worshipped as deities by fishermen." Thus the traditional saying, "Hāna i ka iʻa iki," meaning "Hāna of little fish."*

Rain of Hāna and Koʻolau, Maui

8. The name of the rain or mist at Hāna and at Koʻolau on Maui.

Source: Andrews 265.

Rain of Kaʻuiki, Maui

9. ʻAʻohe nō a he mea like me ka ua Kea
ʻOni a lele Kaʻuiki i ka mālie

There is nothing to compare with the white [Kea] rain
As Kaʻuiki appears to be leaping in the calm

From the song "Ka ua Kea" by Samuel P. Kahoʻomākauliʻi. Source: "Ka ua Kea." *Note: Translator Kimo Alama Keaulana says, "Hāna on Maui is famous for its white rain and Kaʻuiki Hill at Hāna's bay."*

Rain of Puna, Kauaʻi

10. He ua Kea ko Puna
Ke ua maila i Kuaahiahi
He ua hoʻomālie kai no Makaīwa
E ana ana i ka laulā o Kapaʻa

Puna has a white [Kea] rain
Raining now at Kuaahiahi
A rain that quiets the sea of Makaīwa
Measuring the expanse of Kapaʻa

From a makena, or lament, for ʻEmalani Kaleleonālani. Source: Nogelmeier 355.

Keʻehilehua. Rain associated with Koʻiʻālana, Kauaʻi. "Keʻehi lehua" means "to trample the lehua blossoms."

1. Ua Keʻehilehua—Koʻiʻālana, Kauaʻi.

   From a list of rain names and their descriptions. Source: "Names . . . rain."

Kēhau. Same as *Tēhau*; related to *hau*. Dew, mist; associated with Maunakea, Hawaiʻi, and with Kaʻala, Oʻahu, and found in other areas. Also the name of a wind. "Kēhau" is both the name of a specific rain and a generally descriptive term; its various usages are determined by the context.

Rain of Mailehahei, Hawaiʻi

1. Noenoe Mailehahei i ka ua noe
   He noe uahi pōhina ia na ke Kēhau
   He ua hoʻopulu i ka maka o ka Pohu
   Pō aʻela nā uka o Lanihau i ka uahi ē, ahi

   Mailehahei is foggy in the noe rain
   A smoky, hazy fog created by the Kēhau
   A rain that wets the face of the Pohu wind
   The uplands of Lanihau are dark in the smoke, fire

   From a mele given to the Bishop Museum by Thomas E. Cooke of Hilo, Hawaiʻi, who received it from Judge Simon H. Haʻaheo of Kalapana, Hawaiʻi. Hawaiian source: "Mikihilina." English trans. by author. Additional source: Moehonua and Lilikalani.

Rain of Maunakea, Hawaiʻi

2. E aha ʻia ana Maunakea
   Kuahiwi ʻalo pū me ke Kēhau
   . . . .
   Aloha ia uka pūanuanu
   I ka hoʻopulu ʻia e ke Kēhau
   Haʻina ʻia mai ana ka puana
   Pulu ʻelo i ka wai a ka Nāulu

   What is happening with Maunakea
   Mountain that dwells with the [Kēhau] mists
   . . . .

Beloved is that lofty highland, ever cold and damp
Wetted, as it is, with the [Kēhau] mist and dew
Let the tale be known through the telling
Soaked by the water of the sudden Nāulu rains

From a mele mākaʻikaʻi, or travel chant, for ʻEmalani Kaleleonālani. Source: Nogelmeier 114. Additional source: Kaanaana, "He wehi."

Rain of Nāʻilima, Hawaiʻi

3. He loa ke ala ma kai o Nāʻilima
I nā haona ipu wai a ke Kēhau
Ko laila wai ʻauʻau nō ia

Long is the path oceanside of Nāʻilima
In the water receptacles of the Kēhau rain
It is the bathing water of that place

From a mele kālaiʻāina, or political chant, for ʻEmalani Kaleleonālani. Hawaiian source: Nogelmeier 180. English trans. by author.

Mists of ʻUlupalakua, Maui

4. *See Lanipaʻina (#2).*

Rain of Kaʻala, Oʻahu

5. ʻO Kaupono i Kaʻala ke kiʻo wai a ke Kēhau
Ke hiʻi maila e Kokoloea

The pool of the Kēhau is Kaupono atop Kaʻala
Embraced by Kokoloea

From a mele kālaiʻāina, or political chant, for ʻEmalani Kaleleonālani. Hawaiian source: Nogelmeier 176. English trans. by author. *Note: "Kokoloea" is the name of an upland area and plains near Malamanui, Oʻahu.*

6. ʻO Kaʻala, kuahiwi mauna Kēhau, ke ōpū maila lā i Kamaoha. *Kaʻala, mountain hill with cool [Kēhau] rain, rising there perhaps at Kamaoha.*

Source: Pukui and Elbert, 1986, 293. Additional source: Emerson, *Pele* 100.

Rain of Kamaoha, Oʻahu

7. Kapu ka luna o Kamaoha i ke Kēhau
He Kēhau hoʻopili ʻia na ke Kiʻowao
E hoʻopulu ana i pulu ka maka o ka ʻōpua ā

The crown of Kamaoha is sacred in the Kēhau
A Kēhau brought close by the Kiʻowao
Creating moisture so that the face of the clouds is wet

From a mele inoa, or name chant, for Albert Ka Haku O Hawaiʻi by his father, Kamehameha IV. Hawaiian source: Iolani. English trans. by author.

Rain of Lanihuli, Oʻahu

8. Uluwehi nā luna i Lanihuli
Pulu pē i ka nihi a ka ua
. . . .
I laila, liʻa ka manaʻo
Pūkuʻi i ke anu a ka ua
Kolonahe aʻela i ka uka
Me ke kēhau o ka nahele

The heights of Lanihuli are verdant
Soaked by the sweep of the rain
. . . .
There desires are stirred
Crouching in the chill of the rain
As it gently slips through the uplands
With the cool [kēhau] dew of the forest

From the song "Pulu pē nei ʻili i ke anu." Source: Wilcox et al. 232. *Note: Wilcox et al. say, "This traditional song describes beautiful Lanihuli, the peak above the western side of the Pali lookout. Chill and cold often appear in Hawaiian poetry to describe the physical sensations of love and desire. The image of crouching with a loved one in the cool [kēhau] rain of a fragrant forest portrays quite an intimate setting."*

Rain of Malamanui, Oʻahu

9. ʻO kuʻu [hālokoloko] wai lehua
Ke hoʻomaʻū ʻia lā e ke Kēhau
Ua pulu i Malamanui

My pool of lehua nectar
Is being moistened by the Kēhau
It is saturated at Malamanui

From a mele by Kaumealani describing lovemaking. Hawaiian source: Hoʻoulumāhiehie, *Ka Moʻolelo* 151. English trans. by author. Additional source: Hoʻoulumāhiehie, *Epic* 142.

Rain of Kalena, Kauaʻi

10. A ua i Kaliko, pulu ʻelo Kalena
Pōpō ke kapa o ka huakaʻi hele
Mālama o pulu i ka ua Kēhau
Ka ua hoʻopulu maka o ka ʻawapuhi
ʻO ke kuāua ia o Kalenakupu

And raining at Kaliko, Kalena is drenched
The kapa of the travelers are rolled up in a ball
Cared for so as not to be saturated by the Kēhau rain

The rain that drenches the bud of the ʻawapuhi plant
It is the shower of Kalenakupu

From a makena, or lament, for ʻEmalani Kaleleonālani. Hawaiian source: Nogelmeier 353. English trans. by author.

Chilly rain

11. Pulu i ka wai naoa a ke kēhau. *Wet in the chilly water of the dew.*

    Source: Pukui and Elbert, 1986, 262.

Mist

12. A mist; a cold, fine rain floating in the air, mostly in the mountainous regions.

    Source: Andrews 267.

13. *For another example of Kēhau mist, see Nāulu (#67).*

Other examples

14. ʻAuhea wale ʻoe, e ke Kēhau
    E hoʻomaʻū nei i nā pua

    Attend, O Kēhau
    Freshen the flowers

    From the song "He makana" for Helen Henderson, presented on her wedding day by Helen Desha Beamer. Source: Kaaihue 59.

15. Onaona wale ia pua
    I ka mikiʻala mau ʻia
    Hele a nohenohea
    I ka nou a ke kēhau

    This flower is softly fragrant
    And quickly secured
    Lovely
    When pelted by the [kēhau] dew

    From the song "Beautiful ʻilima" for John Alexander Liholiho by his mother, Emma Kanoʻa DeFries. Source: Elbert and Mahoe 37.

16. Nāna i kaomi nā māno
    Hakukoʻi ka wai i nā pali
    Pulu ʻelo i ka ua kēhau
    Ua ʻolu i ka liko o ka lehua

    It put pressure on the water source
    The water rushed over the cliffs
    Soaked by the kēhau rain
    The leaf buds of the lehua are refreshed

From the song "Aia i Alakai" by Kamealoha. Hawaiian source: Holstein 5. English trans. by author.

17. Hāʻule naoa ka wai a ke Kēhau, ke nāʻū lā nā kamaliʻi. *The water of the Kēhau mist falls rippling as the children play nāʻū.*

    From a chant for Kamehameha II. Source: Pukui and Elbert, 1986, 263. *Note: Nāʻū is a children's game played at sunset. The idea is to hold one's breath the longest while saying "nāʻūūūūūūūūūūūū" in an effort to keep the sun from going down.*

18. *For another example of kēhau, see Nāulu (#21).*

Kēhaupua. Misty rain associated with Nuʻalolo, Kauaʻi.

1. He ipu wai ʻala, wai aloha
   Na ke Kēhaupua
   ʻO ke kino ia o ka Haʻikō makani
   Hali ʻala o Nuʻalolo

   A fragrant water bowl, the essence of affection
   By the Kēhaupua misty rain
   It is the embodiment of the Haʻikō wind
   That carries the perfume to Nuʻalolo

   From a makena, or lament, for ʻEmalani Kaleleonālani. Source: Nogelmeier 348.

Kenikeni. Rain associated with Līhuʻe, Kauaʻi.

1. E ka ua Kenikeni o Līhuʻe, ua pau kou hoʻopulu pē ʻana i ka ʻili o kuʻu aloha. *O Kenikeni rain of Līhuʻe, your drenching of my love's skin has ended.*

   From an obituary for Eda Kawaikauomaunahina Kalua. Hawaiian source: Kalua. English trans. by author.

2. Me ka ua Kenikeni o Līhuʻe
   E uē helu mai ʻo Kaapuwai

   With the Kenikeni rain of Līhuʻe
   Kaapuwai wails, recounting your deeds

   From a kanikau, or lament, for Eda Kawaikauomaunahina Kalua. Hawaiian source: Kaapuwai. English trans. by author.

kēwai. Misty rain. Also the name of a wind. "Kēwai" means "watery, misty, dew-laden."

Rain of Hilo Hanakahi, Hawaiʻi

1. Ua ka ua kēwai, kāpili ua o Hanakahi
   Mālie a mao aʻe ka ua

The kēwai rain falls, a rain attached to Hanakahi
Calmness, and the rain subsides

From a mele recited to Kūapāka'a by Kahikuokamoku. Hawaiian sources: Kanepuu, "He moolelo" 11/14/1867; Nakuina, *Moolelo* 47. English trans. by author. Additional source: Nakuina, *Wind* 42.

Rain of Pū'ōpelu, Maui

2. Ke kēwai lā e kahe ana i ke kula o Pū'ōpelu
Kaha ka maka o ke kanaka o Lāhainā

The misty [kēwai] rain is falling on the Pū'ōpelu plain
The eyes of the people of Lāhainā are attracted

From a mele for Lāhainālunа Seminary by Paalua. Source: Fornander, *Fornander* 6: 533.

Rain of Kona, Kaua'i

3. Hanohano Kaua'i i ka mālie
Malino i ke kai o Kona i ka la'i
Ho'oipoipo a ka ipo ahi
I ka ua kēwai lawe mālie
Pulu 'elo ke oho a loke lau li'i
Ka palai moe anu o Maialoa

Kaua'i is glorious in the calm
Calm in the serenity of the sea of Kona
The lovemaking of the fiery sweetheart
In the soft, gentle kēwai rain
The foliage of the small-leafed rose is drenched
The palai fern that rests in the cold of Maialoa

From a mele inoa, or name chant, for Keahi Luahine's great-grandmother. Hawaiian source: Barrère, Pukui, Kelly 88. English trans. by author.

Misty rain

4. Watery, dripping, misty; damp, dew-laden.

Source: Pukui and Elbert, 1986, 145.

5. Wind from a place of rain; a mist connected with rain some distance off. Pili ke kēwai, kūhaluka ka mauna (*The kēwai is close, the mountain is crowded*).

Source: Andrews 269. English trans. by author.

Rain that shakes

6. Kē pū lani ua kēwai
Ke kēwai ua lani kūkū

A striking sound in the heavens, the kēwai rain
The kēwai, the heavenly rain that shakes

From a mele for Kaumuali'i and Kapi'olani. Hawaiian source: *Na Mele Aimoku* 101. English trans. by author.

kiawe. Same as *Kīaweawe.* Rain that streams down gracefully; associated with Kāhili, Maui, and found in other areas.

Rain of Kāhili, Maui

1. Hoa o ka ua, ka lā, ke anu, me ke ko'eko'e
   Huikau loli ka nahele, ka uka o Kāhili
   Loli ka ua kiawe, puehu i luna o ka lā'au
   Ka ua kiawe 'ōnohi kīhe'e 'ula i ke kula

   Companion of the rain, the heat, the cold, and the damp
   The forest, the upland of Kāhili is confused by the change
   The kiawe rain has turned, it is scattered over the trees
   The rainbow-streaked kiawe rain that pours red over the plains

   From a kanikau, or lament, for P. Kamohai. Hawaiian source: J. B. Kawelo. English trans. by author.

Streaming rain

2. Ka ua kiawe i luna o ka lā'au. *The rain streaming down on the tree.*

   Source: Pukui and Elbert, 1986, 146.

Kīaweawe. Same as *kiawe.* Rain that streams down gracefully. Associated with Lele'anae (or Lelea'anae), O'ahu.

1. Ua kaumaha luna o Kōnāhuanui
   Loa'a akula i ka ua
   Kīaweawe o Lele'anae
   Ke ho'okahekahe maila
   I luna o Waipuhia

   The top of Kōnāhuanui is heavy
   Gotten by the rain
   Kīaweawe of Lele'anae
   Pouring down
   Above Waipuhia

   From a chant for Albert Ka Haku O Hawai'i. Hawaiian source: Ami. English trans. by author.

Kiawelehua. Rain associated with Hōpoe, Hawai'i. "Kiawe lehua" means "swaying lehua."

1. He ua Kiawelehua ko Hōpoe
   Popohe i ka lā ka pua wai Ko'olihilihi

Hōpoe has the Kiawelehua rain
Made shapely by the sun is the flower of the Koʻolihilihi water

From a chant for Kawaikini. Hawaiian source: Lohiau, "He mele no Kawaikini." English trans. by author.

2. Ka ua Kiawelehua o Hōpoe. *The Kiawelehua rain of Hōpoe.*

A traditional saying. Hawaiian source: Pukui, *ʻŌlelo* 169. English trans. by author. *Note: Pukui describes the Kiawelehua as "the rain that sets the lehua of Hōpoe to swaying" and says that "when the rain patters down, the lehua of Hōpoe, Puna, gently sway to and fro."*

Kiaweʻula. Similar to *kiawe* and *Kīaweawe.* Rain that streams down gracefully with a faint streak of red, as of a rainbow. "Kiaweʻula" is both the name of a specific rain and a generally descriptive term; its various usages are determined by the context.

Rain associated with the ʻĀpuakea rain

1. *See ʻĀpuakea (#23).*

Rain of the cliffs

2. Kuʻu keiki i ka ua kiawe ʻula i nā pali
Ke hāliʻi maila lā i ke pili
Kuʻu keiki i ke ahe a ka makani he Moaʻe
Kuʻu keiki, kuʻu keiki aloha ē

My darling child in the kiawe ʻula rain on the cliffs
Spreading out over the pili grass
My dear child in the gentle blowing of the wind, the Moaʻe
My child, my beloved child

From a kanikau, or lament, for H. Keowai. Hawaiian source: Kamala. English trans. by author.

Kīhenelehua. Rain associated with Hāmākua and Mahiki, Hawaiʻi, and with the *Pūhenelehua* rain. "Kīhene lehua" means "bundle of lehua blossoms."

Rain of Hāmākua, Hawaiʻi

1. Ka ua Kīhenelehua o Hāmākua. *The Kīhenelehua rain of Hāmākua.*

A traditional saying. Hawaiian source: Pukui, *ʻŌlelo* 169. English trans. by author. *Note: Pukui describes the Kīhenelehua as "the rain that produces the lehua clusters."*

Rain of Mahiki, Hawaiʻi

2. Lei Mahiki i ka ua Kīhenelehua
He ua Pūhenelehua no Makakolohe

I hoʻopalapalaheʻa i ka maka o ka ʻawapuhi
Pakele ʻoe mai hina i ka Puʻukohā

Mahiki is bewreathed with the Kīhenelehua rain
A Pūhenelehua rain that teases the lehua of Makakolohe
That has stained the face of the ginger blossom
You escape, almost tumbling on Puʻukohā, the hill of laughter

From a mele kālaiʻāina, or political chant, for ʻEmalani Kaleleonālani. Source: Nogelmeier 191. *Note: In this rendering, "Puʻukohā" is used instead of "Puʻukoha," which was printed in the source.*

Kīkēhala. Rain associated with Punaluʻu, Oʻahu. "Kīkē hala" means "to break open hala fruit."

1. Ka ua Kīkēhala o Punaluʻu. *The hala-pelting [Kīkēhala] rain of Punaluʻu.*

   A traditional saying. Source: Pukui, *ʻŌlelo* 169.

Kili. Similar to *kili hau, kilihune, kili nahe,* and *kili noe.* A beloved fine, light rain; to rain gently.

Rain of Waiheʻe, Maui

1. He poli aloha ka makua lā
   Ka iki hone o ke kanaka lā
   I ka ua Kili o Waiheʻe lā
   E wiki, e lohi ʻo ia ala lā

   The father has a loving heart
   The soft sweetness of man
   In the drizzling Kili rain of Waiheʻe
   He should make haste before it's too late

   From a mele inoa, or name chant, for Erisapeka. Hawaiian source: Maria. English trans. by author.

kili hau. Also known as kilikili hau. Also the name of a wind. "Kili hau" can refer to being tipsy.

Cold rain

1. Ice-cold shower; cold drizzle; to rain thus. *Fig.*, tipsy.

   Source: Pukui and Elbert, 1986, 151.

2. Ua kilikili hau me ʻoe lā, ʻeā
   Ke noe mai nei i ka nahele lā, ʻeā

   The light, cold kilikili hau is with you
   Misting the forest

From a mele inoa, or name chant, for Ulumāheihei Hoapili by Naonohielua. Hawaiian source: Kanahele 123. English trans. by author.

kilihune. Also known as kilihuna, kilikili huna, kilikilihune, and tilihune. A beloved fine, wind-blown rain.

1. Fine, light rain, wind-blown spray, drizzle; to shower lightly... He wahi lā hoʻokilihune ua kēia (*This is just a day of slight showers*).

   Source: Pukui and Elbert, 1986, 151.

2. Ua haʻalele mai nei iaʻu e ʻū i ke aloha
   I hoa pupuʻu anu i ke kau me ka hoʻoilo
   I hoa peʻe poli no ke ano ahiahi
   No ka noe liʻulā welo i ka ʻilikai
   Me ka ua kilihune i ke kulu aumoe
   Moe ʻoe i laila, i ʻaneʻi ke aloha

   You have left me mourning for love
   For a companion to cuddle with in the cold of summer and winter
   For a bosom companion to hold in the evening
   In the mist of twilight as it settles on the surface of the sea
   And the kilihune rain of midnight
   There you sleep, here is love

   From a kanikau, or lament, for S. W. Kanalu. Hawaiian source: Pilialoha. English trans. by author.

3. *For more examples of kilihune, see Kea (#6) and Lanipōlua (#2).*

kili nahe. A beloved fine, light rain.

1. Light, soft, gentle rain; pleasant patter of gentle rain; to rain gently.

   Source: Pukui and Elbert, 1986, 151.

2. He mele pūʻali inu wai
   No nā keiki lā o Kalehua
   I nanea i ka laʻi o ka hāpuʻu
   I ka ua kili nahe i ka maka o ka lehua

   A temperance league song
   For the children of Kalehua
   Who enjoy the tranquility of the hāpuʻu fern
   In the kili nahe rain on the face of the lehua blossom

   Hawaiian source: Kawahine. English trans. by author.

kili noe. Also known as kilikili noe and kilikilioe. A beloved fine, misty rain.

1. Fine, misty rain, somewhat heavier mist than the kili ʻohu.

   Source: Pukui and Elbert, 1986, 151.

2. ʻIke akula au i ka ua o Koʻiʻālana lā
   Ka ua kili noe i ka maka o ka ʻākōlea
   E wiki ana ka huakaʻi malihini o pulu i ka ua
   O ʻElekeninui, nui maila kā ke anu o ia kuahiwi

   I saw the rain of Koʻiʻālana
   The kili noe rain in the face of the ʻākōlea fern
   The tour is moving quickly to avoid the rain
   Of ʻElekeninui, the cold of that mountain is extreme

   From a makena, or lament, for ʻEmalani Kaleleonālani. Hawaiian source: Nogelmeier 356. English trans. by author. *Note: "ʻElekeninui" is also the name of a stream in Waimea, Kauaʻi.*

kili ʻohu. Fine rain and light mist. Lighter than the kili noe.

Source: Pukui and Elbert, 1986, 151.

Kiliʻoʻopu. Rain associated with Wailuku, Maui. Also the name of a wind.

Rain of Waiheʻe, Maui

1. Kuʻu kāne mai ka ua Kiliʻoʻopu o Waiheʻe
   ʻAuʻau ka ʻuhane i ka wai o Nīʻaukawa

   My dear husband from the Kiliʻoʻopu rain of Waiheʻe
   The spirit bathes in the water of Nīʻaukawa

   From a kanikau, or lament, for Kamakaokalani. Hawaiian source: Wiliama. English trans. by author.

2. He loa Puʻukoaʻe
   He pāpāʻōlelo na ka makani
   Makani lū ʻino i nā lehua o Kaukini
   Polipoli Pūlehu i ka ua Kiliʻoʻopu o Waiheʻe
   Me ka ua nā māmala ʻino a ka wai

   Expansive is Puʻukoaʻe
   A conversation held by the wind
   Wind that violently scatters the lehua blossoms of Kaukini
   Pūlehu is polished by the Kiliʻoʻopu of Waiheʻe
   With the rain come hard strokes of the water

   From a mele ʻāina, or song of the land, for Maui. Hawaiian source: Kanahele 39. English trans. by author. Additional sources: Kualu; Kanahele 54, 57; "Nani Lahaina." *Note: Both Kualu and Kanahele 54, 57 place the Kiliʻoʻopu at Waiehu, not at Waiheʻe.*

Rain of Waikapū, Maui

3. Ua Kiliʻoʻopu—Waikapū, Maui.

From a list of rain names and their descriptions. Source: "Names . . . rain."

Rain of Wailuku, Maui

4. Ua Kiliʻoʻopu—Wailuku, Maui.

From a list of rain names and their descriptions. Source: "Names . . . rain."

kinai. Also known as hoʻokina, hoʻokinakina, kina, kinakina, and kinakinai. Rain associated with Hilo, Hawaiʻi. "Kinai" means "to persist."

1. ʻO Hilo ia lā o ka ua kinai, kinakinai ka ua o Hilo, ka ua mao ʻole o Hilo. *That is Hilo of the constant [kinai] rain, the rains of Hilo go on and on, the rain of Hilo that never clears.*

From a chant. Source: Pukui and Elbert, 1986, 152. Additional source: Fornander, *Fornander* 4: 317.

Kinailehua. Rain associated with Panaʻewa, Hawaiʻi. "Kinai lehua" means "to suppress lehua flowers," "to persist on lehua flowers," or "to quench the thirst of lehua flowers."

1. Ka ua Kinailehua o Panaʻewa. *The Kinailehua rain of Panaʻewa.*

A traditional saying. Hawaiian source: Pukui, *ʻŌlelo* 169. English trans. by author. *Note: Pukui describes the Kinailehua as "the rain that bruises the lehua blossoms" and says, "Both lehua and rain are commonly found in Panaʻewa."*

Kinimakalehua. Name of a mountain rain. *Lit.*, countless lehua blossoms.

Source: Pukui and Elbert, 1986, 153. *Note: "Kinimakalehua" is also the name of a place in Moanalua, Oʻahu (Thrum 213).*

Kiʻoao. Same as *Kiʻowao.* Rain associated with Kahui and Waipuhia, Oʻahu. "Kiʻo ao" means "upland pool."

Rain of Kahui, Central Oʻahu

1. *See ʻUla (#4).*

Rain of Nuʻuanu, Oʻahu

2. *See Hāʻao (#19).*

kīʻohuʻohu. Mist.

1. Noho ana ke akua i ka nāhelehele i ālai ʻia e ke kīʻohuʻohu, e ka ua koko. *The gods dwell in the forest, hidden away by the [kīʻohuʻohu] mists and low-lying [koko] rainbow.*

From a hula prayer. Source: Pukui and Elbert, 1986, 154. Additional source: Barrère, Pukui, Kelly 82.

Ki‘owao. Same as *Ki‘oao.* Rain associated with Nu‘uanu, O‘ahu. Also found on Maui, on other parts of O‘ahu, and on Kaua‘i. "Ki‘o wao" means "upland pool."

Rain of Maui

1. *See ‘Ulalena (#2).*

Rain of O‘ahu

2. The name of the mist or cloud almost always settled on the hills of O‘ahu.

   Source: Andrews 273.

3. Ku‘u kāne mai ka ua Ki‘owao ē
   Mai ka ua Pō‘aihala o Kahalu‘u ē

   My dear husband from the Ki‘owao rain
   From the Pō‘aihala rain of Kahalu‘u

   From a kanikau, or lament, for Ioseba Nawai. Hawaiian source: Noholoa. English trans. by author.

4. The Ki‘owao rain, followed by a lingering mist, is common in Nu‘uanu Valley, but it is also found between Ka‘ala and Mānoa.

   From an email sent to the author by Kimo Alama Keaulana on April 4, 2013.

5. *For more examples of Ki‘owao of O‘ahu, see Maka‘ūpili (#3) and Wa‘ahila (#14).*

Rain of Halemano, O‘ahu

6. Lei Kaua‘ula, kāhiko i Halemano
   Kahe ka wai a ke Ki‘owao
   I ka pāku‘iku‘i ‘ia e ka ua Paliloa, i laila
   ‘Eha ka nahele o Malamanui i ke anu ē

   Kaua‘ula is festooned with a lei, an adornment at Halemano
   The waters of the Ki‘owao rain flow
   Augmented by the Paliloa rain, there
   The forest of Malamanui aches in the cold

   From a song for Lunalilo. Hawaiian source: Ho‘oulumāhiehie, *Ka Mo‘olelo* 252. English trans. by author. Additional source: Ho‘oulumāhiehie, *Epic* 236.

Rain of Honolulu, O‘ahu

7. Mā‘alo ho‘okahi ka ‘uhane i ka la‘i o Honolulu
   E ho‘olohe ana i nā pihe ‘uā lua
   E nānā ana i ka ua Ki‘owao
   Ke hāli‘i maila lā i ke pili

E ho'i e pili ka 'uhane me ke kāwelu
I walea ka 'uhane i ke kui lei kāmakahala

The spirit passes alone through the calm of Honolulu
Listening to the many cries of wailing
Watching the Ki'owao rain
That is covering the pili grass
The spirit returns to touch the kāwelu grass
The spirit enjoys stringing lei kāmakahala

From a kanikau, or lament, for B. Mahune. Hawaiian source: Wailea. English trans. by author.

8. *For another example of Ki'owao of Honolulu, see Kūkalahale (#10).*

Rain of Ka'au, Wai'anae Uka, O'ahu

9. Ke 'ū ala ka ua Ki'owao
Wai ka nahele o Ka'au

The Ki'owao rain groans
The forest of Ka'au is watery

From a chant. Hawaiian source: "Kuluikului." English trans. by author.

Rain of Kahaukomo, O'ahu

10. E kahi a pono ke oho i Kahaukomo
Ke oho o ke kāmakahala pewa i ka ua
He oho 'ōmau līpine 'ia a ke Ki'owao

Comb out the hair properly at Kahaukomo
Like soft petals of the kāmakahala, gathered by the rain
Tresses held in ribbons by the Ki'owao mists

From a mele hō'ihi, or honorific chant, praising the hair of 'Emalani Kaleleonālani. Source: Nogelmeier 53.

11. Pali Waipuhia pahe'e Kahaukomo, 'eā
Komo i nā 'ale a wai a ke Ki'owao, 'eā

Waipuhia is a cliff-face, and Kahaukomo is slippery
Entering the liquid billows of the Ki'owao rain

From a mele māka'ika'i, or travel chant, for 'Emalani Kaleleonālani by Kama'iku'i. Source: Nogelmeier 152.

Rain of Kahui, Central O'ahu

12. Me he kanaka maika'i lā ka ua Ki'owao
Ka pua 'ula i ka uka o Kahui ē, i laila

The Ki'owao rain is like a handsome man
The red blossom in the uplands of Kahui, there

From a mele reflecting on the travels of Hiʻiakaikapoliopele and on her consequent nickname, Wahinepōʻaimoku (Island-encircling woman). Hawaiian source: Hoʻoulumāhiehie, *Ka Moʻolelo* 173. English trans.: Hoʻoulumāhiehie, *Epic* 162.

13. *For another example of Kiʻowao of Kahui, see Kuahine (#2).*

Rain of Kamaoha, Oʻahu

14. *See Kēhau (#7).*

Rain of Keʻehu, Oʻahu

15. ʻAu hoʻolana ʻo Keʻehu wai kēhau
Ka huna wai kēhau a ke Kiʻowao
I kulu kiʻo, hālana i ka pua o ka nēnē

Keʻehu floats in the misty waters
The droplets of dew from the Kiʻowao rain
Flowing into a pool, flooding the flowers of the nēnē

From a mele for Hakaleleponi. Hawaiian source: Fornander, *Fornander* 6: 537. English trans. by author.

Rain of the Koʻolau Mountains, Oʻahu

16. *See ʻUlalena (#23).*

Rain of Lanihuli, Oʻahu

17. Maikaʻi ka ua i Nuʻuanu lā
Kāhiko i ke kāmakahala lā
Nā pua ʻāhihi o Lanihuli lā
I kuia mai e ke Kiʻowao lā
E ka ua Waʻahila i nā pali lā
Ke nihi aʻela i ka nahele lā
Loku ana i ka lau kukui lā
I ke oho lau liʻi o ke koa lā
Malu ai nā hale i ka uka lā
Ka uka i Kahālauaola lā

The rain at Nuʻuanu is splendid
Adorning the kāmakahala blossoms
The ʻāhihi blossoms of Lanihuli
Strung into lei by the Kiʻowao
By the Waʻahila rain on the cliffs
Softly moving through the forest
Showering heavily upon the kukui leaves
Upon the little leaves of the koa trees
That shelter the houses in the upland
The upland at Kahālauaola

From a mele hōʻihi, or honorific chant, for ʻEmalani Kaleleonālani by Wahineikaili Kamaʻikuʻiʻōpio. Hawaiian source: Nogelmeier 34. English trans. by author.

Rain of Luakaha, Oʻahu

18. *See Kūkalahale (#11).*

Rain of Māeaea, Oʻahu

19. ʻO ka noe mai a ka ua Kiʻowao lā
Me ke one wali o Māeaea lā

The mist of the Kiʻowao rain
And the smooth sands of Māeaea

From the song "Kuu hoa i Halemano" by Mrs. Eliza Holt. Hawaiian source: Holstein 118. English trans. by author.

Rain of Nuʻuanu, Oʻahu

20. Ua Kiʻowao—Nuʻuanu; hard rain that comes down the middle of the valley.

From a list of rain names and their descriptions. Source: "Names . . . rain."

21. ʻO Nuʻuanu ua huli i ka makani
Ua lālau nā lima o ka ua Kiʻowao
E ʻāʻumeʻume ana me Leleaʻanae

Nuʻuanu has turned into the wind
The hands of the Kiʻowao rain have taken hold
Struggling with Leleaʻanae

From a mele for Kanaʻina. Hawaiian source: Fornander, *Fornander* 6: 540. English trans. by author.

22. ʻO Nuʻuanu kuahiwi kiʻekiʻe i luna
Me he mea ala i hana ʻia ka nani e ka ua Kiʻowao

Nuʻuanu is the high mountain region above
As though the beauty was crafted by the Kiʻowao rain

From a mele kālaiʻāina, or political chant, for ʻEmalani Kaleleonālani. Source: Nogelmeier 177–78.

23. Kū ʻalawa nā maka i uka
Nānā ʻo ka ua Kiʻowao
Paʻū maila i ka nahele
Hoʻokio maila i nā pali
ʻUhane hele lau lāʻau
Kui ʻāhihi o Nuʻuanu

Stand with the eyes glancing inland
Observing the Kiʻowao rain

Soaking the forested regions
Whistling among the cliffs
A ghostly spirit traversing the leaves
Stringing the ʻāhihi blossoms of Nuʻuanu

From a kanikau, or lament, for ʻEmalani Kaleleonālani. Source: Nogelmeier 324.

24. Kapu Nuʻuanu i ka ua Kiʻowao lā
ʻO ke alo hāliu i Heʻeia lā
Nānā i ka ua Mololani lā

Nuʻuanu in the Kiʻowao rain is sacred
Its front turned toward Heʻeia
Observing the Mololani rain

From a mele inoa, or name chant, for Albert Ka Haku O Hawaiʻi. Hawaiian source: Haui. English trans. by author.

25. Ua nani ʻo Nuʻuanu
I ka lau o ke kāwelu
Ua hālāwai akula
Me ka makani nui
E ʻiʻini ana ka manaʻo
E ʻike iā Kahuailana
ʻO koʻu hoa nō ia
ʻO ka ua Kiʻowao

Beautiful is Nuʻuanu
With its blades of kāwelu grass
Having been met
With strong winds
One desires
To see Kahuailana
This is indeed my friend
The Kiʻowao rain

From a mele inoa, or name chant, for Kamehameha IV. Hawaiian source: Luahine. English trans. by author. Additional source: "He kanikau no ka moi i aloha nuiia." *Note: Kahuailana is Kahuailanawai, a pool in upper Nuʻuanu.*

Rain of Nuʻuanu, Oʻahu, and of Alakaʻi, Kauaʻi

26. Cool mountain rain accompanied by wind and fog, sometimes associated with Alakaʻi Swamp on Mt. Waiʻaleʻale, Kauaʻi, as well as Nuʻuanu Valley, Oʻahu.

Source: Pukui and Elbert, 1986, 154. *Note: Pukui and Elbert, 1986, 509, say that "Kiʻowao" means "upland."*

Rain of the forest

27. Ua Kiʻowao nō ka ua mau i ka nahele, i ka wao kele, wao akua, i ka moku lāʻau, pewa lāʻau, noho mau ka ua ma ia wahi, no laila i kapa ʻia aku ai ka inoa o kēia ua, ua Kiʻowao.

    The Kiʻowao is the rain that is always in the forest, in the upland forest, in the uninhabited mountain region, in the grove of trees and the tree clumps. The rain is constant in such places; therefore, the name of this rain is called the Kiʻowao rain.

    From a letter to the newspaper *Kuokoa* regarding famous rains. Hawaiian source: Kalaaukumuole, "Manao wehewehe." English trans. by author. Additional source: Kalaaukumuole, "Thought explanations."

Other examples

28. Kohāhā ka leo ka ua Kiʻowao. *Loud sounds the rain Kiʻowao.*

    From a chant. Source: Pukui and Elbert, 1986, 158.

29. Kiliʻopu mai nei e ke anu
    Kilikili hau i kuʻu poli
    Konini kalaʻihi ka ʻōpua
    I ke kuwai ʻia e ke Kiʻowao

    Delighted by the cold
    Sprinkling upon my heart
    The stormy cloud moves to and fro
    Rubbed by the Kiʻowao rain

    From the song "He wahi kilihau anu" by Leleiōhoku. Hawaiian source: Holstein 28. English trans. by author.

30. ʻO ka ua Kiʻowao. I hea lā kahi kūpono e kapa aku ai i kēia ua, he Kiʻowao? No ka mea, ʻehā nō mau wao i mōakāka leʻa iā kākou, ʻo ia hoʻi kēia: 1. Wao akua. 2. Wao nahele. 3. Wao kele. 4. Wao kanaka. ʻO ka ua lā i ka wao hea, ka ua i kapa ʻia aku ai pēlā ka inoa? I ka wao kele paha, i ka wao nahele paha, i ka wao akua paha? E haʻi mai ka mea ʻike i ke ʻano.

    The Kiʻowao rain. Where is the place from which this rain gets the name Kiʻowao? The reason being, there are four regions (wao) that are well known to us. They are: 1. Wao akua (distant mountain region). 2. Wao nahele (inland forest region). 3. Wao kele (rainforest region). 4. Wao kanaka (upper agricultural region). From which wao does the rain derive its name? Perhaps in the wao kele, maybe in the wao nahele, or perhaps in the wao akua? May those who know reply.

    From a letter to the newspaper *Kuokoa* regarding rain, wind, and sun names. Hawaiian source: Kanuha, "Mau mea." English trans. by author. Additional source: Kanuha, "Some things."

Kīpēhala. Rain associated with ʻEleao, Hawaiʻi. "Kīpē hala" means "to pelt hala trees."

1. Ua mao ʻole ka ua o ʻEleao
No ʻEleao ka ua Kīpēhala
Ua hoʻopumehana o Hanakahi

The rain of ʻEleao is unceasing
The Kīpēhala rain belongs to ʻEleao
Rain that warms Hanakahi

From a chant for Hilo in the legend of Pākaʻa and Kūapākaʻa. Hawaiian source: Kanepuu, "He moolelo" 11/14/1867. English trans. by author. Additional sources: Nakuina, *Moolelo* 48; Nakuina, *Wind* 43.

Kīpehipehipuahala. Rain associated with Pohukaina, Oʻahu. "Kīpehipehi pua hala" means "to pelt hala fruit."

1. *See Kūkalahale (#19).*

Kīpehipualehua. Rain associated with Mākaulele (or Mōkaulele), Hawaiʻi. "Kīpehi pua lehua" means "to pelt the lehua blossoms."

1. He kanikau aloha no Mary Haliilua
Kuʻu wahine mai nā pali huli lua o Hilo
Mai ka ua Kīpehipualehua o Mākaulele
Ke nihi aʻela ma kai o Panaʻewa

This is a chant of mourning with affection for Mary Haliilua
My dear wife from the cliffs of Hilo that face each other
From the Kīpehipualehua rain of Mākaulele
Silently creeping along the oceanside of Panaʻewa

From a kanikau, or lament, for Mary Kaawaloa i Haliilua. Hawaiian source: Keliihaleole. English trans. by author.

Kīpū. Same as *ʻĀpuʻupuʻu*, *Kīpuʻu*, and *Kīpuʻupuʻu*. Rain and wind associated with Waimea, Hawaiʻi. "Kīpū" means "to bundle up" or "to remain," as mist or rain. It may also be a shortened form of "Kīpuʻu." "Kīpū" is both the name of a specific rain and a generally descriptive term; its various usages are determined by the context.

1. Ua Kīpuʻupuʻu o Waimea ("ua Kīpū" ka pono): I uka nō ka ua e kīpū mai ai, ʻo ka wai ke huʻe ana i kai o Waiʻauwia me Waihaka, he hiki nō i ke kanaka ke hele i kahi e makemake ai me ke kīpū nō i ke kapa nui, no laila mai kēia inoa, ua Kīpū o Waimea.

Kīpu‘upu‘u rain of Waimea ("Kīpū rain" is the proper term): Inland is where the rain remains [kīpū]; the water washes out below Wai‘auwia and Waihaka. People can go wherever they please, wrapped tightly [kīpū] in large kapa; that is how it got this name, Kīpū rain of Waimea.

From a letter to the newspaper *Kuokoa* regarding famous rains. Hawaiian source: Kalaaukumuole, "Manao wehewehe." English trans. by author. Additional source: Kalaaukumuole, "Thought explanations."

2. Hana Waimea i ka ‘upena a ka makani
I ka ‘alihi pīkoi a ke Kīpū
A ka ua Leikoko‘ula i ke pili
Me he ‘ahu‘ula ala i luna o ka lā‘au

Waimea creates a net with the wind
With the upper cord formed by the Kīpū wind-rain
And the Leikoko‘ula rain over the pili grass
Like a feather cape above the trees

From a mele māka‘ika‘i, or travel chant, for ‘Emalani Kaleleonālani. Hawaiian source: Nogelmeier 121. English trans. by author.

3. Aloha ‘ino nō ku‘u kaikamahine, ku‘u lei ‘ā‘ī, ku‘u hoa hele o ke alanui ē

. . . .

E ho‘i kākou i ke anu o Waimea, makani Kuehulepo o Līhu‘e, i ka ua Kīpū i nā pali ē

Pity for my daughter, the lei that adorns my neck, my traveling companion on the road

. . . .

Let us return to the cold of Waimea, to the Kuehulepo wind of Līhu‘e, in the Kīpū rain on the cliffs

From a mele ‘uhane, or lament, for Kealohapauole, presumably by Kaikamahineole. Hawaiian source: "He uhane." English trans. by author.

Kīpu‘u. Same as *‘Āpu‘upu‘u*, *Kīpū*, and *Kīpu‘upu‘u*. Rain and wind associated with Waimea, Hawai‘i. "Kīpu‘u" refers to the bumpiness of the skin in the cold, wind-driven rain. It may also refer to the hilly land of Waimea.

1. E ho‘i aku nō wau e pili
Me ka ua Kīpu‘u o ka nahele

And I will return to be close
With the Kīpu‘u rain of the forest

From the song "By and by ho‘i mai ‘oe." Source: Lili‘uokalani 93–94. Additional source: Holstein 23. *Note: The editors of Lili‘uokalani say that the abbreviated name*

"Kīpuʻu" in this song is "the only name that clearly links the song to the island of Hawaiʻi," as it is a cold wind and rain of Waimea. In Holstein, the song is credited to "Kuahine," possibly referring to Queen Liliʻuokalani.

Kīpuʻupuʻu. Same as *ʻĀpuʻupuʻu*, *Kīpū*, and *Kīpuʻu*. Rain and wind associated with Waimea, Hawaiʻi, and found in other areas. "Kīpuʻupuʻu" refers to the bumpiness of the skin in the cold, wind-driven rain. It may also refer to the hilly land of Waimea.

Rain of Hilo and Waimea, Hawaiʻi

1. The ua Kīpuʻupuʻu is a rain with a wind that beats against you. It is known in Hilo as well as Waimea and lasts but a few minutes.

   From a list of rain names and their descriptions. Source: Kelsey, "Rains."

Rain of Mahiki, Hawaiʻi

2. I kēia manawa i hao mai ai ua kupua Mahiki nei i kona mana, ʻo ia hoʻi, nou maila ʻo ia i ka ua kīkoni ʻili o ka ulunahele o Mahiki, pō ka ululāʻau mai ʻō a ʻō.

   I kēlā wā i neʻe aku ai ʻo Hiʻiaka i mua, a ʻo ke aikāne aku nō ma hope ona, a i ia wā i ʻōlelo aʻe ai ʻo Hiʻiaka i ke kahu, iā Pāʻūopalaʻā, "ʻAuhea ʻoe, e Pāʻūopalaʻā!"

   Ō maila nō hoʻi ke kahu me ka nīnau ʻana mai, "He aha ia, e kuʻu haku?"

   "E malu aʻe hoʻi kō kinolau ma luna o māua, i ʻole hoʻi māua e maka ʻeleʻele i ka ua pehi maka a Mahiki."

   He manawa ʻole ia, malu ana nā lau palaʻā me ke ʻamaʻu ma luna o lāua nei, hele ka ua me ke koʻekoʻe ma waho. Oi hele lāua nei a mao ka ua, i nānā mai ka hana o Mahiki, e hele aku ana nō lāua nei, me kona manaʻo loa ua make lā ʻo Hiʻiaka mā i ke anu a me ka ʻōpili o ke kuāua āna i hoʻokuʻu mai ai. . . .

   Lohe maila ua Mahiki nei i kēia kau a Hiʻiaka, pane maila ʻo ia me ka huhū, "ʻAʻole ʻoe e ola iaʻu, e kēnā wahine hoʻokano. E make ana ʻoe iaʻu i nei lā. Hala aʻela ka make a kuʻu kino ua Kīpuʻupuʻu, eia mai nō ua make ou."

   Then Mahiki employed his powers, pelting them with the piercing rain of Mahiki Forest, darkening the entire woodland from edge to edge.

   Hiʻiaka moved forward with her companion behind her, as she spoke to her attendant, Pāʻūopalaʻā, "Are you listening, Pāʻūopalaʻā?"

   The attendant responded, asking, "What is it, my chiefess?"

   "Make your body forms into a shelter above us, so we are not blinded by the eye-piercing rain of Mahiki Forest."

In no time, fronds of palaʻā and ʻamaʻu fern made a shelter for the two of them as the icy rain poured down outside. They carried on until the rain diminished, and when Mahiki looked, he saw they were still going, while he had been certain that Hiʻiaka and her fellows had died in the cold and cramping chill of the torrents of [kuāua] rain he had unleashed. . . .

Mahiki heard this chant by Hiʻiaka and snarled back, "I shall not spare you, you arrogant woman. You will die by my hand this day. You escaped my rain form, the Kīpuʻupuʻu downpour, but here comes the death of you."

From the legend of Hiʻiakaikapoliopele. Hawaiian source: Hoʻoulumāhiehie, *Ka Moʻolelo* 114. English trans.: Hoʻoulumāhiehie, *Epic* 107–8. *Note: "Mahiki" is both the name of a place in Waimea and the name of a male demigod of the Mahiki Forest.*

3. A mai laila mai a mahiki ke ama o ka waʻa i ka ua Kīpuʻupuʻu, a hōʻea nō hoʻi i nā puʻu kūlua o Kohala.

   With a mahiki, a leap, like the leaping outrigger of a canoe, we reached Mahiki in the Kīpuʻupuʻu rain of Waimea, and then we made it to the companion hills of Kohala.

   From Hiʻiakaikapoliopele's account of her journey to a steersman as they crossed the Pailolo Channel on their way to Oʻahu. Hawaiian source: Hoʻoulumāhiehie, *Ka Moʻolelo* 141. English trans.: Hoʻoulumāhiehie, *Epic* 133.

4. Kīmō ka ʻalā i ke poho o Mahiki
   He hana lalahū ia na ke Kīpuʻupuʻu

   The ʻalā stone pounds the hollows of Mahiki
   A stiffening brought on by the Kīpuʻupuʻu wind and rain

   From a mele by Hiʻiakaikapoliopele as she noticed her husband's arousal upon seeing her and Wahineʻōmaʻo draw up their skirts. Hawaiian source: Hoʻoulumāhiehie, *Ka Moʻolelo* 357. English trans. by author. Additional source: Hoʻoulumāhiehie, *Epic* 332.

5. Lei Mahiki i ka ua me ka makani
   Nanai nā ʻale nani a ke Kīpuʻupuʻu
   He nalu kāhiko ʻia na Kawelowelo
   Ke mahalo ʻia maila e Kawiliwahine, i laila
   I laila nō māua me ke koʻekoʻe
   E nonoho ai i ka ua Kīpuʻupuʻu
   Me oʻu hoa Mālanaʻeʻelekoa

   Mahiki wears a lei of rain and wind
   The glorious billows of the Kīpuʻupuʻu strut
   Waves festooned by Kawelowelo
   Admired by Kawiliwahine, there
   There we were in the chill

Dwelling in the Kīpuʻupuʻu rain
With my friend Mālanaʻeʻelekoa

From a mele recalled by Hoʻoulumāhiehie as he described the fine physiques of Hiʻiaka-ikapoliopele and her beautiful companions. Hawaiian sources: Hoʻoulumāhiehie, *Ka Moʻolelo* 115; "He akua." English trans. by author. Additional source: Hoʻoulumāhiehie, *Epic* 109. *Note: In this rendering, "Mālanaʻeʻelekoa" is used instead of "mālana ʻeʻelekoa," which was printed in Hoʻoulumāhiehie,* Ka Moʻolelo *115.*

Rain of Puʻuepa, Hawaiʻi

6. Pakele ʻoe mai hina i ka Puʻukohā
Kawiliwahine i Kahuenahā
Me he pahi lā e ʻō mai ana ka ua Kīpuʻupuʻu i ke kula o Puʻuepa

You escape, having almost fallen because of the Puʻukohā
Kawiliwahine at Kahuenahā
Like a knife that pierces is the Kīpuʻupuʻu rain on the plains of Puʻuepa

From a mele kālaiʻāina, or political chant, for ʻEmalani Kaleleonālani. Hawaiian source: Nogelmeier 191. English trans. by author. *Note: In this rendering, "Puʻukohā" is used instead of "Puʻukoha," "Kawiliwahine" instead of "Kāwili wāhine," and "Kahuenahā" instead of "ka hua nahā," which were printed in the Hawaiian source.*

Rain of Uli (or ʻŌuli), Hawaiʻi

7. Kuʻu hoa i ke anu o Waimea
Mai ka ua Kīpuʻupuʻu anu o Uli
E nanahu ana i ka ʻili o ke kanaka a ʻeha

My companion in the cold of Waimea
From the cold Kīpuʻupuʻu rain of Uli
Biting the skin of a man till it hurts

From a mele inoa, or name chant, for Albert Ka Haku O Hawaiʻi. Hawaiian source: Keelikolani, "He mau inoa." English trans. by author.

8. Miki ke kai a ka ua Kīpuʻupuʻu
Hele hoʻokano i ka nahele o Uli
Hālāwai aʻela me ka ua Paliloa
ʻO nā kūkā ʻana ia me ke anu
Ua paʻa nā manaʻo me ke Kīpuʻupuʻu

Receding like the sea, the Kīpuʻupuʻu rain pulls away
Strutting to the forest of Uli
Meeting with the Paliloa rain
Consulting with the cold
The thoughts are fixed on being with the Kīpuʻupuʻu

From a mele that describes the smiting of several young aliʻi of Niʻihau with the incomparable and unattainable beauty of Puakaʻōhelo. Hawaiian source: "He kaao . . . no Puakaohelo" 3/17/1894. English trans. by author. Additional source: "Miki ke kai."

Rain of Waimea, Hawaiʻi

9. Ka ua Kīpuʻupuʻu o Waimea. *The Kīpuʻupuʻu rain of Waimea.*

A traditional saying. Source: Pukui, *ʻŌlelo* 169. *Note: Pukui says that this phrase is "an expression often used in songs of Waimea, Hawaiʻi. When Kamehameha organized an army of spear fighters and runners from Waimea, they called themselves Kīpuʻupuʻu after the cold rain of their homeland."*

10. Ke Kīpuʻupuʻu hoʻānu ʻili o Waimea. *The skin-chilling Kīpuʻupuʻu of Waimea.*

A traditional saying. Hawaiian source: Pukui, *ʻŌlelo* 188. English trans. by author.

11. Waimea i ka ua Kīpuʻupuʻu. *Waimea in the Kīpuʻupuʻu rain.*

A traditional saying. Hawaiian source: Pukui, *ʻŌlelo* 319. English trans. by author. Additional source: Elbert and Mahoe 50. *Note: Pukui says, "Waimea, Hawaiʻi, is famed in old mele for its cold Kīpuʻupuʻu rain."*

12. Name of a chilly wind and rain at Waimea, Hawaiʻi.

Source: Pukui and Elbert, 1986, 155.

13. Ua Kīpuʻupuʻu—Waimea, Hawaiʻi; like hail.

From a list of rain names and their descriptions. Source: "Names . . . rain."

14. I kēia wahi, huli maila ʻo Hiʻiaka a ʻōlelo maila i ke aikāne, "Kokoke kāua e hōʻea i Hāmākua a mai laila aku, hiki kāua i ka ua Kīpuʻupuʻu o Waimea, a mai ia wahi aku hōʻea kāua i nā Kohala."

Here, Hiʻiaka turned and spoke to her intimate companion, "We will soon reach Hāmākua, and from there, we will arrive at the Kīpuʻupuʻu rain of Waimea, and from there, we will reach the Kohala lands."

From a description of the journey ahead of Hiʻiakaikapoliopele and Wahineʻōmaʻo as they travel to fetch Lohiʻau, Pele's lover. Hawaiian source: Hoʻoulumāhiehie, *Ka Moʻolelo* 112. English trans. by author. Additional source: Hoʻoulumāhiehie, *Epic* 106.

15. "Ua pono kou mau ʻau," wahi a Hiʻiaka i pane aku ai i kēia wahi kanaka, me kona hoʻomau ʻana mai nō hoʻi i ke kamaʻilio ʻana, "ʻO nā ʻau nō kēnā ou, hole nō Waimea i ka ihe a ka makani. He hewa ia lā?"

Pane maila nō hoʻi ua wahi kanaka nei me ka ʻōlelo ʻana mai, "Na wai hoʻi ka hewa o kēia mea maikaʻi āu i hana maila, e Hiʻiaka? ʻAʻole nō paha e nele ana koʻu ʻike hou i ka ua Kīpuʻupuʻu o kuʻu ʻāina hānau."

"Your stalks are fine," said Hiʻiaka to this man, then adding, "Those are now your stalks, stripped just like Waimea was stripped by the spears of the wind. Is anything wrong now?"

The man spoke in reply, saying, "Who could fault this fine thing you have done, Hiʻiaka? Now, for certain, I will again see the Kīpuʻupuʻu rain of my homeland."

From an account of Hiʻiakaikapoliopele's restoration of two ghosts to life by inserting kī stalks into their skin. Hawaiian source: Hoʻoulumāhiehie, *Ka Moʻolelo* 118. English trans.: Hoʻoulumāhiehie, *Epic* 111.

16. Iā ia nei e kū ana i waena o nā kānaka, ua lilo kēia i mea nānā nui ʻia mai e ua poʻe kamaʻāina nei o "Hole Waimea i ka ihe a ka makani, i ka hao mai a nā ʻale a ke Kīpuʻupuʻu."

. . . .

"Ua pono," wahi a ke aliʻi wahine o Waimea i pane mai ai me ka nīnau ʻana mai hoʻi, "ʻAʻohe naʻe paha ou makemake e komo i ka mokomoko a ko mākou poʻe kānaka ikaika e leʻaleʻa mai nei?"

"Na wai hoʻi ka ʻole? He mea maikaʻi nō hoʻi paha ia āu, e ke aliʻi, e nīnau maila. He aha auaneʻi hoʻi, e leʻaleʻa malihini aku nō hoʻi paha au me nā keiki o ka ua Kīpuʻupuʻu nei," i pane aku ai ʻo Kauakahiapaoa.

. . . .

I ka lohe ʻana hoʻi o ʻĀinakō, ke koa, ka ikaika a me ka lawakua o Waimea, he keiki Kauaʻi kona hoa paio, ua lelele aʻela ʻo ia i luna, a hoʻokīkakaha ihola kona mau kīpoʻohiwi, me ka paha ʻana iho.

ʻO Waimea kēia
O ka ua Kīpuʻupuʻu
Puʻupuʻu wela a ʻĀinakō
Haʻahaʻa Waiʻaleʻale

A maopopo ko Kauakahiapaoa kū ʻana e mokomoko me ko ka ua Kīpuʻupuʻu keiki ikaika a kāʻeʻaʻeʻa hoʻi ma ia hana, pane maila ke aliʻi wahine, i mua ʻo Kauakahiapaoa.

As he mingled among them, he was much noticed by the people of "Waimea, stripped by the spears of the wind and struck by the wind-driven billows of the Kīpuʻupuʻu rain."

. . . .

"Good, then," said the chiefess of Waimea, asking, "Would you care to join the grappling match our athletes are enjoying?"

"Who could refuse? It is good of you, Chiefess, to ask. It seems I shall enjoy myself as a visitor with the sons of the Kīpuʻupuʻu rain," replied Kauakahiapaoa.

. . . .

When ʻĀinakō, the brave, the strong, the giant of Waimea, heard that a Kauaʻi youth would be his opponent, he jumped up and flexed his shoulders, uttering this taunt.

This is Waimea
Of the pummeling Kīpuʻupuʻu rain
With ʻĀinakō's fiery fists
Waiʻaleʻale will be humbled

Once it was set for Kauakahiapaoa to fight the strongest young champion of this land of the Kīpuʻupuʻu rain, the chiefess told Kauakahiapaoa that he should first go to the men's eating house.

From an account of a visit to Waimea, Hawaiʻi, by Kauakahiapaoa, a Kauaʻi chief. Hawaiian source: Hoʻoulumāhiehie, *Ka Moʻolelo* 400–2. English trans.: Hoʻoulumāhiehie, *Epic* 374–75.

17. "ʻAʻole i nele ʻo Waimea nei lā i ka ʻaʻahu hau nani e pau ai ka wela ikiiki o nā kānaka i koʻele maila mai Kohala mai a hōʻea ihola i ka ua Kīpuʻupuʻu nei. Nānā nō hoʻi ka maka, a wae nō hoʻi i ka ʻono a nā mea e makemake ai, a laila, he mea ʻōʻō wale ʻia nō i ka nahele o Mahiki, a ʻo ia ka wā e ʻike ʻia ai ka hao mai a nā ʻale a ke Kīpuʻupuʻu. Pehea ia manaʻo iā ʻoe, e ke aliʻi?"

. . . .

"Aia hoʻi a hewa koʻu waha i ka ʻai āu, e ke aliʻi, a laila, hōʻike aku au iā ʻoe i kaʻu mau pane no kāu kāua iaʻu e kaʻulua iki i Waimea nei i kēia pō, i ʻike au i ka ʻiniki a ka hau anu o ka ua Kīpuʻupuʻu nei," wahi a Kauakahiapaoa i pane aku ai i ke aliʻi wahine.

"Waimea always has a beautiful cloak of the [hau] dew to ease the heat and discomfort of those who have trudged from Kohala to reach this land of the Kīpuʻupuʻu rain. Look about, choose the flavor you desire, and let your spear be thrust into the forest of Mahiki, for only then will you really come to know the surging billows of the Kīpuʻupuʻu. Does that thought appeal to you, O Chief?"

. . . .

"Once I have eaten my fill of your fare, Chiefess, I will respond to your offer for me to linger in Waimea tonight and experience the cool nip of the Kīpuʻupuʻu rain," Kauakahiapaoa replied to the royal woman.

From a conversation between the chiefess of Waimea, Hawaiʻi, and Kauakahiapaoa. Hawaiian source: Hoʻoulumāhiehie, *Ka Moʻolelo* 404. English trans.: Hoʻoulumāhiehie, *Epic* 377.

18. Aloha ka ua Kīpuʻupuʻu o Waimea
Mea mai ana ʻoe i kou ʻiʻi

Huʻea kou ʻanoʻi mai loko a waho
Eia au lā, ua kānāwai

Beloved indeed is the Kīpuʻupuʻu rain of Waimea
You speak to me of your desire
Your desire pouring forth from within
Yet here I am, bound by oath

From a mele of Kauakahiapaoa refusing a proposition made by the chiefess of Waimea, Hawaiʻi. Hawaiian source: Hoʻoulumāhiehie, *Ka Moʻolelo* 404. English trans.: Hoʻoulumāhiehie, *Epic* 378.

19. Huli aʻela ʻo Kalaniʻōpuʻu i ka mano pā pōhaku mai o nā aliʻi o Hawaiʻi... ʻo Kānekoa hoʻi, ka ua Kīpuʻupuʻu o Waimea.

Kalaniʻōpuʻu turned to the many stone walls [guardians] from among the chiefs of Hawaiʻi... and to Kānekoa, the Kīpuʻupuʻu rain of Waimea.

From the story of Kamehameha I. Hawaiian source: Kamakau, *Ke Kumu Aupuni* 22. English trans. by author. Additional source: Kamakau, *Ruling* 86.

20. Aloha maila ʻo Kamehameha iā Kānekoa, i kona hoʻomanaʻo ʻana aʻe i ko Kānekoa hāʻawe hoʻomanawanui ʻana iā ia i ke kula o Kawaihae—i ka piʻina ikiiki o Lanimaomao—ʻalo aku ʻo ka ua Kīpuʻupuʻu o Waimea. Ua iho ka ua Kīpuʻupuʻu a lana ka wai i ke kua o Kānekoa, ʻo Kamehameha nō ma hope. A ke piʻi akula lāua ʻo ka piʻina o Uli, e hoʻokūnihi aʻe ana ʻo Kānekoa iā ia ma ka ʻaoʻao. ʻAe, he lei ʻāʻī ʻo ia na kona poʻe makua kāne.

Kamehameha thought fondly of Kānekoa, recalling how Kānekoa patiently carried him on his back over the plains of Kawaihae, and over the stifling rise of Lanimaomao, enduring the Kīpuʻupuʻu rain of Waimea. The Kīpuʻupuʻu rain descended and the water rested on Kānekoa's back, with Kamehameha behind. And as they climbed the ascent of Uli, Kānekoa kept him safe on the side along the precarious ridge. Yes, he was a beloved one to his uncles.

From the story of Kamehameha I. Hawaiian source: Hoʻoulumāhiehie, "He moolelo" 2/22/1906. English trans. by author. Additional sources: Kamakau, *Ke Kumu Aupuni* 76; Desha, *He Moʻolelo* 1: 242; Kamakau, *Ruling* 124; Desha, *Kamehameha* 170.

21. ʻO Waimea i ka ua Kīpuʻupuʻu ma ka ʻohi ʻia ʻana o nā kānaka koa, a ʻo Nāpoʻopoʻo hoʻi kekahi o nā ʻāina aliʻi i paʻa ai ke aupuni o Kamehameha.

Waimea in the Kīpuʻupuʻu rain was where warriors were drafted, and Nāpoʻopoʻo was one of the royal lands that solidified Kamehameha's dominion.

From the story of Kamehameha I. Hawaiian source: Desha, *He Moʻolelo* 1: 279. English trans. by author. Additional source: Desha, *Kamehameha* 194.

22. Ku'u makuahine ho'i
Ku'u makuahine o ka ua Kīpu'upu'u anu o Waimea
Mai ka makani anu lā he Waikōloa

My beloved mother indeed
My dear mother of the cold Kīpu'upu'u rain of Waimea
From the chilly wind, a Waikōloa

From a kānaenae, or chant of eulogy, for H. Kamahiai. Hawaiian source: Makuaole. English trans. by author.

23. Aia lā, aia lā, 'o ka makani kū honua
Makani ua Nāulu ko Kawaihae
He ua anu hō'eha 'ili ka makani
Ua Kīpu'upu'u o Waimea

There it is, there it is, the wind that strikes the land
Kawaihae has the Nāulu wind-rain
The wind is a cold rain that hurts the skin
Kīpu'upu'u rain of Waimea

From a wind chant in the legend of Pāka'a and Kūapāka'a, probably referring to winds that are accompanied by rains of the same name. Hawaiian source: Kanepuu, "He moolelo" 11/14/1867. English trans. by author. Additional sources: Nakuina, *Moolelo* 54–55; Nakuina, *Wind* 48.

24. Hole Waimea i ka ihe a ka makani
Hao mai nā 'ale a ke Kīpu'upu'u
He lā'au kala'ihi na ke anu
I 'ō'ō i ka nahele o Mahiki

Waimea strips the spears of the wind
Waves of the Kīpu'upu'u rain bluster
A rigid tree fashioned by the cold
Into a digging stick in the forest of Mahiki

From a mele inoa, or name chant, for Kamehameha I. Hawaiian source: *Nā Leo Hawai'i Kahiko,* Record I, Side A, Band 9. English trans. by author. Additional source: Moehonua and Lilikalani. *Note:* Nā Leo Hawai'i Kahiko *says, "Kamehameha needed more spear fighters, and having heard of a company of twelve hundred young men of Waimea, Hawai'i, who were trained runners, he went to see for himself. He was pleased with their swiftness and knew that they would make excellent spear fighters. He appointed Nānu'uakalani'ōpu'u to train and lead them. They called themselves the Kīpu'upu'u after the icy cold rain of their homeland. The first thing they did was to go to Mahiki forest to make spears. It was there that the young men thought of composing a chant in honor of their chief, Kamehameha I. The first composition was criticized by several expert poets and hula masters. It began with 'Hole Waimea i ka ihe a ka makani, hala kika i ka pu'ukolu.'*

(ʻWaimea is pierced by the spear-like blasts of the wind; slipping and sliding over the triple hills.ʼ) It was the slipping and sliding that was objected to. With the few changes, the chant was completed to the satisfaction of all and presented as a gift to the ruler by the newly trained warriors of Waimea. It was first chanted as an oli and later, as a hula. This was one of the most popular chants of Kamehameha's day and was heard wherever his armies moved."

25. Kauna kula ʻo ka ua Kīpuʻupuʻu
Ua ʻalohi lua i ka ʻEʻelekoa

. . . .

Kuʻu wahine i ka ua Kīpuʻupuʻu
Mai ka ua noe anu a ka Mālana
E lala ai i ke ahi ka māʻeʻele
He kapa ia no ke kini o Waimea ē!

The Kīpuʻupuʻu rain settles on the plains
Sparkling brightly in the ʻEʻelekoa rain

. . . .

My darling wife in the Kīpuʻupuʻu rain
From the cold, misty rain of the Mālana
The numbness will be warmed over the fire
It is a kapa for the multitudes of Waimea!

From a kanikau, or lament, for Kamehameha IV. Hawaiian source: Moehonua and Lilikalani. English trans. by author.

26. Aloha ʻino nō kuʻu wahine

. . . .

Kuʻu wahine mai ke kai hāwanawana o Kawaihae
Mai ka ua Kīpuʻupuʻu o Waimea

What great pity for my beloved wife

. . . .

My dear wife from the whispering sea of Kawaihae
From the Kīpuʻupuʻu rain of Waimea

From a kanikau, or lament, for Luakauwahine. Hawaiian source: Pawai. English trans. by author.

27. Waimea i ka ua, i ka ua Kīpuʻupuʻu. *Waimea in the rain, the pelting Kīpuʻupuʻu rain.*

From the song "Hilo Hanakahi" by Keola Nālimu. Source: Wilcox et al. 65. Additional source: Elbert and Mahoe 50.

28. Kilakila nā rough riders
Me ka ua Kīpuʻupuʻu

Me ka nani aʻo Puʻuokalani
Me ka hae o ka lanakila

Majestic are the rough riders
And the Kīpuʻupuʻu rain
And the beauty of Puʻuokalani
And the flag of victory

From the song "Hawaiian Rough Riders." Hawaiian source: Elbert and Mahoe 43. English trans. by author. *Note: Elbert and Mahoe say that "the cold Kīpuʻupuʻu rain is associated with Waimea and was the name of a cowboy club at the Parker Ranch. Puʻuokalani (Hill of the royal chief) is a nearby place."*

29. Kaulana ʻIkuā me Kaʻauʻa lā
Nā ʻeu kīpuka ʻili
Nā āiwaiwa o ʻEuropa lā
No Waimea e ka ʻeu
I ka ua Kīpuʻupuʻu
I kahua Waiomina

Famous are ʻIkuā and Kaʻauʻa
The lively ropers
The wondrous ones of Europe
From Waimea comes such spirit
From the Kīpuʻupuʻu rain
To the arenas of Wyoming

From the song "Waiomina" by Helen Lindsey Parker. Source: Wilcox et al. 276.

30. Koʻiʻi ka lehua i ke anu
Kīpuʻupuʻu aʻo Waimea

Refreshed is the lehua in the chill
Kīpuʻupuʻu of Waimea

From the song "Ke aliʻi hulu mamo" for Elizabeth Kahanu Kalanianaʻole. Source: Kaaihue 72.

31. He lā nui a he lā hauʻoli
Waimea o ka ua Kīpuʻupuʻu
Eia ke kamalani ua hiki mai
Waimea o ka ua Kīpuʻupuʻu

. . . .

E ola e ke kama nona nei lei
O ka ua Kīpuʻupuʻu ē
Likeke, lei aloha ʻoe a Kahilu

O ka ua Kīpuʻupuʻu ē
He puana kēia i aloha ʻia
O ka ua Kīpuʻupuʻu ē
ʻO Kaleiokū kou inoa
O ka ua Kīpuʻupuʻu ē

An important day, a joyous day
Waimea of the Kīpuʻupuʻu rain
Here is the royal child, he has arrived
Waimea of the Kīpuʻupuʻu rain

. . . .

Life to the child to whom this lei belongs
Of the Kīpuʻupuʻu rain
Likeke, beloved child of Kahilu
Of the Kīpuʻupuʻu rain
This is a refrain of love
Of the Kīpuʻupuʻu rain
Kaleiokū is your name
Of the Kīpuʻupuʻu rain

From the song "Kaleiokū" for Richard Smart by Helen Desha Beamer. Hawaiian source: Kaaihue 79. English trans. by author.

32. Maikaʻi ke anu i Waimea lā
Kāhiko i ke Kīpuʻupuʻu lā
I ke kapa a ka ʻEʻelekoa lā
I kuku ʻia e ka Waiʻahulu lā
E ka ua ʻUla i ke pili lā
I pilia mai e Moʻolau lā

Splendid is the chill in Waimea
Adorned by the Kīpuʻupuʻu
By the kapa of the ʻEʻelekoa
That is beaten by the Waiʻahulu
By the ʻUla rain over the pili grass
Clung to by Moʻolau

From a mele mākaʻikaʻi, or travel chant, for ʻEmalani Kaleleonālani by Kupahīnano Papa. Hawaiian source: Nogelmeier 108. English trans. by author. Additional source: Andrews 504.

33. Kuʻu kāne mai ke kai hāwanawana lā o Kawaihae
Kuʻu kāne mai ke kula pili o Kawaihae Uka
Kuʻu kāne hoʻi mai ka ua Kīpuʻupuʻu o Waimea

My dear husband from the whispering sea of Kawaihae
My dear husband from the pili grass plain of Kawaihae Uka
My dear husband, indeed, from the Kīpuʻupuʻu rain of Waimea

From a kanikau, or lament, for Kamakaokalani. Hawaiian source: Wiliama. English trans. by author.

34. He aloha nō paha ʻoe
E ke anu o Waimea
E ka ua Kīpuʻupuʻu
Leikokoʻula i ke pili
Hāliʻi maila i luna
I ka welelau o ke kuahiwi

Greetings to you
O chill of Waimea
O Kīpuʻupuʻu rain
Leikokoʻula rain over the pili grass
Spreading above
To the top of the mountain

From the song "Pua o ka he-i." Hawaiian source: Holstein 163. English trans. by author.

35. Kuʻu kāne, e kuʻu kāne hoʻi
Kuʻu kāne mai ka ua Kīpuʻupuʻu anu o Waimea
E nanahu ana i ka ʻili o ke kanaka
He ahi ke kapa e mehana ai ko laila kini ē
He kini Peʻepākaiaulu, pā makani

My beloved husband, oh, my dear husband indeed
My dear husband from the cold Kīpuʻupuʻu rain of Waimea
Biting and nipping the skin of a person
A fire is the blanket that warms the people of that place, oh
A multitude of the Peʻepākaiaulu, a blowing of the wind

From a kanikau, or lament, for Kamehameha IV by his wife, ʻEmalani Kaleleonālani. Source: Nogelmeier 338.

36. *For another example of Kīpuʻupuʻu of Waimea, see Mālana (#1).*

Rain of Waoʻala, Oʻahu

37. Ua pulu ʻelo i ka ua
Kīpuʻupuʻu o Waoʻala
E hoʻomau ana i ka liko
Onaona lua i ka maka

Drenched by the rain
Kīpuʻupuʻu of Waoʻala
Continuing upon the leaf buds
So gentle and sweet to the eyes

From the song "Liko o Waoala." Hawaiian source: Holstein 20. English trans. by author.

Rain of Maunahina, Kauaʻi

38. Aia i Maunahina
I ka ua Kīpuʻupuʻu
Kō mai ana i ka ʻili
I ka poli o ka ipo ahi

At Maunahina
In the Kīpuʻupuʻu rain
Pulling at the skin
At the bosom of the impassioned lover

From a mele kuʻi, or chant accompanied by hula kuʻi, in which dancers stamp their feet and make vigorous gestures. Hawaiian source: Kawelaakawai. English trans. by author. Additional source: Barrère, Pukui, Kelly 47.

Kiu. Same as *Tiu*. Rain associated with Maui and Kauaʻi. Also the name of a wind.

Rain of East Maui

1. Ka lī anu haʻukeke a ka ua Kiu
I kahi a māua e nonoho ai
Me kuʻu wahine mai ka ua Hāmākualoa ē

The cold, shivering chill of the Kiu rain
Where we lived
With my darling wife from the Hāmākualoa rain

From a kanikau, or lament, for Batesepa Puhia, the wife of Davida Malo. Hawaiian source: Malo, "Eia." English trans. by author.

Rain of Makawao, Maui

2. *See ʻUlalena (#13).*

Rain of Makaweli, Kauaʻi

3. ʻAʻala ka ʻiliahi o Makaweli
ʻAʻala ke kiele me ke kupukupu
E aʻoaʻo ana ke ʻala
Me ka lehua maka noe i ke anu
Anu māʻeʻele i ka noe a ke Kiu
ʻAʻala ka hala o Māpuana i ke kai

The ʻiliahi tree of Makaweli is fragrant
The kiele plant and ferns are fragrant
The fragrance confers
With the lehua maka noe shrub in the cold
Numb with cold in the mist of the Kiu
The hala trees of Māpuana by the sea are fragrant

From a mele inoa, or name chant, for Wahinealoha. Hawaiian source: Savia. English trans. by author.

Rain of Mānā, Kauaʻi

4. E Kū, e Lono, e Kāne, Kanaloa
ʻAkahi ʻoe a ʻike i ka mole wai
I nā mole wai pūhae a ka makani
I nā lile wai ʻono kau i ka pali
I nā muliwai loloa a ka ua Kiu
ʻOlolī ka wai ʻolokeʻa i Mānā
Uhalu ʻole ke kaha ʻōkolo i ka helu

Kū, Lono, Kāne, Kanaloa
You are just now seeing the source of water
The water sources torn by the wind
The sparkling, delicious water placed on the cliffs
The long streams created by the Kiu rain
Narrow are the waters crisscrossing at Mānā
Innumerable are the places across which they crawl

From a mele for Haili, the daughter of Kaumualiʻi. Hawaiian source: Pukui, *Nā Mele Welo* 38. English trans. by author.

Koʻapuaiʻa. Rain associated with Kauaʻi.

Rain of Hāʻena, Kauaʻi

1. He wai mākaʻikaʻi na ko laila kupa
Ke nānā lā i ke Koʻapuaiʻa
ʻIke aku i ke Kiuwailehua
ʻAʻala ka ihu o Hāʻena ua makani
Kuʻu mea aloha ē, ma ʻaneʻi mai ā

Waters visited by the people of that area
Observing the Koʻapuaiʻa
Seeing the Kiuwailehua wind
The nose of Hāʻena is scented with fragrance; it is windy
My beloved, come here

From a mele for Albert Ka Haku O Hawaiʻi by Keoni Ana, or John Young. Hawaiian source: Ana. English trans. by author.

Rain of Mānā, Kauaʻi

2. Makemake au i ke inu wai o lalo
I ka hoʻonani mai a ke Koʻapuaiʻa
Pāpaʻanā kōʻeleʻele Mānā
ʻEleu nō i ke kaha o Nohomalu ē, i laila

I wish to sip of the waters below
Enhanced by the Koʻapuaiʻa showers
Mānā shudders and clamors in haste
Rushing to the sheltered strands of Nohomalu, yes, there

From a mele recalled by Hoʻoulumāhiehie as he described the fine physiques of Hiʻi-akaikapoliopele and her companions. Hawaiian source: Hoʻoulumāhiehie, *Ka Moʻolelo* 73. English trans.: Hoʻoulumāhiehie, *Epic* 70. Additional source: *Na Mele Aimoku* 169.

Koeʻula. Rain associated with ʻAhulili, Hawaiʻi.

1. Aloha ka ua Koeʻula a ka ʻuhane e kanikau nei ē
I lani nui ʻoe, e ke aloha, naʻu e hiʻi nei ē
Hiʻi ka lā wela i ke alo o ʻAhulili
E hahai a holo ana me ke anu o Pōhakuloa

Love to the Koeʻula rain, which the spirit mourns over
You are a great chief, my beloved, whom I embrace
The hot sun is cradled by ʻAhulili
As it follows and travels with the cold of Pōhakuloa

From a kanikau, or lament, for M. E. Manu. Hawaiian source: Hoapili. English trans. by author.

koʻiawe. Also known as koʻiaweawe. Light, moving rain; to shower.

Rain of the cliffs

1. Ka ua koʻiawe ko nā pali. *The cliffs have the koʻiawe rain.*

From a mele aloha, or song of love, for the newspaper *Ka Hoku o ka Pakipika.* Hawaiian source: Keukahi. English trans. by author.

2. *For another example of the koʻiawe of the cliffs, see Pōʻaihala (#5).*

Koʻilipilipi. Same as *Poʻolipilipi*. Rain associated with Kalihi, Oʻahu. "Koʻi lipilipi" means "sharp adze" or "adze that tapers."

1. Ka ua Koʻilipilipi o Kalihi. *The Koʻilipilipi rain of Kalihi.*

A traditional saying. Hawaiian source: Pukui, *ʻŌlelo* 169. English trans. by author. *Note: Pukui says that the Koʻilipilipi is "a pouring rain that lasts for days. A story is told of a couple who slept while the rain poured. The rain lasted so long that when they awoke, the sides of their heads were flattened, as though cut away by an adze."*

Koʻiʻula. Reddish- or rainbow-colored rain, mist, or cloud; associated with Nāpili, Maui.

1. Ua Koʻiʻula, he ua kēia ma ke kula, i hui pū ʻia me ka ua Koko, papā ihola ka lā, piʻo ke ānuenue; hui pū kēia mau mea ʻelua, me ka ua i ke kula; ka ʻula, e like me ka ua Koko i ka moana. No laila mai kēia inoa i kapa ʻia ai ua Koʻiʻula, ua nōweo ʻula o Nāpili.

   Koʻiʻula rain, this is a rain on the plains joined together with the Koko rain. The sun shines and the rainbow arches; these two combine with the rain on the plains. The redness is like that of the Koko rain over the ocean. Therefore the name Koʻiʻula rain was given, a bright, scarlet rain of Nāpili.

   From a letter to the newspaper *Kuokoa* regarding famous rains. Hawaiian source: Kalaaukumuole, "Manao wehewehe." English trans. by author. Additional source: Kalaaukumuole, "Thought explanations."

Koko. Similar to and possibly the same as *koko ʻula*; probably related to *Leikoko-ʻula*. Rain with rainbow, often considered a symbol relating to royalty or divinity; heavy rain that turns streams red-brown from runoff; associated with Maui and found on other islands. "Koko" means "blood." It is both the name of a specific rain and a generally descriptive term; its various usages are determined by the context.

Rain of Kāʻanapali, Maui

1. *See Paʻūpili (#4).*

Rain of Kaupō, Maui

2. Ua Koko—He ua hāliʻi i ka ʻilikai me ka piʻo pū o ke ānuenue. He hōʻailona o ka pilikia no kekahi mea ʻano nui ke ʻike ʻia kēia ua.

   . . . .

   I ke alanui aku, hālāwai mākou me kekahi kamaʻāina nō o Kaupō, a naue like mākou me ke keaka pū ʻana i nā mea i loaʻa iā mākou. Iā mākou e hele nei, aia nō ka nānā ʻana o ka maka o ka mea kākau ma ʻō a ma ʻaneʻi o kēia ʻāina, a ʻike akula au i ka ua kualau i ke kaʻahele aʻe i ka moana, a me nā ao e hoʻopuni aʻe ana i ke kuahiwi Maunakea, a hoʻomanaʻo wale aʻela nō i koʻu ʻāina hānau ʻo Kona. ʻAʻole i pau ka ua kualau, kū ana nō ka ʻōnohi ānuenue i ka moana kai a hoʻokahi ko mākou huli like ʻana a ʻōlelo akula au iā Keneki, "ʻO kēlā ka ua Koko e like me ka hoʻākāka i loaʻa iā kākou i kēia ʻauinalā."

ʻAʻole kēia he hōʻailona pilikia, no ka mea he maikaʻi ko mākou ala hele ma nā wahi a pau o Kaupō, a hiki hou nō i Honolulu nei.

Koko rain—A rain that blankets the surface of the sea and is accompanied by an arching rainbow. When this rain is seen, it is a sign of trouble for a rather important person.

. . . .

Along the road we met with another native of Kaupō, and we traveled together, chatting about what we had gotten. While we were going, the eyes of this writer were looking here and there across this land, and I saw the kualau rain shower moving on the ocean and the clouds encircling the mountain Maunakea, and I thought back to my birthplace, Kona. Before the kualau had ceased, a patch of rainbow appeared over the deep sea, and we all turned at once, and I said to Kenneth, "That is the Koko rain, like the explanation we got this afternoon."

This was not a sign of trouble, as our paths throughout Kaupō and our return to Honolulu were pleasant.

From an account of East Maui rains and their descriptions. Hawaiian source: Maunupau, *Huakai* 27, 29. English trans. by author.

Rain as a sign

3. ʻO kēia kekahi o nā hōʻailona aliʻi. He ua ʻula kekahi ʻōlelo ʻana o ka ua koko, inā ma ka ʻāina, a inā ma ka moana. Inā e ʻike ʻia kekahi ua koko ma ka ʻāina a ma ka moana paha, a laila, penei i ʻōlelo ʻia e kekahi kahuna. "ʻO wai lā ke aliʻi e make ana i kēia ua koko?" He nui nā ʻano o kēia hōʻailona e pili ai i nā aliʻi, inā he aliʻi make, a i ʻole, he aliʻi hānau paha, a i ʻole he aliʻi e hiki mai ana paha, a i ʻole e hele aku ana paha. ʻO kēia mau hōʻailona aliʻi, ua like nō ia me ke kuʻi ʻana o ka hekili, a ma ke ʻōlapa ʻana o ka uila, a me ke kualau ma ka moana, he mau hōʻailona aliʻi lākou a pau, inā he aliʻi make, aliʻi hele paha i ka huakaʻi, a i ʻole, he aliʻi hānau paha. ʻAʻole nō e hiki wale ana kēia mau hōʻailona i nā lā a pau, aia nō a hiki i ka manawa e kūpono ai no ia mau hōʻailona, a laila ʻike ʻia akula.

   This is one of the signs of royalty. The rainbow-hued koko rain is also called the ʻula (red) rain, whether it is on land or over the ocean. If a koko rain is seen on land or over the ocean, then a priest would say, "Who is the chief who will die by this koko rain?" There are many interpretations of this sign that pertain to chiefs—it could mark the death of a chief, the birth of a chief, or it could mark the arrival or departure of a chief. These chiefly signs, such as the crashing of thunder, the flashing of lightning, the kualau rain shower with wind over the ocean, are all signs of royalty, whether for the death of a chief, the voyage of a chief, or the birth of a chief. These signs do not appear every day, only when the time is right for the sign; then it is seen.

From research by S. N. Haleole. Hawaiian source: Fornander, *Fornander* 6: 109. English trans. by author.

4. Blood-red rain: This is given in the Hawaiian text as "ua koko," literally, "blood rain." Hawaiians used the expression to refer to rainbow-colored mists and clouds, prismatic effects seen in sun- and rain-filled valleys; also to distinguish a heavy rain that turns streams rust-red from the wash of hillsides. All these associations are relevant, always with the pervasive idea that rainbows and the color red, the blood color, are manifestations of divinity.

   Source: Pukui and Korn 184.

5. Ua koko—literally bloody rain, a term applied to a rainbow when lying near the ground, or to a freshet-stream swollen with the red muddy water from the wash of the hillsides. These were important omens, claimed as marking the birth of tabu chiefs.

   Source: Emerson, *Unwritten* 258. Additional source: Pukui and Elbert, 1986, 362, 509.

6. Ma ʻaneʻi e kamaʻilio ʻuʻuku no ka wahine a Makaliʻi. ʻO kēia wahine ʻo Malanaikuaheahea, he wahine maikaʻi a me ka nani loa, ʻo kona kino ke nānā aku, ua like me ka ʻula o ke ahi ka ʻula. I kona kū ʻana a hele mai kona hale noho mai, ua pākū ʻia ka mālamalama o ka lā, a pouli, neʻe maila ka ua koko, uhi maila ka ʻohu, kokolo maila ka noe, hāliʻi maila ke ʻawa, kahe maila ka wai ʻula, ʻōlapa ka uila. Ma hope o kēia mau mea, ʻike ʻia maila ke kino o Malanaikuaheahea, e hele mai ana ma luna o ka welelau o ka lima o nā kānaka, ua nani loa me ka hiwahiwa ua wahine lā ke nānā aku, no ka mea, ua hoʻohinuhinu loa ʻo Makaliʻi i kāna wahine, a ua nani loa.

   Here we will speak a little about the wife of Makaliʻi. This wife, Malanaikuaheahea, was a fine woman with great beauty. To look upon her body, it was like the red of fire. When she rose and went out of her house, the light of the sun was blocked and it became dark, the koko rain moved in, the ʻohu fog covered, the noe mist crawled in, the ʻawa rain spread forth, the red water flowed, and lightning flashed. After these things occurred, the form of Malanaikuaheahea would be seen traveling atop the tips of the hands of men. That woman was very beautiful and precious to look upon because Makaliʻi really dressed up his wife, and she was gorgeous.

   From the legend of ʻAukelenuiaiku. Hawaiian source: Fornander, *Fornander* 4: 79, 81. English trans. by author. *Note: Fornander,* Fornander *4: 80, says, "Tributes of the elements to rare beauty and high birth are characteristically the Hawaiian ideal, not restricted to legendary lore. Heavy rains, or an electric storm on the death or funeral of one of high station, or the red rain—ua koko—or rainbow presence are acknowledged as signs of royal recognition."*

7. *For more examples of koko as a sign of royalty and divinity, see kīʻohuʻohu (#1) and ʻUlalena (#14).*

Rain associated with the Koʻiʻula rain

8. *See Koʻiʻula (#1).*

Rain over the sea

9. ʻO ʻoe ka ʻuhane i ka ua koko, i ke ānuenue piʻo i ke kai ē

   . . . .

   Aloha ka ua Koko, ua kīhei kilika i ke kai

   You are the spirit in the low-lying koko rainbow, in the arching rainbow over the sea

   . . . .

   Love to the Koko rain, a silk cloak over the sea

   From a mele ʻuhane, or lament, for Kealohapauole, presumably by Kaikamahineole. Hawaiian source: "He uhane." English trans. by author.

Rainbow-hued rain

10. *See kualau (#16).*

koko ʻula. Or kōkō ʻula. Similar to and possibly the same as *Koko*; probably related to *Leikokoʻula*. Rain with rainbow; associated with royalty. "Koko ʻula" means "red blood." "Kōkō ʻula" means "network of red color," as of a spreading rainbow.

Rain of Waimea, Hawaiʻi

1. Ua hiki ua anu Waimea lā
   Ka ua koko ʻula i ke pili lā

   Having arrived, Waimea is cold
   The koko ʻula rain is over the pili grass

   From a mele inoa, or name chant, for Lele. Hawaiian sources: Kalai; "Pā ka leo" 20. English trans. by author.

Rain of Waolani, Oʻahu

2. Aia i Waolani kuʻu lei ua koko ʻula
   He lei na ka ua noe i kui a lawa

   There at Waolani was my lei of koko ʻula rain
   A lei sewn to completion by the misty rain

   From a mele by Paliuli as love for her brother welled up in her. Hawaiian source: Manu, *Keaomelemele* 41. English trans. by author.

Rain as a sign of royalty

3. Lamalama i luna ka ʻōnohi lā
Kāhiko ua koko ʻula lā
Ka hōʻailona kapu o ke kama lā
He ēwe mai nā kūpuna

The display of a rainbow illuminates above
An adornment with the blood red [koko ʻula] rain
This is the sacred sign of the princess
The lineage passed down from the ancestors

From the song "He inoa no Kaʻiulani." Source: Liliʻuokalani 127. Additional sources: Elbert and Mahoe 44; Holstein 101.

Rain associated with rainbows

4. Kōkō ʻula—network of red color, as of a spreading rainbow.

Source: Pukui and Elbert, 1986, 162.

Rain over the ocean

5. *See Nāulu (#60).*

Koloawao. Same as *Kolowao.* Rain associated with Kaʻala, Oʻahu.

1. ʻO lākou nō kai luna o Kaʻala—kāua
Ua pulu paha i ka ua Koloawao—kāua
Kilikili hau anu i Malamanui—kāua

They are indeed the ones atop Kaʻala—the two of us
Probably drenched by the Koloawao rain—the two of us
A cold, icy drizzle at Malamanui—the two of us

From a mele inoa, or name chant, for Albert Ka Haku O Hawaiʻi. Hawaiian source: Kapooloku. English trans. by author.

Kolowao. Same as *Koloawao.* Rain associated with Kaʻala, Oʻahu. "Kolo wao" refers to creeping in the mountains.

1. Ka ua Kolowao o Kaʻala. *The Mountain-creeper [Kolowao] rain of Kaʻala.*

A traditional saying. Source: Pukui, *ʻŌlelo* 169. *Note: Pukui says, "This rain is accompanied by a mist that seems to creep among the trees."*

2. Eia aʻe ka ua ke hele mai nei
He huakaʻi ʻimi kāoʻo loa na ka makani
A makani Ihuanu luna o Kaʻala
Ua hoʻopua haʻu ʻia e ka Waikōloa
He mia ala na ka ua Kolowao

ʻO ka ua puni ke aloha i haʻi ai
Ua hoʻonuʻu i ka waiwai āukauka
He mai, eia hoʻi

Here comes the rain
On a journey forced along by the wind
A bitterly cold [Ihuanu] wind up on Kaʻala
Billowed out by the cold Waikōloa wind
Poured upon by the Kolowao rain
The rain we speak of is surrounded with love
It has gathered wealth until it is rich
Come, it is here

From a mele of welcome. Source: "Eiae ka ua."

Kona. Heavy leeward rain and wind. "Kona" means "leeward."

Rain of Lānaʻi

1. *See Nāulu (#25).*

Rain usually with wind

2. ʻŌʻiliʻili lupea nā hōkū māhana ʻelua
   He ua Kona me ka makani

   The striking appearance of twin stars
   A Kona rain with wind

   From a mele for Kūaliʻi. Hawaiian source: Fornander, *Fornander* 4: 376–77. English trans. by author. Additional source: *Na Mele Aimoku* 85.

3. Ua Kona—a rainstorm from the south; generally with wind.

   From a list of rain names and their descriptions. Source: "Names . . . rain."

4. Rain accompanying the southwest wind, a leeward rain of Kona.

   Source: Kent 378. Additional source: Andrews 293.

5. He ua Kona, he ua nui loa ia. *Kona rain, this is an extremely heavy rain.*

   Hawaiian source: Malo, *Ka Moolelo* 19–20. English trans. by author. Additional source: Malo, *Hawaiian* 14.

Kuahine. Same as *Tuahine*. Rain associated with Mānoa, Oʻahu, and found on other parts of Oʻahu.

Rain of Kaʻau, Waiʻanae Uka, Oʻahu

1. Eia au, e ka ua Kuahine
   E ua nei i ka nahele o Kaʻau

Here I am, O Kuahine rain
Raining upon the forest of Kaʻau

From a mele for Hakaleleponi. Hawaiian source: Fornander, *Fornander* 6: 537. English trans. by author.

Rain of Kahui, Central Oʻahu

2. He aha lā ka mea lena i ka uka o Kahui?
He Kuahine lāua me ke Kiʻowao

What is expanding in the uplands of Kahui?
The Kuahine and the Kiʻowao

From a mele inoa, or name chant, for chiefs. Hawaiian source: Kameʻeleihiwa 65. English trans. by author.

Rain of Kānewai, Oʻahu

3. E *Ka Hae Hawaii*—aloha ʻoe
Ko mākou makamaka, mai ka hale malu a ke ao
Mai ka ua Kuahine o Mānoa
Ke hoʻopiha aʻela a piha ʻo Kānewai

O Ka Hae Hawaii—*greetings*
Our intimate friend, from the peaceful dwelling in the clouds
From the Kuahine rain of Mānoa
That fills Kānewai to capacity

From a farewell song for the newspaper *Ka Hae Hawaii.* Hawaiian source: Lohiau, "He mele no ka nupepa." English trans. by author.

Rain of Kanoenoe, Oʻahu

4. Noho nō i Kanoenoe ke Kuahine
Hele a haiamū ka ua i ke kula

The Kuahine rain dwells in Kanoenoe
The rains gather in the plains

From a mele reflecting on the travels of Hiʻiakaikapoliopele and on her consequent nickname, Wahinepōʻaimoku (Island-encircling woman). Hawaiian source: Hoʻoulumāhiehie, *Ka Moʻolelo* 173. English trans.: Hoʻoulumāhiehie, *Epic* 162.

Rain of Kōnāhuanui, Oʻahu

5. I luna o Kōnāhuanui
He kapa no Kalaʻihauola
No ka wahine o ka mauna
I walea paha i ke koʻekoʻe
I ka makani anu he Kēhau
Pupuʻu i ka ua Kuahine

Ako i ka hale lau Moaʻe
Malu i ka hale lau maiʻa

Above Kōnāhuanui
A cape for Kalaʻihauola
For the woman of the mountains
Who perhaps takes delight in the chill
In the cold wind, a Kēhau breeze
Crouching there in the Kuahine rain
Thatching a house with the [Moaʻe] trade wind
Taking shelter in the house of banana leaf

From a kanikau, or lament, for Kamehameha IV by his wife, ʻEmalani Kaleleonālani. Source: Nogelmeier 324. *Note: Pukui,* ʻŌlelo *251, says that "Kalaʻihauola" is the name of one of two supernatural women whose stone forms guarded the Nuʻuanu trail near the gap. "It was around Kalaʻihauola that the umbilical cords of babies were hidden to ensure their good health. When the new road over the Nuʻuanu Pali was made, these stones were destroyed."*

Rain of Kulaokahuʻa, Oʻahu

6. Kuʻu kāne hoʻi
Kuʻu kāne aloha i ke Kulaokahuʻa ē
Mai ka ua Kuahine hāliʻi pili
I pili mao ʻole ʻia ke anu me ke koʻekoʻe

My dear husband indeed
My beloved husband of Kulaokahuʻa
From the Kuahine rain that covers the pili grass
Forever joined by cold and dampness

From a makena, or lament, for Kamehameha IV by his wife, ʻEmalani Kaleleonālani, upon the king's passing. Hawaiian source: Nogelmeier 331. English trans. by author.

Rain of Mānoa, Oʻahu

7. Ka ua Kuahine o Mānoa. *The Kuahine rain of Mānoa.*

A traditional saying. Source: Pukui, *ʻŌlelo* 169–70. Additional source: Pukui and Elbert, 1986, 169. *Note: Pukui says, "This rain is famed in the songs of Mānoa. According to an old legend, Kuahine was a chiefess, the wife of Kahaukani. Their daughter Kahalaopuna was so beautiful that rainbows appeared wherever she was. Once, two gossiping men claimed they had made love to her. This so angered her betrothed husband that he beat her into unconsciousness. She was revived by an owl god, but after hearing more gossip, her betrothed killed her. In grief, her mother became the Kuahine rain. Her father adopted two forms—the wind Kahaukani and a hau tree. It was said that this tree moaned in grief whenever a member of royalty died."*

8. Huli aku ke alo i Mānoa lā
I mehana i ka ua Kuahine lā

The front turns to Mānoa
To be warmed by the Kuahine rain

From a mele inoa, or name chant, for Manoanoa. Hawaiian source: Lihau. English trans. by author.

9. He ua Kuahine, ua nō i ke kakahiaka a awakea mālie, hoihoi ka mana'o i ka pau koke 'ana o ka ua, 'oli'oli ke kanaka ma laila i kapa 'ia aku ai kēia inoa, ua Kuahine o Mānoa, kaena ai ko Mānoa po'e kahiko, penei: "E ho'i ka u'i o Mānoa ua ahiahi, ka ua 'Āpuakea." No ka ho'iho'i 'ana o ka po'e kahiko i ka ua i luna pono o Pōhākea e ua ai, ma ka 'ōlelo 'ana aku penei: "Ma uka ka ua, ma kai ka ua; kiola, kiola ka ua i Pōhākea." Ua ho'olohe nō ka ua i ka 'ōlelo 'ana, ua ho'okō nō. No laila i kapa 'ia ai kēia inoa, ka ua 'Āpuakea o Mololani.

A Kuahine rain, a morning rain that moderates at noon, the mind is joyous at the early ceasing of the rain, man is happy whereby [he has] given it this name, Kuahine rain of Mānoa, boasted by the old people of Mānoa, thusly: "The comely youth of Mānoa goes home it is evening, the 'Āpuakea rain." For the old people turned the rain right over Pōhākea, it is spoken of in this way: "Above is the rain, below is the rain, throw, throw the rain to Pōhākea." The rain hears the saying and obeys, therefore is given this name, the 'Āpuakea rain of Mololani.

From a letter to the newspaper *Kuokoa* regarding famous rains. Hawaiian source: Kalaaukumuole, "Manao wehewehe." English trans.: Kalaaukumuole, "Thought explanations." *Note: There are several references in this example that relate to the legend of Kahalaopuna, the beautiful chiefess of the uplands of Mānoa, who was murdered by her betrothed. The phrase quoted above, "E ho'i ka u'i o Mānoa ua ahiahi," is an old saying that refers to Kahalaopuna, who would descend every day to Kalehuawehe in Waikīkī to surf, and would return to her home in the evening ("He moolelo kaao Hawaii no Kahalaopuna"). "Pōhākea" in this context refers to a peak in the Wai'anae Mountains, which, in one version of the legend, is where Kahalaopuna was taken to be killed by her jealous betrothed (Fornander,* Fornander *5: 188–91). Like Kahalaopuna, who faithfully listened to and followed her betrothed, despite his violent plans (Thrum 121, 123–25), the 'Āpuakea rain of Mololani is said to listen and obey. "Mololani" is the name of the crater at the end of the Mōkapu Peninsula in Kāne'ohe Bay and a name for Kāne'ohe Bay itself (Nogelmeier 99). It is also a term that means "well-kept, well-nursed" (Pukui and Elbert, 1986, 253), as was Kahalaopuna, which is why, in a version of the mo'olelo, it was difficult for her to follow Kauhi as they hiked up the mountain to where he would murder her (Thrum 123). In the same version, "Kauakuahine" (The Kuahine Rain) is the name of Kahalaopuna's mother (ibid., 118).*

10. Ho'oipo i Mānoa ka ua Kuahine, 'eā
He ua lī 'a'e lehua no Kaho'iwai, 'eā

Romancing in Mānoa is the Kuahine rain
A rain of Kaho'iwai that brings a chill over the lehua

From a mele māka'ika'i, or travel chant, for 'Emalani Kaleleonālani by Kuhea. Source: Nogelmeier 154.

11. Na ka ua Kuahine o Mānoa
I kono mai ia'u e naue aku
'Ike i ka nani lā o ia uka
Kāhiko 'ia e ka uhiwai

It was the Kuahine rain of Mānoa
That invited me to go
See the beauty of this upland
Adorned by the uhiwai mist

From the mele "Kupaa ka manao me Liliu," also known as "Kupaa oiaio me ka lahui," for "ka poe i aloha i ka aina" (the loyal people who love the land). Hawaiian source: Olepau. English trans. by author. Additional source: *Buke Mele* 13–14.

12. Aia i ka luna a'o Mānoa
Ke aloha 'āina e luana nei
'Oiai mākou e walea ana
Hone ana e ka ua Kuahine
Me he lā nō a e 'ī mai ana
'Imi 'ia e ka pono o ka 'āina

There in upper Mānoa
Lovers of the land are at leisure
As we relax
The Kuahine rain sounds softly
As if to say
Seek the well-being of the land

From the song "Hele malihini ia Manoa" by J. W. Kamali. Hawaiian source: *Buke Mele* 23. English trans. by author.

13. Ua naka nā hulu o ke kōlea
Ka uahi poepoe pūpanapana
Nāna i kiani aku e ka lono
Hiolo ai ho'i nā 'enemi
Lilo ai lākou a i mea 'ole
I ka ua Kuahine a'o Mānoa

The feathers of the kōlea quivered
Round smoke from a pistol
Signals the news

The enemies have been defeated
They have become insignificant
In the Kuahine rain of Mānoa

From the patriotic mele inoa, or name chant, "Wehi ohu no Wilikoki" for Robert Wilcox by J. P. Kina. Hawaiian source: *Buke Mele* 33. English trans. by author.

14. Ua nani Mānoa i ka'u 'ike
I ka loku a ka ua Kuahine

Beautiful is Mānoa in my view
In the downpour of the Kuahine rain

From the song "Hawaii i ka ehuehu" by Home Kauwila. Hawaiian source: *Buke Mele* 73. English trans. by author.

15. A waho o ka nuku a'o Māmala
Hū mai ke aloha i ka 'āina
Huli aku e ke alo a i Mānoa
Ka nene'e mai a ka ua Kuahine

Outside the harbor entrance of Māmala
Love for the land wells up
Turning to face Mānoa
The Kuahine rain creeps forth

From the song "Ma'i korela ma Honolulu" by S. Kanalu. Hawaiian source: *Buke Mele* 74. English trans. by author.

16. Ka ua Kuahine o Mānoa ē, me ka ua Līlīlehua o Pālolo, a me ka ua Kūkalahale ē, me nā ua kaulana o ua 'āina nei a ka 'e'epa nui wale, ua pau ko 'oukou ho'opulu hou 'ana i ka 'ili waianuhea o Aunty.

The Kuahine rain of Mānoa, the Līlīlehua rain of Pālolo, the Kūkalahale rain, and the famous rains of this extraordinary land, you will never again moisten Aunty's sweet skin.

From a eulogy for Mrs. Kaiawe Helelani Punohu. Hawaiian source: Craig. English trans. by author.

17. Ku'u hoa o ka i'a lauahi lima o Kālia
Hoa nānā i ka ua Kuahine o Mānoa
Mai ka ua Hāli'ipili o 'Ualaka'a
Auē ku'u kāne ē

My companion of the fish of Kālia that is caught by quick hands
Companion who observes the Kuahine rain of Mānoa
From the Hāli'ipili rain of 'Ualaka'a
Pity for my dear husband!

From a kanikau, or lament, for J. Henry. Hawaiian source: Kahinawe, "He kanikau." English trans. by author.

18. Hoʻohihi aku au e ʻike
Ia wai māpunapuna
Ua Kuahine piʻo ānuenue
O ia uka ʻiuʻiu

I am entranced to see
That bubbling well spring
Kuahine rain with its arching rainbow
Of those lofty heights

From the song "Moanikeʻala" by Leleiōhoku. Source: Wilcox et al. 176. Additional source: Holstein 15. *Note: Wilcox et al. say, "Kuahine, or Tuahine, is the misty curtain of rain of Mānoa Valley."*

19. *For more examples of Kuahine of Mānoa, see Kūkalahale (#11) and Makaʻūpili (#3)* .

Rain of Nuʻuanu, Oʻahu

20. Ke Kona o Nuʻuanu, ka ua Kuahine
ʻUla i ka ua ka pali o Malailua

The Kona wind of Nuʻuanu, the Kuahine rain
The cliff of Malailua is red from the rain

From a mele for Kaumualiʻi. Hawaiian source: Fornander, *Fornander* 6: 475. English trans. by author. *Note: Fornander says that the Kuahine is a mountain rain.*

Rain of Waʻahila, Oʻahu

21. Na ka ua Kuahine o Waʻahila e noho haʻanipo lā i ka wao. *By the Kuahine rain of Waʻahila (Mānoa, Oʻahu) that dwells in love with the uplands.*

From a chant. Source: Pukui and Elbert, 1986, 45.

22. *For an example of Kuahine of Waʻahila Ridge, see Waʻahila (#26).*

Rain of Wahiawā, Oʻahu

23. He nui nā ʻoihana a ka Waiʻōpua
He ʻoihana nō ia na ke Kuahine
Hoʻomaikaʻi paʻa pono ihola i ke kula
None, pāʻani i ka Waikōloa
Pāʻani leʻaleʻa i Wahiawā

The Waiʻōpua wind has many tasks
An undertaking by the Kuahine rain
Bringing long-lasting pleasure to the plains
Teasing, playing in the Waikōloa wind
Pleasurable fun at Wahiawā

From a mele used to describe how Hiʻiakaikapoliopele paid her debt to Kaʻanahau. Hawaiian source: Hoʻoulumāhiehie, *Ka Moʻolelo* 151. English trans. by author. Additional source: Hoʻoulumāhiehie, *Epic* 142.

Other examples

24. He lei ua puīa
I ke ʻala me ke aloha
Palai lau lipolipo
I pulu i ka ua Kuahine

A lei permeated
With perfume and love
Palai with deep green fronds
Drenched by the Kuahine rain

From the song "He lei na ke aloha." Hawaiian source: Holstein 173. English trans. by author.

25. ʻAʻole nō au e hopo
I ke kīhāpai lehulehu
A nā koa e ʻalo nei
Hopo ʻole i ka ua Kuahine
I ka leo mali a ke kāhuli

I do not fear
The crowded garden
That the koa trees endure
Unafraid of the Kuahine rain
Of the flattering words of the land shell

From a mele inoa, or name chant, for Pekipeki. Hawaiian source: Kanahele 110. English trans. by author.

26. He nui nō wau a he hiwahiwa
He punahele na ka ua Kuahine
Haʻina ʻia mai ana ka puana
No ke ʻala o ka mea pohihihi

I am important and beloved
A favorite of the Kuahine rain
Tell the refrain
For the fragrance of the mysterious one

From the song "Ke ala o ka mea pohihihi." Hawaiian source: Mauuhilo. English trans. by author. Additional source: *Buke Mele* 103.

kuakualau. Same as *Kualau.* Shower over the sea, accompanied by wind.

1. *For an example of kuakualau, see pūnohu (#2).*

Kualau. Same as *kuakualau.* Shower over the sea, accompanied by wind; associated with Hilo, Hawaiʻi, and with Niʻihau. Also found in other areas. "Kualau" is both the name of a specific rain and a generally descriptive term; its various usages are determined by the context.

Rain of Hilo, Hawaiʻi

1. *See Kanilehua (#37).*

Rain over the sea near Hilo, Hawaiʻi

2. Kuʻu kaikuahine o ka ua Kualau
   O ka ua hāliʻi mai i ka ʻili o ke kai

   My darling sister of the Kualau rain
   Of the rain that blankets the surface of the sea

   From a kanikau, or lament, for Mrs. Maleka Ii. Hawaiian source: Kaapa. English trans. by author.

Rain of Kapaʻahu, Hawaiʻi

3. A ʻo kekahi hōʻailona, ʻo ia hoʻi, ʻo ke kualau, ʻo ia hoʻi, ka ua kaʻi i ka moana. A lohe au i ka ʻōlelo ʻia, "kualau ʻōhua," ʻo ia hoʻi ka manawa e pae mai ai ka ʻōhua.

   And another thing, another sign is the kualau, that is, the rain that forms a line out at sea. And I heard it said, "kualau ʻōhua," that is, the time that the ʻōhua (fish) appear.

   From a story by Emma Kapūnohuʻulaokalani Kauhi about portents. Source: Kauhi 59, 124.

Rain of Kīlauea, Hawaiʻi

4. ʻO ka hāʻule ihola nō ia o ka ua kualau, a piha loa aʻela ʻo ka lua i ka wai, a hū ma waho, ʻaneʻane e pio ke ahi, ʻuʻuku wale nō koe ma loko o ka poli o Lonomakua kahi i hoʻopumehana ai.

   The kualau rain fell, and the volcanic pit became so full of water that it overflowed. The fire was almost extinguished, but a little remained in the bosom of Lonomakua, where it was kept warm.

   From the legend of Pele in the newspaper series "Ka hoomana kahiko." Hawaiian source: Kaawa. English trans. by author.

Rain of Leleiwi, Hawaiʻi

5. Kuʻu kāne mai ka ua Kualau i ka moana
   Ke hoʻolei aʻela i uka o Leleiwi

My dear husband from the Kualau rain over the ocean
That is cast upon the shore of Leleiwi

From a kanikau, or lament, for S. N. Wahine. Hawaiian source: Kaneheleloa. English trans. by author.

Rain over the ocean off of Hāna, Maui

6. Ku'u kāne mai ka makani Lau'awa o Hāna
Mai ka ua kualau i ka moana

My beloved husband from the Lau'awa wind of Hāna
From the kualau rain over the ocean

From a kanikau, or lament, for J. Henry. Hawaiian source: Kahinawe, "He kanikau." English trans. by author.

Rain of O'ahu

7. 'Auhea wale 'oe, e ke Kualau
E ka ua nihi hele ma ka moana
Hō'ike a'e 'oe i kou nani
I ka mālamalama 'oi kelakela

Where are you, O Kualau shower
O cautious rain that goes over the ocean
Display your beauty
And penetrating radiance

Part of a chant from an account of Mr. and Mrs. J.W.K. Kihe's tour of O'ahu. Hawaiian source: Mokumaia. English trans. by author.

Rain of Kualoa, O'ahu

8. Aia au ke 'ike lā i ka nui o ka 'ino, a hiki mai, ke pā lā ka makani me ka lauwili, aia ke kualau ke 'oni a'ela ma mua, a mea akula au i ua mau malihini nei, "'Eā, ua nui nō ka hanohano o ko 'olua lā?" "Ke aha lā ho'i?" "E nānā a'e nō ho'i 'olua lā, aia ka Nāulu, ke 'oni i mua o Kualoa; hele nō ke kualau, hele nō ka Nāulu a ka makani e hāpaipai mai ana; ua 'ike 'ia kekahi o 'olua," a hū a'ela ko mākou 'aka.... Kā'alo a'ela 'o Ka'a'awa, aia 'o Kahana ke ho'opulu 'ia maila e ka Nāulu; ke 'ike lā au i ka 'ūkele o ke alanui.

I am seeing the strength of the storm, and since it has arrived, the wind has been blowing from all directions, the kualau rain has been moving ahead, and I mentioned to the visitors, "Isn't this day of yours receiving great honor?" "How so?" "Look over there, there is the Nāulu shower, moving before Kualoa; the kualau rain travels, the Nāulu travels, borne on the wind; one of you has been recognized," and we burst into laughter.... Passing Ka'a'awa, there is Kahana being drenched by the Nāulu rain; I can see mud on the road.

From an account of Mr. and Mrs. J.W.K. Kihe's tour of O'ahu. Hawaiian source: Mokumaia. English trans. by author.

Rain of Kawainui, Kaua'i

9. Wā Hanalei ē
Wā i nā mākāhā i'a
A Kawainui, ma'oeā
Ka 'apa a ka ua Kualau

Hanalei is grumbling
Grumbling at the fish inlet
At Kawainui sluggishly
Lingers the Kualau rain

From a song recited before starting the guessing game called pū kaula. Source: Fornander, *Fornander* 6: 214–15.

Rain of Ni'ihau

10. No laila, ke kāhea nei au iā 'oukou, e ka po'e o Ni'ihau i hele i Kaua'i (ka po'e makamaka 'ole) e hele mai lākou ma O'ahu nei, ma Pālolo nei, e hana ai, e mahi ai, e ma'au'auwā ai, e kūkulu hale ai, e hana ai i nā hana pono, e noho loa ai a make. Ua pau ka minamina, ke aloha 'ana i ko 'oukou 'āina komohana, i ka ua Kualau o Ni'ihau, a me ka ua lele makani o Lehua, a me ka wai huna a ka pāo'o.

And so, I am calling to you, the people of Ni'ihau who went to Kaua'i (those without friends), they should come here to O'ahu, to Pālolo to work, farm, peddle, build houses, to do moral work, to live long until death. Over is the sorrow, the grieving for your land in the west, in the Kualau rain of Ni'ihau, and the windblown rain of Lehua, and the hidden water of the pāo'o fish.

From a letter to the editor offering a place for those evicted from Ni'ihau to live and work on O'ahu. Hawaiian source: Kanepuu, "Eia." English trans. by author.

Rain over the ocean

11. He ua kualau, ma ka moana ia ua. *A kualau rain, this rain occurs at sea.*

Hawaiian source: Malo, *Ka Moolelo* 20. English trans. by author. Additional source: Malo, *Hawaiian* 14.

12. A strong wind; a gale of wind; wind and rain upon the ocean; he ua me ka makani ma ka moana (*a rain and wind over the ocean*).

Source: Andrews 298. English trans. by author.

13. Ua kualau—the rain that comes from the sea.

From a list of rain names and their descriptions. Source: "Names . . . rain."

14. Shower accompanied by sea wind . . . ʻAuhea wale ʻoe, e ke Kualau, ka ua nū hele ma ka moana (*Listen, O Kualau shower, rain moaning over the open sea*).

From a chant. Source: Pukui and Elbert, 1986, 170.

15. Hū mai ke aloha no Kahoʻoluhi
Kāpena wiwo ʻole o ka moana
Nāna i lula mai nā ʻale
Nā makani paio o ka moana
Hea mai, e ka ua Kualau
"I hea ʻoe, e Kahoʻoluhi?"

Love swells for Kahoʻoluhi
The courageous captain of the ocean
He ruled the waves
And the blustering winds of the ocean
Call out, O Kualau rain
"Where are you, Kahoʻoluhi?"

From a mele for Joseph Nāwahīokalaniʻōpuʻu. Hawaiian source: Sheldon, *Ka Puke* 243. English trans. by author. Additional source: Sheldon, *Biography* 326–27.

16. E ui aku ana au iā ʻoe
Aia i hea ka wai a Kāne?
Aia i kai, i ka moana
I ke kualau, i ke ānuenue
I ka pūnohu, i ka ua koko

I ask you
Where is the water of Kāne?
There toward the sea, in the ocean
In the kualau shower, in the rainbow
In the pūnohu rainbow, in the rainbow-hued koko rain

From a mele for the god Kāne. Hawaiian source: Emerson, *Unwritten* 258. English trans. by author.

17. Ma kēia wahi ua ʻike aku mākou i ka uahi o ke Kīlauea, e hele pū mai ana me ka ua kualau i ka moana. *We saw the smoke of the* Kīlauea *from here and with it the kualau rain at sea.*

From an account by Thomas K. Maunupau of an expedition through Kaupō, Maui. Source: Maunupau, *Huakai* 88, 154. *Note:* "Kīlauea" *was the name of a steamer.*

18. Ke kualau pua ia mai lalo
Inā i ke kai o Palepalemoana

The kualau shower appears from the west
If at the sea of Palepalemoana

From a mele kālaiʻāina, or political chant, for ʻEmalani Kaleleonālani. Hawaiian source: Nogelmeier 206. English trans. by author. *Note: Nogelmeier says, "A kualau is a rain cloud and column of rain moving inland from over the sea."*

19. Kanikau lā, he aloha nou e Kamakaokalani
    Kuʻu kāne mai ka ua kualau i ka moana

    A dirge, an expression of love and sorrow for you, Kamakaokalani
    My dear husband from the kualau rain over the ocean

    Hawaiian source: Wiliama. English trans. by author.

20. *For more examples of kualau over the ocean, see Koko (#2, #3), Lanipōlua (#3), and Nāulu (#65).*

Kuaʻoʻe. Rain associated with Hawaiʻi and Oʻahu. The rain name appears as "Kua-o-e" in an 1862 Hawaiian-language newspaper (Pawai).

Rain of Hilo, Hawaiʻi

1. *See Kanilehua (#32).*

Rain of Mokuola, Hawaiʻi

2. ʻAʻohe hoa o ka ua Kuaʻoʻe
   Ka ua kāhiko lā maka o ka lehua
   O nā lehua maka ʻehu kai o Mokuola

   The Kuaʻoʻe rain is friendless
   The rain that adorns the face of the lehua bloom
   The sea-sprayed faces of the lehua of Mokuola

   From a kānaenae, or chant of praise, for Aberahama K. Kalana from his kūpuna. Hawaiian source: "He kanaenae aloha." English trans. by author.

Rain of Mololani, Kāneʻohe, Oʻahu

3. Kuʻu wahine mai ka ua Kuaʻoʻe o Mololani
   Mai ke koʻa mokumoku o Heʻeia
   Aloha ia wahi a kāua i hele ai

   My beloved wife from the Kuaʻoʻe rain of Mololani
   From the broken coral beds of Heʻeia
   Fondness for these places where we traveled

   From a kanikau, or lament, for Luakauwahine. Hawaiian source: Pawai. English trans. by author.

kuāua. Shower; to shower.

1. The kuāua is a rain without wind extending over a small area—a shower.

   From a list of rain names and their descriptions. Source: Kelsey, "Rains."

2. He ʻaloʻalo kuāua no kuahiwi. *Facing the mountain showers (said of a brave person).*

   Source: Pukui and Elbert, 1986, 171.

3. Penei ka hoʻomaka ʻana i ka mahi ʻai, ua puhi ke kanaka i ke ahi, waiho a hāʻule ke kuāua mua, helu mua ka mahi ʻai, ʻakahi kuāua, hoʻokahi kuāua i koe, maʻū kuʻu wahi wela, a hāʻule hou ke kuāua, a ahiahi o ia lā, hele i ka ʻako lau, ka inoa o nā lau e ʻako ai, he hualani, he hōkeo, he lapa, he kawelo, he likolehua, a nui aku, a ʻope a paʻa, waiho a kakahiaka o kekahi lā, piʻi aku a hiki i ka wela āna, a kanu a pahu mua i ka mākālua, a nui nō hoʻi ka mākālua, a laila, kanu ka lau, a paʻa ka wela, pau ke kanu ʻana.

   This is how farming would begin. A man would set a fire and leave it until the first kuāua shower fell. The farmer would count, "First kuāua, one kuāua more and my new field will be damp." And when the kuāua fell again, that evening he would go to cut sweet potato slips—the names of which are hualani, hōkeo, lapa, kawelo, likolehua, and many more—and then he would tie them securely in a bundle and wait until morning of the next day, head inland to his new field, and dig and make holes in the ground. And when there were many holes, then the slips would be planted, the new field would be complete, and the planting would be finished.

   From an article on ancient farming practices. Hawaiian source: Kealakai, "He wahi moolelo." English trans. by author. Additional sources: Kealakai, "Cultivation"; Solis 22.

4. *For more examples of kuāua, see Nāulu (#62) and Waʻahila (#26).*

Kuiʻilima. Same as *Kuilima.* "Kui ʻilima" means "to string ʻilima flowers."

1. Name of a Honolulu rain.

   Source: Pukui and Elbert, 1986, 174.

Kuilima. Same as *Kuiʻilima.* Rain associated with Nuʻuanu, Oʻahu. "Kui lima" means "to lock arms."

Rain of Honolulu, Oʻahu

1. Kuʻu hoa i ka ua Kuilima
   Ua Kūkalahale o Honolulu

   My companion in the Kuilima rain
   The Kūkalahale rain of Honolulu

   From a mele inoa, or name chant, for Albert Ka Haku O Hawaiʻi by his uncle, Kamehameha V. Hawaiian source: Alii Iolani. English trans. by author.

Rain of Nuʻuanu, Oʻahu

2. Aloha Nuʻuanu i ke kāmakahala lā
   Ke kuia maila e ka ua noe ē
   Ka ua Kuilima o Nuʻuanu lā
   Ke hehi [maila] i ka maka o ka ʻāhihi ē

   Beloved Nuʻuanu with its kāmakahala plants
   Strung together by the misty rain
   The Kuilima rain of Nuʻuanu
   Treading on the face of the ʻāhihi flowers

   From a mele inoa, or name chant, for Kaluaihaleola. Hawaiian source: Lewaina. English trans. by author.

Kūkalahale. Rain and wind associated with Honolulu and the larger Kona district of Oʻahu. "Kū kala hale" means "standing under the eaves of the house" or "striking the house gables." "Kūkala hale" means "announcing to the homes."

Rain of Oʻahu

1. Nākolo, nakulu, paʻapaʻaʻina kaupoku o Kākuhihewa ē
   Hewa ka iʻa a ʻUmiamaka i ka lili a ka ua Kūkalahale

   The roofs of Kākuhihewa are roaring, rumbling, and crackling
   The fish of ʻUmiamaka was wronged by the jealousy of the Kūkalahale rain

   From a dirge for Joseph Nāwahīokalaniʻōpuʻu by people of Oʻahu. Hawaiian source: Sheldon, *Ka Puke* xxii. English trans. by author. Additional source: Sheldon, *Biography* 017. *Note: The second line of this example is similar to the traditional saying, "Hewa ka iʻa a ʻUmiamaka, he ōkea loko," which means, "The fish of ʻUmiamaka was defective; it had sand within." This would be said in reference to something that was defective or wrong, or when someone had been deceived (Pukui,* ʻŌlelo *105).*

2. Kuʻu lei ka makani ʻŌlauniu
   Hoa pili o ka ua Kūkalahale

   My lei, the ʻŌlauniu wind
   Close friend of the Kūkalahale rain

   From the song "Nanimauloa" by F. S. Hawaiian source: *Buke Mele* 103. English trans. by author.

Rain of Honolulu, Oʻahu

3. He ua Kūkalahale (strike the house gables) ko Honolulu. *Honolulu has a Kūkalahale rain.*

   From a textbook on the Hawaiian language. Hawaiian source: Kahananui and Anthony 108. English trans. by author.

4. Ka ua Kūkalahale o Honolulu. *The Kūkalahale rain of Honolulu.*

   A traditional saying. Source: Pukui, *ʻŌlelo* 170. *Note: Pukui says that the Kūkalahale, which is often mentioned in songs, is "the rain that announces itself to the homes by the pattering it makes on the roofs as it falls."*

5. A i ia wā i uhau aʻe ai ʻo Hiʻiaka i kēia kau, ʻoiai nō ka ua Kūkalahale o Honolulu nei e hoʻokawewe ana i ka lau lāʻau a me nā hākala hoʻi o nā hale pili o ia mau lā aloha o ka nohona o ko kākou mau kūpuna.

   Then Hiʻiaka offered up this chant, while the Kūkalahale rain of Honolulu pattered upon the leaves of the trees and the thatched roofs of those beloved days of our ancestors.

   From the legend of Hiʻiakaikapoliopele. Hawaiian source: Hoʻoulumāhiehie, *Ka Moʻolelo* 441. English trans.: Hoʻoulumāhiehie, *Epic* 411.

6. I ia wā i haʻalei aku ai ʻo Kauakahiapaoa i kāna kilu.... "He mau wahi hua mele paoa loa kā hoʻi kaʻu i oli aʻe nei. E ʻeha ana hoʻi au i ke kaikamahine o ka ua Kūkalahale."

   I kēlā wā, lālau ihola ʻo Peleʻula i kāna ipu, a oli aʻela nō hoʻi.

   . . . .

   Eia au ē, ka ua Kūkalahale
   A pā!

   At that point, Kauakahiapaoa tossed his kilu.... "Those were certainly unlucky lines I just chanted. I shall be wounded by the girl of the Kūkalahale rains."

   Whereupon Peleʻula reached for her gourd marker and chanted.

   . . . .

   Here am I, the Kūkalahale rain
   Let it strike!

   From an episode in the story of Hiʻiakaikapoliopele, when Peleʻula and Kauakahiapaoa are playing the game of kilu. Hawaiian source: Hoʻoulumāhiehie, *Ka Moʻolelo* 446–47. English trans.: Hoʻoulumāhiehie, *Epic* 417.

7. Kuʻu kaikunāne o kēia wahi e noho ʻia nei
   Haʻalele mai nei ʻoe iaʻu i ka pili a kāua
   Kuʻu hoa o ka ua Kūkalahale o Honolulu

   My beloved brother of this place of residence
   You have left me and our closeness
   My dear companion of the Kūkalahale rain of Honolulu

   From a kanikau, or lament, for Keawehunahala by his sister. Hawaiian source: Kaaipaka. English trans. by author.

8. He aloha nō ʻo Honolulu i ka ua Kūkalahale
   Ka nuku aʻo Māmala ʻau aʻe nei ma hope

Dearly loved is Honolulu in the Kūkalahale rain
The entrance of Māmala Bay fares on behind

From the song "He aloha nō ʻo Honolulu" by Lot Kauwē. Source: Wilcox et al. 50.

9. Kuʻu kāne mai ka ua Kūkalahale
Mai ka ʻai nānā i ka ʻai ke ola

My beloved husband from the Kūkalahale rain
From the food that is looked upon to sustain

From a kanikau, or lament, for J. Henry. Hawaiian source: Kahinawe, "He kanikau." English trans. by author.

10. ʻO Honolulu ke ahupuaʻa. . . . ʻO ka ua Kūkalahale ka ua mau; ka ua Kiʻowao noho mai i uka, kilihēhē nō i ka pua kāmakahala.

Honolulu is the ahupuaʻa. . . . The regular rain is the Kūkalahale; the Kiʻowao rain stays inland, soaking the kāmakahala flowers.

From a historical account of Hawaiʻi. Hawaiian source: Kamakau, "Ka moolelo o Hawaii nei" 6/22/1865. English trans. by author. Additional source: Kamakau, *Tales* 6.

11. E nā keiki o ka ua Kūkalahale, noho ua Kiʻowao o Luakaha—no ʻoukou ka ʻōlelo ʻia: "Hōʻā ke ahi, kōʻala ke ola, ʻo nā hale wale nō kai Honolulu; ʻo ka ʻai nō kai Nuʻuanu, ʻo ka iʻa nō kai laila." E kiu, e holo, e hoʻi ka uʻi o Mānoa, ua hāliʻi ka ua Līlīlehua i nā kahawai, ke ōpū maila ka ua Kuahine i ka pua o ka ʻōhiʻa—he lualaʻi lua Kahoʻiwai na Kanaloahoʻokau. Hoʻowaha kamaliʻi o Pālolo, ua pulu ʻelo i ka ua Puanaiea, he mau wāhine noho i ka lā ʻo Kuʻialauahi me Huewa.

Children of the Kūkalahale rain, dwelling in the Kiʻowao rain of Luakaha—it is for you that it is said: "Light the fire; in cooking there is life. There are only houses in Honolulu; the food is in Nuʻuanu, even the fish are there." Look ahead, get going, return home, young beauty of Mānoa. The Līlīlehua rain has spread over the valleys; the Kuahine rain is opening the flowers of the ʻōhiʻa trees. Kahoʻiwai is a splendid repose for Kanaloahoʻokau. The children of Pālolo talk excessively and are drenched by the Puanaiea rains. Kuʻialauahi and Huewa are women who dwell in the sun.

From an article by Samuel Mānaiakalani Kamakau about his publication of "Ka moolelo o Kamehameha I," the story of Kamehameha I. Hawaiian source: Kamakau, "Aloha." English trans. by author. Additional source: Kamakau, "Beloved."

12. The Kūkalahale rain falls in the area bounded by Punchbowl and River Street, and Vineyard Boulevard and the sea.

From an email sent to the author by Kimo Alama Keaulana on April 4, 2013.

13. Ku'u hoa mai ka 'ilihune kapa 'ole
Ku'u hoa mai ka ua Kūkalahale o Honolulu

My dear companion from destitute times when we had no kapa covering
My dear companion from the Kūkalahale rain of Honolulu

From a kanikau, or lament, for H. Haumea. Hawaiian source: David. English trans. by author.

14. *For more examples of Kūkalahale of Honolulu, see Kanilehua (#17) and Kuilima (#1).*

Rain of Kapu'ukolo, O'ahu

15. 'O ia 'o lāua me ka ua Kūkalahale
E kūkahekahe ana i Kapu'ukolo

The twosome is like the Kūkalahale rain
Gossiping away at Kapu'ukolo

From a mele māka'ika'i, or travel chant, for 'Emalani Kaleleonālani, credited to P. Nāhaolelua. Source: Nogelmeier 138. *Note: Pukui, Elbert, and Mookini 90 say that "Kapu'ukolo" is the name of an old section of Honolulu that is bounded, in part, by the Nu'uanu Stream and Honolulu Harbor.*

Rain of Kona, O'ahu

16. Contemporary sources with traditional knowledge suggest that "Kūkalahale" may be translated as "to stand under the eaves." This translation comes from Theodore Kelsey (1891–1987), an expert in Hawaiian language and culture, who taught it to M. Puakea Nogelmeier. Nogelmeier explains that the Kūkalahale is a passing rain, so people do not have to go inside to wait for the rain to pass; they can just stand under the eaves. "The orographic rain that O'ahu gets is stalled on the windward side until the clouds lift enough to pass the Ko'olau Range, at which point the clouds, and rain, drift right over the Kona district and out to sea," Nogelmeier says. This was confirmed for him at a gathering of Native Hawaiian speakers in Wai'anae, where a heavy rain set in. One of the mānaleo said that, unlike the Kūkalahale, this rain would require them to stay indoors—"'A'ole kēia he Kūkalahale; ma ka hale kākou e pa'a ai."

Drawn from an email sent to the author by M. Puakea Nogelmeier on October 14, 2012.

17. *For more examples of Kūkalahale of Kona, see Kuahine (#16) and Wa'ahila (#4, #5).*

Rain of Mānoa, O'ahu

18. Ua Kūkalahale—Mānoa Valley, Honolulu, O'ahu; a rain that blows under the eaves of the houses.

From a list of rain names and their descriptions. Source: "Names . . . rain."

Rain of Pohukaina, Oʻahu

19. Aloha ke hoa pili o ka ua Kūkalahale
O ka ua Kīpehipehipuahala o Pohukaina

Love to the close companion of the Kūkalahale rain
Of the Kīpehipehipuahala rain of Pohukaina

From a kanikau, or lament, for Albert Ka Haku O Hawaiʻi. Hawaiian source: Kahinalua. English trans. by author.

Kukupaʻu. Or Kukupaʻū or Kukupau. Rain associated with Hawaiʻi and Kauaʻi. "Kukupaʻu" means "to do with zest." "Kuku paʻū" means "to beat overlaid kapa." "Kuku pau" means "to beat completely," as kapa.

Rain of Hilo Hanakahi, Hawaiʻi

1. Kemu Waimea i ka ua Kukupaʻu o Hanakahi. *Waimea absorbs the pouring [Kukupaʻu] rain of Hanakahi.*

Source: Pukui and Elbert, 1986, 144.

Rain of Koʻolau, Kauaʻi

2. I kumu wai hoʻi Neki na ka wai
Ua i Pueo, ke kumu o ka ua
Hohola ihola i luna o Hā[l]aulani
Ka [paneʻe kū] a ka ua i ka lāʻau
Ka ua Kukupau i luna o ka lau hala
Ka ua Kanikaukū, me he kanaka lā
Ka ua hahi i ke kai o Manolau

The headwaters at Neki are fed by the water
Raining at Pueo, the origin of the rain
Spreading out above Hālaulani
The rain forces its way through the trees
The Kukupau rain atop the hala leaves
The Kanikaukū rain, seemingly a person
The rain that walks on the sea of Manolau

From the makena, or lament, "He kumu lewa no Kaleleonalani" for ʻEmalani Kaleleonālani. Hawaiian source: Nogelmeier 353. English trans. by author. *Note: According to Soehren, "Neki" is the name of a peak at the Waiʻoli/Waipā/Lumahaʻi border in Haleleʻa, Kauaʻi, and "Manolau" is an area along the seaward border of Hanalei and Waiʻoli. Keaweamahi says that "Hālaulani" is the name of a peak above the ahupuaʻa of Waiakalua, which has groves of hala trees. It is also the name of an area on the seaward side of Anahola.*

Kulaʻikanaka. Rain associated with Kawelowelo, Hawaiʻi. "Kulaʻi kanaka" means "to knock people down."

1. He kanaka ka ua me ka lā
   He ua Kulaʻikanaka ko Kawelowelo

   The rain and the sun are like people
   Kawelowelo has a Kulaʻikanaka rain

   From a chant for Haili, daughter of Kaumualiʻi. Hawaiian source: Pukui, *Nā Mele Welo* 44. English trans. by author.

kulu paka kahi. Related to *kulu pākakahi*. Rain associated with Niʻihau and the month of Hilina, which corresponds to November there. "Kulu paka kahi" means "to fall in single drops."

1. Aloha kuʻu wahine mai ka ua kulu pākakahi
   Mai ka ua kulu paka kahi o Hilina

   Love for my wife from the scattered rain
   From the rain of Hilina that falls in single drops

   From a kanikau, or lament, for Rode Koleiki. Hawaiian source: Kahale. English trans. by author.

kulu pākakahi. Related to *kulu paka kahi*. Rain associated with Niʻihau and the month of Hilina, which corresponds to November there. "Kulu pākakahi" means "to fall scattered," as light rain.

1. *See kulu paka kahi (#1).*

kūnihi. Rain controlled by a seer of Kamāʻoa, Hawaiʻi. "Kūnihi" means "to turn aside."

1. Ka ua kūnihi a Kaʻupena. *The [kūnihi] rain of Kaʻupena that turns aside.*

   A traditional saying. Source: Pukui, *ʻŌlelo* 170. *Note: Pukui says, "Kaʻupena was a seeress of Kamāʻoa Plain, in Kaʻū. Whenever rain approached, she called it to come to her home and to leave the homes of her neighbors alone so that their crops would not be ruined by a too-early rain. The rain obeyed."*

Kūpunikapa. Rain associated with Lanakila, Maui. "Kūpuni kapa" means "to clutch kapa."

1. Ka ua Kūpunikapa o Lanakila. *The Kūpunikapa rain of Lanakila.*

   A traditional saying. Hawaiian source: Pukui, *ʻŌlelo* 170. English trans. by author. *Note: Pukui says, "The rain of Lanakila, Maui, is so cold that it makes one clutch and hold his clothing close to his body."*

L

Lanaʻula. Rain associated with Haʻikū, Hawaiʻi.

1. Kū ana i Māhoe iā ia
   ʻO Haʻikū i Maunaloa
   A ola mai ka ua Lanaʻula

   Standing at Māhoe towards her
   Hāʻikū at Maunaloa
   The Lanaʻula rain lives on

   From a pule hula, or hula prayer. Source: Nāmakaokeahi, *History* 164 (Hawaiian section) and 155 (English section).

Lanihaʻahaʻa. Rain associated with Hāna, Maui. Also found at Hilo, Hawaiʻi. "Lani haʻahaʻa" means "low sky."

Rain of Hilo, Hawaiʻi

1. The Lanihaʻahaʻa rain is known in Hilo as well as Maui.

   From a list of rain names and their descriptions. Source: Kelsey, "Rains."

Rain of Hāna, Maui

2. ʻO Hāna kēia, ʻakahi o loaʻa
   ʻO Hāna, ʻāina ua Lanihaʻahaʻa

   This is Hāna; finally here
   Hāna, land of the Lanihaʻahaʻa rain

From the lament "He kanikau no Peʻapeʻa." Hawaiian source: Fornander, *Fornander* 6: 428. English trans. by author.

3. Hāna ua Lanihaʻahaʻa—ʻo ke ʻano o kēia ua, he haʻahaʻa ke ao ke hele mai ka moana mai, a hāʻule ka ua a kapa ʻia iho kēia inoa. He uʻi maoli nō ka inoa o ka ua o kēia ʻāina kaulana nui wale, ʻaʻole paha he ua o kekahi mau wahi ʻē aʻe, e like me kēia, a na wai nō e ʻōlelo wale ia ʻo Maui nō e ka ʻoi.

   Hāna of the Lanihaʻahaʻa rain—this rain is as such: a low cloud moves in from the ocean, and then the rain falls. That is how it got its name. The name of the rain of this extremely famous place is very pretty. There is probably no other rain elsewhere like this, and who would argue, Maui is truly the best.

   From an account of East Maui rains and their descriptions, as told to Thomas K. Maunupau by Josua Ahulii of Kaupō, Maui. Hawaiian source: Maunupau, *Huakai* 27. English trans. by author.

4. Pā kēia i ka ʻāina ua Lanihaʻahaʻa, a ʻo ko ia nei hoʻokele pololei akula nō ia i kona wahi waʻa no ka lae kaulana ʻo ʻUpolu ma Kohala, ʻoiai, ʻo ia nō paha ka ʻāina kokoke loa mai o Hawaiʻi i Maui.

   The breeze breathed upon the land of the [Lanihaʻahaʻa] rain of the low sky, and the voyager steered his canoe straight for the famous point of ʻUpolu at Kohala, as perhaps that is the land where Hawaiʻi is the closest to Maui.

   From the story of Kalaeone's journey to Hawaiʻi Island to tell Kamehameha I about the death of Kahekili. Hawaiian source: Desha, *He Moʻolelo* 2: 123. English trans.: Desha, *Kamehameha* 353.

5. "Haʻahaʻa," literally, hanging low. I am reminded of an old song uttered, it is said, by a hero from the top of Kaʻuiki Hill in Hāna, Maui: "ʻĀina ua Lanihaʻa-haʻa." Land of rain, where the heavens hang (ever) low.

   From a note by Nathaniel B. Emerson regarding the word "haʻahaʻa," which appears in a chant uttered by Lohiʻau in the Hiʻiakaikapoliopele legend. Source: Emerson, *Pele* 210.

6. Lana mālie ka houpo kūʻono o Kaʻuiki e nūnē kaniʻuhū ana
   Ua kau akula i ke alo o nā maka nā hoʻomanaʻo hōʻīloli—ʻo ia kā!
   ʻO ka manu kaʻupu hālō ʻale o ia moana ʻoehu i ʻAlenuihāhā
   Ke hāhā lā ke aloha o ka hoʻokipa iā Hāna ua Lanihaʻahaʻa ē

   Stilled, the deep heart of Kaʻuiki is in sorrowful wonder
   Placed before the eyes are memories that evoke strong emotions—is it so?
   The kaʻupu bird peers over the waves of that blustery ocean in the
   ʻAlenuihāhā Channel
   The affection of the host reaches for Hāna of the Lanihaʻahaʻa rain

   From a lament for Joseph Nāwahīokalaniʻōpuʻu by people of the lands of Kama, where he was warmly hosted during his trip to the legislature. Hawaiian source: Sheldon, *Ka Puke* xxi. English trans. by author. Additional source: Sheldon, *Biography* 015.

7. ʻOiai lākou ma Hāna . . . Ua hoʻokipa pumehana ʻia me ke ohohia nui ʻo Nāwahī, ʻoiai ʻo ia ma ua ʻāina kaulana nei o ka ua Lanihaʻahaʻa.

   While they were at Hāna . . . Nāwahī was warmly hosted amid great excitement while he was in that famous land of the Lanihaʻahaʻa rain.

   From the biography of Joseph Nāwahīokalaniʻōpuʻu. Hawaiian source: Sheldon, *Ka Puke* 120. English trans.: Sheldon, *Biography* 195.

8. Ka ua Lanihaʻahaʻa o Hāna. *The Lanihaʻahaʻa rain of Hāna.*

   A traditional saying. Hawaiian source: Pukui, *ʻŌlelo* 170. English trans. by author. *Note: Pukui says that this phrase "refers to Hāna, Maui. Once, the young warrior chief Kāʻeokūlani ran to a banana grove to escape a sudden squall. As he stood safe and dry in the shelter of the banana leaves, he lifted his spear. It accidentally pierced through the leaves, and a trickle of water came through. He remarked that the sky where he stood was so low he had pierced it."*

9. Lani haʻahaʻa. *Low sky, of the low sky, a poetic reference to Hāna, Maui, and its rains.*

   Source: Pukui and Elbert, 1986, 193.

10. Makemake ke ʻike aku
    I ka ua Lanihaʻahaʻa
    Me ka Waiʻānapanapa
    ʻO Maui nō lā ka ʻoi

    I want to see
    The Lanihaʻahaʻa rain
    And Waiʻānapanapa
    Maui indeed is the best

    From the song "Maui no ka oi" by Rev. S. Kapu. Hawaiian source: Holstein 127. English trans. by author.

11. ʻAʻohe ou loa, e Hawaiʻi
    I ka haulani a ka huelo poki
    Lula ʻia nā ʻale o ka moana
    A hiki i ka ua Lanihaʻahaʻa

    Your expanses are nothing, Hawaiʻi
    To the speeding of the whaleboat
    The ocean's billows are dominated
    To Hāna of the Lanihaʻahaʻa rain

    From a mele for Joseph Nāwahīokalaniʻōpuʻu. Hawaiian source: Sheldon, *Ka Puke* 364. English trans.: Sheldon, *Biography* 462.

Lanikea. Rain associated with Wai'ōhi'umalo, Maui. "Lani kea" means "white sky."

1. Nani wale ka nohona a Lili'umoku
   I ka malu 'olu o Kauikalani
   'O ka nihi a ka ua Lanikea
   I uka o Wai'ōhi'umalo

   It is splendid aboard Lili'umoku
   In the cool shade of Kauikalani
   The creeping of the Lanikea rain
   Above Wai'ōhi'umalo

   From the song "Wai o Kanewai." Hawaiian source: Holstein 106. English trans. by author. *Note: There were two schooners named* Lili'u. *One sailed between Hawai'i, Maui, and O'ahu from approximately 1867 to 1877 carrying passengers and cargo; the other also provided interisland service and operated from 1878 to 1882.*

Lanipa'ina. Rain associated with 'Ulupalakua, Maui. "Lani pa'ina" means "crackling sky."

1. Ka ua Lanipa'ina o 'Ulupalakua. *The Sky-crackling [Lanipa'ina] rain of 'Ulupalakua.*

   A traditional saying. Source: Pukui, *'Ōlelo* 170.

2. E 'olu'olu 'oe a me kou kāpena e ho'okomo iho i nā mea 'ano hou i hana 'ia ma 'Ulupalakua i ka lā 4 iho nei o Iulai, 'oiai ho'i ka ua Lanipa'ina e kilihune ana i luna o "Prospect Hill."

   Aia ma nā hora hu'ihu'i o ke aumoe, i lohe 'ia aku ai nā leo nahenahe o ka pū'ali "Ka Ua Lanipa'ina Club" e mele ana ma loko o Rose Ranch i nā mele ho'oipo a ke aloha, 'oiai ho'i nā pua lili'i o ka magnolia e mōhai ana i ka nani wehi o ka lehua maka noe; pēlā ihola i ho'omana'o 'ia ai 'o ka lā 4 ia, 'oiai nō ho'i ka pū'ali mele e mele ana me ka hau'oli, e hāli'i mai ana ho'i ka 'ohu i ka nahele o ka uka, e kulu ana ho'i nā kēhau anu o ia pō, pēlā e ho'omana'o mau 'ia ai ka lokomaika'i piha o J. J. Halstead me Jas. K. Smyth Jr. i nā boys o ia pū'ali.

   Aia ho'i i ka hiki 'ana aku i ka hora 12 awakea, i ka wā ho'i a nā pua e luhe mālie mai ana i ka nani, a e kilihune kili hau ana ka ua Lanipa'ina i ka liko o ka pua, pēlā i 'ike 'ia aku ai ka pū'ali koa, "Ka Ua Lanipa'ina Military Company" e paikau ho'okahakaha mai ana ma lalo o ke alaka'i maiau 'ana a ko lākou Kāpena R. W. Wilcox. . . .

   Ma hope o ka pau 'ana o kēia mau le'ale'a, ua kāhea 'ia mai ka pū'ali "Ka Ua Lanipa'ina" e komo aku ma kahi o Jas. Makee e ho'olele ahikao pū me ia ma ia pō, 'a'ohe nō i li'uli'u, ua 'ike 'ia akula nā ahikao e lele mai ana i ka lewa me nā waiho'olu'u like 'ole, ke hele ala a pipi'o i ka lewa.

ʻO ka pau ihola kēia o nā mea ʻano nui o ua lā 4 nei; no laila, e hoʻōki au ma ʻaneʻi, me ka hoʻomanaʻo mau loa i nā leo nahenahe me nā paikau koa ʻana a ka pūʻali "Ka Ua Lanipaʻina."

You and your captain, please add the new events that took place at ʻUlupalakua this past Fourth of July while the Lanipaʻina rain was drizzling over "Prospect Hill."

It was during the chilly midnight hours that the melodious voices of the soldiers from the "Ka Ua Lanipaʻina Club" could be heard singing romantic songs of love inside Rose Ranch, while the dainty magnolia flowers were presenting offerings to the adorning beauty of the lehua maka noe; that was how the day was recognized as the Fourth. And as the song group was singing joyously, a light ʻohu cloud was covering the upland forest and the chilly kēhau mists were drizzling that night; that is how the great generosity of J. J. Halstead and Jas. K. Smyth Jr. toward the boys of this group will always be remembered.

When 12 noon came, indeed the period when flowers gently droop in their beauty, and the Lanipaʻina rain was coming down in a cold drizzle over the buds of the flowers, that was when the group of soldiers, "Ka Ua Lanipaʻina Military Company," could be seen parading back and forth under the expert leadership of their captain, R. W. Wilcox. . . .

After these amusements ended, the "Ka Ua Lanipaʻina" soldiers were called to Jas. Makee's place to shoot fireworks with him that evening. In no time the fireworks could be seen flying through the sky with various colors, arching in the upper heavens.

This was the conclusion of the events of the Fourth; therefore, I will stop here, forever remembering the melodious voices and marching soldiers of the "Ka Ua Lanipaʻina" group.

From a letter to the newspaper *Ka Lahui Hawaii* describing the Fourth of July celebrations at ʻUlupalakua. Hawaiian source: Makaikai. English trans. by author.

Lanipili. Heavy rain that lasts for days or a cloudburst; associated with Hilo, Hawaiʻi. Also found on Oʻahu and Kauaʻi. "Lani pili" means "clinging sky."

Rain of Hilo, Hawaiʻi

1. Ua Lanipili

   Aia a ʻōmalumalu mai nā ao, a e hāʻule mai ana nā paka ua nunui ma ʻō a ma ʻaneʻi o ua ʻāina nei, a e hoʻomau ʻia ana ka noke mao ʻole a nā paka ua, a e pohāpohā mai ana paha nā māpuna leo hekili i kekahi wā, a e kolo māʻeʻele mai ana ke anu, a e hāʻule mai ana nā huahekili ma luna o ka honua, a e ʻaneʻane ana e uihā i kā ia nei noke mau, a e hōʻea mau mai ana nā hoʻomanaʻo ʻana no nā hoa aloha a me nā makamaka, me he mau māpuna wai lā, a laila, e puana ʻia mai auaneʻi kona inoa, ʻae, "Kuʻu hoa mai ka ua Lanipili o Hilo. He ua hakihaki lehua no Panaʻewa." I kekahi manawa e ʻoi ana ma mua o hoʻokahi pule ka lōʻihi o kona ua ʻana, a he uihā nā malihini ke hiki mai nēia ʻano ua.

Lanipili Rain
When clouds cast a heavy shade, and large raindrops fall everywhere on this land, and the raindrops do not clear up but continue to persist, and perhaps the explosive sound of thunder cracks at times, and a numbing cold crawls in, and hail drops to the earth, and you are about to grow weary of its persistence, and the recollections of beloved companions and friends continue to surface like bubbling springs, then by and by its name will be uttered, yes, "My companion from the Lanipili rain of Hilo. It is a rain of Panaʻewa that breaks lehua flowers." Sometimes this rain lasts for more than one week, and visitors get weary when this kind of rain comes.

From a description of rains by Lilia H. Richards and company, who traveled to the uplands of Puna in 1876. Hawaiian source: Richards et al. English trans. by author.

2. Ka ua Lanipili is a rain that falls for days and nights. . . . The ua Lanipili is the continuous rain of Hilo.

From a list of rain names and their descriptions. Source: Kelsey, "Rains."

3. Iā Makaliʻi lau ʻawaʻawa ka uka o Puna
Hala ka wai ma uka o Kapāpala
Lanipili o Hilo ē

During Makaliʻi there are many disasters in the uplands of Puna
The waters pass by the uplands of Kapāpala
Lanipili rain of Hilo

From a mele by Kamapuaʻa to Pele. Hawaiian source: Kahiolo, *He Moolelo* 57. English trans. by author. *Note: "Makaliʻi" is the name of a Hawaiian lunar month.*

4. Kuʻu kaikuahine o ka ua Lanipili
O ka ua kinakinai mao ʻole lā o Hilo

My darling sister of the Lanipili rain
Of the constant, endless rain of Hilo

From a kanikau, or lament, for Mrs. Maleka Ii. Hawaiian source: Kaapa. English trans. by author.

Rain of Honolulu, Oʻahu

5. Kuʻu kāne mai ka ua Lanipili o ke kaona
I pili wale ʻia nō e ka ua o Honolulu

My husband from the Lanipili rain of town
That was joined by the rain of Honolulu

From a kanikau, or lament, for Kupa Maila. Hawaiian source: Maila. English trans. by author.

Rain of Koaeʻa, Kauaʻi

6. Liʻiliʻi ka uka o Koaeʻa
Nānā i ka ua Lanipili
Ka ʻōʻō, manu leʻa o ka nahele

Tiny is the upland of Koaeʻa
Notice the Lanipili rain
The ʻōʻō, merry bird of the forest

From a mele danced with an ʻulīʻulī, or gourd rattle. Hawaiian source: Emerson, *Unwritten* 111. English trans. by author.

Downpour

7. ʻO ka loku o ka ua Lanipili
Pili ai māua i laila
ʻAʻohe loa o ke kualono
I ke koi a Welikanoe

The downpour of the Lanipili rain
Brought us together there
The ridge is not long
When Welikanoe urges

From the song "Waipuilani" by Mrs. Emma Kapena. Hawaiian source: Holstein 62. English trans. by author.

Rain of the forest

8. Ia noi hoʻi āu a ʻike wale aku
I ka ua Lanipili i ka nahele
Hoʻoipo mau nō ke ʻala anuanu
Me ka pua ʻaʻala o ka ʻiu

That request of yours is to see
The Lanipili rain in the forest
The cool scent is romancing
With the fragrant blossom of the heights

From the song "Kiss goodnight" for ʻEmalani Kaleleonālani. Source: Nogelmeier 13.

Rain of the uplands

9. E ake wale au e ʻike lā
I ka ua Lanipili o uka
Hoʻoipo mau nō ke ʻala huʻihuʻi
I ka pua o ke ʻaʻaliʻi

I desire to see
The Lanipili rain of the uplands
The cool perfume continues to woo
The flower of the ʻaʻaliʻi plant

From the song "Sweet moonlight" by A. K. Kunuiakea. Hawaiian source: Holstein 75. English trans. by author.

Other examples

10. Ua laʻi nō hoʻi ʻoe, e ka ua Lanipili
Pili ʻia e māua ka ua me ke koʻekoʻe
I mehana hoʻi au i kō leo, e ke hoa
Kuʻu hoa o ka ʻanoʻi o ke kulu aumoe

You have become peaceful, O Lanipili rain
The rain and the chill joined us
But I am warmed by your voice, O friend
My companion of midnight's desire

From the song "Sweet Violet." Hawaiian source: Holstein 47. English trans. by author.

11. E Lono lau ʻai nui
E ua mai ka Lanipili
Ka ua hoʻūlu ʻai
Ka ua hoʻūlu kapa
Pōpōkapa wai lehua
A Lono i ka lani

O Lono of the abundant food crops
May the Lanipili rain fall
The rain that makes crops flourish
The rain that makes kapa plants flourish
Rainbow-colored waters of the Pōpōkapa rain
Of Lono in the heavens

From a pule hoʻouluulu ʻai or pule hoʻomau, or prayers to increase crops, to the god Lono. Hawaiian source: Malo, *Hawaiian* 177. English trans. by author.

12. ʻO kekahi ia o kāna kapa ʻana i kāna keiki i ka lapuwale, no ka mea, ua noho ʻoluʻolu loa ia me kāna keiki; ua aʻo aku iā ia i ka naʻauao, a me ka pono, a me ka ʻoiaʻiʻo; ʻo kāna hoa noho ia o ka hale mehameha, hoa pupuʻu anu o nā pō kēhau, a me ka ua Lanipili o ka hoʻoilo—hoa ʻaʻe ʻale o nā kai.

This was the one time she called her child "lapuwale" (idiotic), for she had lived very contentedly with her son; she taught him knowledge and morals and honesty; he was her companion in a lonely home, her huddling partner in the cold of the dewy evenings and of the Lanipili rain of winter, her companion in treading over the billows of the channels.

From a story about a woman who rebuked her son for telling the truth about her wrongdoing. Hawaiian source: "Lanakila." English trans. by author.

Lanipōlua. Rain associated with Hilo and Hōpoe, Hawaiʻi, and with Koʻolau, Oʻahu. "Lani pōlua" means "very dark sky."

Rain of Hilo, Hawaiʻi

1. Ua Lanipōlua
   Aia a ʻomamalu mai nā ao pōhina, a noho mai i luna o nā ululāʻau, me he mau ʻohu wai lā, ʻaʻole hoʻi he ʻano ua ikaika loa, akā, he mau kilihune noe naʻe e noho paʻa mau ana me ka hoʻomao ʻole aʻe, a e ʻano pōhina mai ka ʻike ʻana aku i nā ululāʻau, ʻaʻole nō hoʻi e ʻike ʻia ka mālamalama o nā kukuna o ka lā, a ʻoia noho ʻōmalumalu wale iho nō. A laila, e haʻi ʻia mai auaneʻi kona puana ʻana, "Kuʻu hoa mai ka ua Lanipōlua o Hilo; no Hilo ka lehua nuʻa i ke one o Hanakahi; ʻakahi ke aloha lā, ua hiki mai nō ē."

   E noho paʻa mau mai nō nā ʻohu i luna o Panaʻewa, a moe lōʻihi aʻe ma nā ululāʻau o Paliuli, a piʻo pōhai aʻe a hiki i nā lehua o Haili. He ua ʻano pohu mālie kēia, ʻaʻole nō e pau koke, a ma kona hopena, he ʻehu wai wale nō, a ʻo ka hope loa, he wā mālie e hiki mai ana.

   Lanipōlua Rain
   When a heavy shade is cast by gray clouds that rest above the trees as watery mists—not a very strong rain, but a misty, light rain that usually settles down without clearing up, and visibility in the forests becomes dim; the sparkle of the sun's rays cannot be seen, only the shade—then, by and by, its saying would be uttered, "My companion from the Lanipōlua rain of Hilo; the lehua flowers piled onto the sands of Hanakahi belong to Hilo; love has finally arrived, you are here."

   The mist continues to settle atop Panaʻewa and lies long over the forests of Paliuli; then it arches and gathers, making its way to the lehua flowers of Haili. This is a quiet and gentle type of rain. It does not stop quickly, and at its conclusion, it is only a spray of water, and at the very end, a calm period will arrive.

   From a description of rains by Lilia H. Richards and company, who traveled to the uplands of Puna in 1876. Hawaiian source: Richards et al. English trans. by author.

Rain of Hōpoe, Hawaiʻi

2. Aloha ʻoe, e Hon. Iosepa Nāwahī
   Ka ʻoi, ke kilohana o nā mokupuni
   Kuʻu hoa i ka ua Lanipōlua
   ʻElua kāua i ʻalo aku ai
   I ka pō ua nui aʻi Hōpoe

I ka lehua popohe mai i ke pili
E pili ke aloha me ʻoe
Me ka ua kilihune o uka
ʻO ke aloha ʻāina kūpaʻa mau

Aloha to you, Hon. Joseph Nāwahī
The finest, the most cherished of the islands
My companion in the Lanipōlua rain
We two faced it together
The night of pouring rain at Hōpoe
Amid the beautiful lehua that bloomed among the pili
May the love always be with you
Like the gentle [kilihune] rains that adorn the highlands
The ever-loyal love for the land

From an expression of condolence for Joseph Nāwahīokalaniʻōpuʻu. Hawaiian source: Sheldon, *Ka Puke* 216. English trans.: Sheldon, *Biography* 297.

Rain of Punahoa, Hawaiʻi

3. Kuʻu wahine mai ka bele kuʻi lua o Haili
Nākolo aʻela i kai o Punahoa
Kuʻu hoa pili i ka ua Lanipōlua
Mai ka ua kualau ma ka moana

My wife from the booming bell of Haili
Reverberating below Punahoa
My close companion in the Lanipōlua rain
From the kualau shower over the ocean

From a kanikau, or lament, for Mary Kaawaloa. Hawaiian source: Keliihaleole. English trans. by author.

Rain of Koʻolau, Oʻahu

4. Ua Lanipōlua, he mea hoʻomaikaʻi i ka mea i aloha ʻia, kāne a wahine paha, ua lilo aku i ka mea ʻē, puka maila ke aloha. Uē helu akula, "Kuʻu hoa o ka ua Lanipōlua; me he ipo lā no ke aumoe o Koʻolau, kuʻu hoa o ka malu kukui, malu kuawa ē"; ma laila mai kēia inoa, ua Lanipōlua, i kapa ʻia aku ai e ka poʻe kahiko, "ua Lanipōlua."

The Lanipōlua rain, used to express admiration to someone who is loved, be it man or woman, who has been lost to another. Love issues forth, a wailing call is made, "My dear companion of the Lanipōlua rain, like a lover for the late nights of Koʻolau. My companion of the shade of kukui trees and the shelter of the valley." Hence this name, Lanipōlua rain, which was coined by the ancient people, "ua Lanipōlua."

From a letter to the newspaper *Kuokoa* regarding famous rains. Hawaiian source: Kalaaukumuole, "Manao wehewehe." English trans. by author. Additional source: Kalaaukumuole, "Thought explanations."

5. E ō, e ka wahine hele lā ʻo Kaiona
Alualu waili'ulā o ke kaha pua ʻōhai
ʻO ka ua Lanipōlua, pō anu o ke Koʻolau
Kuʻu hoa o ka malu kī, malu kukui o Kahoʻiwai

Answer, Kaiona, woman who travels in the sunshine
Following mirages on the plain of ʻōhai blossoms
The Lanipōlua rain on a cold night of Koʻolau
My companion of the kī and kukui shade of Kahoʻiwai

From the song "He inoa no Pauahi," also known as "Wahine hele lā o Kaiona," for Bernice Pauahi Bishop and attributed to Leleiōhoku. Hawaiian source: *I Mua e nā Pōkiʻi*. English trans. by author. Additional source: Holstein 7.

6. Kuʻu haku i ka ua Lanipōlua, me he pō lā kā ke anu o ke Koʻolau. *My lord in the dark, pouring [Lanipōlua] rain, like the night and the cold of the Koʻolau (the lord protecting one from darkness and cold).*

From a chant for Bernice Pauahi Bishop. Source: Pukui and Elbert, 1986, 193.

Rain associated with darkness

7. Kuʻu hoa pili o ka ua Lanipōlua. *My constant companion in the [Lanipōlua] rains from the deep, dark heavens.*

Source: Pukui and Elbert, 1986, 339.

8. No ka mea, kuhi ihola nō au i loko o koʻu mau manaʻo hūhewa, ʻo ke ʻano lā o ka ua Lanipōlua, penei: He ua lā ia no nā lā ʻelua, ʻekolu paha, a nui aku; he ua lā e ʻeloʻelo ai ka ʻāina, kū ka hālelo, moku kahawai i ka naoa lā e nā paka ua kulu koʻikoʻi, me ka paʻa mau lā o ka lewa i nā ao ua i ke ao a me ka pō, a no laila mai ke ʻano o kēlā huaʻōlelo pōlua; he pouli mau o ka pō a me ke ao ke ʻano. Akā, i koʻu manawa i nīnau aku ai i kekahi mau kahiko no kēia mau manaʻo oʻu, hōʻole maila lākou, ʻaʻole pēlā.

Because I supposed in my confused thought, the kind of Lanipōlua rain was this: it was a rain of two days, perhaps three, or more; a rain in which the land was drenched, the rocks stood out jagged, streams were broken by excessive rainfall, with the rain clouds prevailing day and night, whereby is obtained that word "pōlua" (two nights); prevailing darkness of the day and night was its kind. But when I asked some old people of these views of mine, they refuted it as not so.

From a letter to the newspaper *Kuokoa* regarding rain, wind, and sunlight names. Hawaiian source: Kanuha, "Mau mea." English trans.: Kanuha, "Some things."

Rain associated with thunder

9. Ku'u makuahine ho'i
   Ku'u makuahine i ka ua Lanipōlua
   Mai ka ua hekikili i luna o ka lā'au

   Oh, my dear mother!
   My beloved mother in the Lanipōlua rain
   From the thundering rain atop the trees

   From a kānaenae, or chant of eulogy, for H. Kamahiai. Hawaiian source: Makuaole. English trans. by author.

Lau'awa. Gentle rain and wind associated with Hāna, Maui.

Rain of Hāna, Maui

1. Lei Hāna i ka makani ua Lau'awa
   Ke mahalo 'ia maila e ka Paiolopawa

   Hāna is bedecked with a lei of wind and rain of the Lau'awa
   Admired by the Paiolopawa

   From a mele lei, or song presenting a lei, originally composed for Lunalilo and inherited by Kalākaua. This portion of the mele was composed by M. Kekauonohi. Hawaiian source: *Na Mele Aimoku* 159. English trans. by author. Additional source: Reichel, *Ke'alaokamaile.*

Rain of Ka'uiki, Maui

2. Maika'i Ka'uiki hemolele i ka mālie
   E kilohi ana i ka hono o nā moku
   Ke 'alawa iho 'oe he nani, he maika'i
   Ke one i ka hehia e ka Lau'awa
   A ka ua 'Āpuakea i ka moana
   I hehua mai e ke anu me ka Ho'olua

   Beautiful is Ka'uiki, peaceful in the calm
   Overlooking the circle of islands (Lāna'i, Moloka'i, Kaho'olawe)
   When one glances about, there is a charm, a beauty
   In the sands trodden down by the bitter [Lau'awa] rains
   The white, misty ['Āpuakea] rain out at sea
   Scattered along by the cold and the Ho'olua rain

   From the chant "Maika'i Ka'uiki." Source: Pukui, *Nā Mele Welo* 164–65.

Rain of Mokuhano, Maui

3. Komohana i luma'ina e ka Lau'awa
   Nonō ka puka ihu o Mokuhano
   Oī i ka pā 'ia e Koholālele

The west is engulfed by Lau'awa rains
Mokuhano gives a gasping snuffle
Sounding at the touch of the Koholālele wind

From a mele recalled by Ho'oulumāhiehie. Hawaiian source: Ho'oulumāhiehie, *Ka Mo'olelo* 417. English trans.: Ho'oulumāhiehie, *Epic* 389.

Drenching rain

4. Ilihia wale ho'i au
I ka ho'opē a ka Lau'awa
Me he lā 'o ku'u ipo
Hō'alu kīpuka 'ili

I am thrilled
With being drenched by the Lau'awa rain
Which is like my sweetheart
Who loosens the rawhide lasso

From the song "Hoalu kipuka ili" by Lolena. Hawaiian source: Holstein 149. English trans. by author.

Lauhīnano. Rain associated with Makahanaloa, Hawai'i. "Lau hīnano" refers to hīnano bracts.

1. Kui Makahanaloa ē i ka ua Lauhīnano
He ua kui pua lei na ka Moani

Makahanaloa strings the Lauhīnano rain
A rain that strings the lei flowers of the Moani breeze

From a mele inoa lei, a name chant worn as a lei, for 'Emalani Kaleleonālani. Hawaiian source: Nogelmeier 104. English trans. by author.

Laukoa. Same as *Makalaukoa*. Rain associated with upper Nu'uanu, O'ahu. "Lau koa" means "koa leaf."

1. Eia au lā ua anu
Ua pulu i ka ua Laukoa
I ke kulu Ki'owao i ka uka
Pulu 'elo ka maka o ka lehua
Puīa 'oko'a i ke onaona
Onaona 'auli'i ka nahele
Me ka pe'a hala i Moelana

Here I am, cold
Soaked by the Laukoa rain
By the drops of the Ki'owao in the uplands

The buds of the lehua are drenched
Completely imbued with perfume
Delicate is the fragrance of the forest
With the hala border at Moelana

From a mele inoa, or name chant, for Sarah Kukapu. Hawaiian source: Kawailuna. English trans. by author.

Laupua. Misty rain. Also the name of a wind.

1. Ano wale mai nō ka hali'a
I ke ehu wai 'ala no lehua
Kilihune i ka noe Laupua
I ka liko māhie o ka lehua

The fond memories are awe-inspiring
In the sweet ehu spray of lehua blossom perfume
Spraying the Laupua mist
And the delightful leaf buds of the lehua

From a mele hō'ihi, or honorific chant, for 'Emalani Kaleleonālani. Hawaiian source: Nogelmeier 42. English trans. by author.

Lēhei. Rain associated with Makawao, Maui. "Lēhei" means "to leap."

1. Kaualēheiomakawao—a lua fighting stroke. (Also known as Ualēhei-omakawao.)

Source: Pukui and Elbert, 1986, 134. *Note: "Kaualēheiomakawao" means "the Lēhei rain of Makawao."*

Lehua. Rain associated with Hawai'i, Maui, and O'ahu. Also the name of a wind and of the 'ōhi'a lehua tree and its blossoms.

Rain of Haili, Hawai'i

1. Ua Lehua i Haili
I lu'ulu'u i ke onaona
Ua like me ka hau anu
'Ōiwi poli o ka lama

Lehua rain at Haili
Weighed down with perfume
Like the cold hau dew
Bosom that glows in health

From the song "E iala hookuli." Hawaiian source: Holstein 92. English trans. by author.

Rain of Kaiʻaiki, Maui

2. Kuʻu wahine hoa pili i ka ua Lehua
Mānai lehua, makani Kiliʻoʻopu
Makani Kilipoe anu o Kaiʻaiki
Māua me kuʻu wahine e noho ai

My wife, my intimate companion in the Lehua rain
Stringing lehua blossoms, Kiliʻoʻopu wind
Cold Kilipoe wind of Kaiʻaiki
Where we would dwell

From a kanikau, or lament, for Doreka Richardson. Hawaiian source: Richardson. English trans. by author.

Rain of Kailua, Maui

3. Ua ua Lehua, he Lehua hala
Ua i ka lehua o Kailua
Lehua maka kōnunu i ka wai

The Lehua rain showers, a passing Lehua
Raining upon the lehua blossoms of Kailua
Face of the lehua rounded in the water

From a mele by Keaulumoku expressing his love for the Hāmākua districts of Maui, where he had lived with Kamehamehanuiʻailūʻau and Kahekili. Hawaiian sources: Kamakau, *Ke Kumu Aupuni* 66; Kamakau, *Ruling* 113. English trans. by author.

Rain of Mānoa, Oʻahu

4. Punihei hoʻi au iā ia ala lā
I ka leo o ke kai leo nui lā
Ke wā maila i Kālia lā
Alia kāua e naue lā
I ka ua Lehua i nā pali lā
Ke noe maila i Mānoa lā

I am entranced by it
By the sound of the loud-voiced sea
Roaring at Kālia
Let us wait before moving on
As the Lehua rain is over the cliffs
Misting over Mānoa

From a mele inoa, or name chant, for Erisapeka. Hawaiian source: Kaholomoku. English trans. by author.

Leiha'a'ākōlea. Rain associated with O'ahu.

1. Lu'ulu'u wale au i kō aloha ē
   . . . .
   Huhuhui, lele lewa kāua
   E 'e'eha i ka ua Leiha'a'ākōlea
   E ō mai ana ho'i 'oe

   I am weighed down by love of you
   . . . .
   Brought together as one, we shall soar
   To share our pain in the Leiha'a'ākōlea rains
   You will offer your response

   From a mele by Hi'iakaikapoliopele as she played kilu, hoping to hear a response from Lohi'au. Hawaiian source: Ho'oulumāhiehie, *Ka Mo'olelo* 436. English trans.: Ho'oulumāhiehie, *Epic* 407.

Leiha'akolo. Rain associated with Kawelowelo, Hawai'i.

1. Hui ka lewa luna me ka lewa lalo i Pu'umoe'awa
   Hui kai 'ō lā'au o Kawelowelo
   Huhuhui, lelelewa
   'Eha i ka ua Leiha'akolo

   The upper and lower heavens meet at Pu'umoe'awa
   Meeting with the probing sea of Kawelowelo
   Joining together, swaying back and forth
   Pained by the Leiha'akolo rain

   From a mele by Hi'iakaikapoliopele as she noticed her husband's arousal upon seeing her and Wahine'ōma'o draw up their skirts. Hawaiian source: Ho'oulumāhiehie, *Ka Mo'olelo* 357. English trans.: Ho'oulumāhiehie, *Epic* 332.

Leikoko'ula. Or Leikōkō'ula. Probably related to *Koko* and *koko 'ula*. Rain associated with Waimea, Hawai'i, and found in other areas. "Lei koko 'ula" and "lei kōkō 'ula" can refer to a lei of red color or a rainbow-hued lei.

Rain of Mahiki, Hawai'i

1. Kuleana i Mahiki ka ua Leikōkō'ula
   He ua Leikōkō'ula pua i ke kula
   I pua nō, i mohala i luna o ka lā'au
   He wai 'ahulu na ka ua Paliloa
   Na ka ua Mālana'e'elekoa i 'Ōuli

   The Leikōkō'ula rain belongs in Mahiki
   A Leikōkō'ula rain blossoming on the plains

Blooming, spreading atop the trees
Dirty water caused by the Paliloa rain
By the Mālanaʻeʻelekoa rain at ʻŌuli

From a mele kālaiʻāina, or political chant, for ʻEmalani Kaleleonālani, credited to F. K. Hipa. Hawaiian source: Nogelmeier 212. English trans. by author. *Note: In this rendering, "Mālanaʻeʻelekoa" is used instead of "malana ʻeʻelekoa," and "i ʻŌuli" instead of "iō Uli," which were printed in the Hawaiian source.*

2. Lei Mahiki i ka ua Leikokoʻula
   He ua lei holehole ʻula i ke pili

   Mahiki is garlanded with the Leikokoʻula rain
   A red lei of rain that strips the pili grass

   From a kanikau, or lament, for Kamehameha IV. Hawaiian source: Moehonua and Lilikalani. English trans. by author.

Rain of Waimea, Hawaiʻi

3. Ua Leikokoʻula—Waimea, Hawaiʻi; red rain.

   From a list of rain names and their descriptions. Source: "Names . . . rain."

4. *For more examples of Leikokoʻula of Waimea, see Kīpū (#2) and Kīpuʻupuʻu (#34).*

Rain of Kekaʻa, Maui

5. *See Paʻūpili (#6).*

Rain of Haleʻauʻau, Oʻahu

6. *See Nāulu (#31).*

Rain of Lēʻahi, Oʻahu

7. Onaona wale Lēʻahi
   I pilia Kaimanahila
   I ke kai pākuʻikuʻi
   E kaunu nei i ka lipo
   Eia mai nō au
   ʻO ka ua Leikokoʻula

   So fragrant is Lēʻahi
   Kaimanahila clings
   To the pounding sea
   Making love in the dark
   Here I come
   The Leikokoʻula rain

   From a mele inoa, or name chant, for Miki. Hawaiian source: Kawahamana, "He mele." English trans. by author.

Rain of Luanui, Oʻahu

8. *See Noe (#15).*

Rain of Waitūʻauhoe, Kauaʻi

9. ʻO ka uhiwai paha ko Kaʻala
ʻO ka ʻehu o ke kai, kai ka moana
Ua kohu pūnohu ʻula i ke kai
ʻAne like me ka ua Leikōkōʻula
Ke nihi hele aʻela i nā pali
Ma ka lihi o ka Waitūʻauhoe

Perhaps Kaʻala has a heavy uhiwai fog
The spray of the sea is upon the ocean
Like a pūnohu ʻula mist rising at sea
Almost like the Leikōkōʻula rain
Creeping along the cliffs
Along the edge of Waitūʻauhoe

From a mele hula, or hula chant, by Wahineikeouli Paa [Wahinekeouli Pa] for her birthplace, Kalalau on Kauaʻi. Hawaiian source: Pukui, *Nā Mele Welo* 156. English trans. by author.

Rain associated with the *ʻUlalena* rain

10. He hoa noʻu i ka ua Leikokoʻula
I hoʻoluʻu kamalena ʻia e ka ua ʻUlalena

A companion for me in the Leikokoʻula rain
Dyed yellow by the ʻUlalena rain

From a mele inoa, or name chant, for Albert Ka Haku O Hawaiʻi by his father, Kamehameha IV. Hawaiian source: Iolani. English trans. by author.

Leilehua. Rain associated with Mākaulele (or Mōkaulele), Hawaiʻi. "Lei lehua" means "lei of lehua blossoms."

1. *See Kanilehua (#18).*

Leimāʻohu. Misty rain associated with Hawaiʻi and Oʻahu. "Lei māʻohu" means "lei of mist."

Rain of Hinakahua, Hawaiʻi

1. Hoʻomanaʻo maila au i nā ʻai o Hinakahua
Ke ua maila nō ka ua Leimāʻohu
ʻO ka ʻohu nō ia o ke kanaka

I recall the wagers at Hinakahua
The Leimāʻohu rain falls
A person's breath on a cold morning

From a mele inoa, or name chant, for Victoria Kamāmalu. Hawaiian source: Hanaia. English trans. by author. *Note: Pukui, Elbert, and Mookini 46 say that Hinakahua was formerly a site for dancing and playing games, including maika (stone rolling), mokomoko (boxing), and kilu (a sexual game).*

Rain of Waiānuenue, Hawai'i

2. Ka ua Leimā'ohu o Waiānuenue. *The [Leimā'ohu] rain of Waiānuenue that is like a wreath of mist.*

   A traditional saying. Source: Pukui, *'Ōlelo* 170. Additional source: Pukui and Elbert, 1986, 240. *Note: Pukui says, "Waiānuenue ('Rainbow water') in Hilo, Hawai'i, is now known as Rainbow Falls. On sunny days a rainbow can be seen in the falls, and on rainy days the rising vapor is suggestive of a wreath of mist."*

Rain of Nu'uanu, O'ahu

3. Lana Nu'uanu i ka ua Leimā'ohu
   E ha'aheo ana i luna o ka 'āhihi

   Nu'uanu floats in the Leimā'ohu rain
   Strutting atop the 'āhihi blossoms

   From a mele māka'ika'i, or travel chant, for 'Emalani Kaleleonālani by her aunt, Kama'iku'i Young. Hawaiian source: Nogelmeier 99. English trans. by author.

Leinahua. Rain associated with Kā'anapali, Maui.

Rain of Kā'anapali, Maui

1. Ka ua Leinahua o Kā'anapali. *The [Leinahua] rain of Kā'anapali that leaps and produces fruit.*

   A traditional saying. Source: Pukui, *'Ōlelo* 170.

Rain of Kahana, Maui

2. Rain on the Kahana side of this island, "Ka ua Leinahua" or "the rain that leaps and produces fruit."

   Source: Sterling 97.

lelehuna. Also known as lelehune. Fine windblown rain spray, dust, mist; to fall as fine rain.

Source: Pukui and Elbert, 1986, 202.

lele ua. Windblown rain.

Source: Pukui and Elbert, 1986, 202.

Lena. Probably related to *'Ulalena* and *'Ulalenalena*. Rain associated with Maui and Kaua'i. "Lena" means "yellow."

Rain of Liliko'i, Maui

1. Ke 'ike aku i ka poli o ka mau'u
He nu'a wale lā no ka 'awapuhi
I ka hehia e ka ua Lena o Liliko'i ē hē ā

Upon seeing the bosom of the grass
A lush clump for the 'awapuhi plant
Trodden upon by the Lena rain of Liliko'i

From "Mehe [*sic*] hālau lā ka malu o ke kukui," a mele aloha, or love song, about a man ignorant in lovemaking. Hawaiian source: Pukui, *Nā Mele Welo* 198. English trans. by author. Additional sources: Nogelmeier 258; Ilihia 5/19/1894.

Rain of 'Ulalena, Maui

2. 'O ka ua Lena 'Ulalena
Kiumoenahele o Liliko'i

The Lena rain of 'Ulalena
The Kiumoenahele wind of Liliko'i

From a mele inoa, or name chant, for Kamehameha III. Hawaiian source: *Na Mele Aimoku* 207. English trans. by author. Additional source: Kinau.

Rain of Maui and of Hanalei, Kaua'i

3. Name of a yellow-tinted rain famous at Hanalei, Kaua'i, and on Maui.

Source: Pukui and Elbert, 1986, 203.

Rain of Lono

4. Keiki 'ehu, kama 'ehu a Kanaloa
Ua mokuhia kāmakamaka ka lino
Ua ka'a 'ia e ka ua Lena a Lono ē

Child with reddish-tinged hair, ruddy child of Kanaloa
Severed, tie the braid securely
Twisted by the Lena rain of Lono

From a mele inoa, or name chant, for 'Aikanaka. Hawaiian source: *Na Mele Aimoku* 3–4. English trans. by author. Additional source: Tatar 181. *Note: "Lono" refers to the god Lono.*

Līhau. Gentle, cool rain; associated with Pāhoa, Hawai'i, and found in other areas. Also the name of a wind and of a mountain in Lāhainā, Maui. "Līhau" means "cool, moist, fresh." It is both the name of a specific rain and a generally descriptive term; its various usages are determined by the context.

Rain of ʻŌlaʻa, Hawaiʻi

1. Kuʻu kāne mai ka malu ʻōhiʻa loloa o Panaʻewa
   Mai ka ua Līhau anu o uka o ʻŌlaʻa
   Pūkuʻi aku mākou i ke anu o ka mauna
   Mahana aku au i ka poli o kuʻu kāne
   Auē kuʻu kāne

   My dear husband from the shade of the tall ʻōhiʻa of Panaʻewa
   From the cold Līhau rain of the uplands of ʻŌlaʻa
   We huddle in the cold of the mountain
   I am warm in the bosom of my husband
   Pity for my beloved husband!

   From a kanikau, or lament, for S. N. Wahine. Hawaiian source: Kaneheleloa. English trans. by author.

Rain of Pāhoa, Hawaiʻi

2. He kupa ʻo ia, a kulaiwi, no kēia hapa o Puna paia ʻala i ka hala, a me ka hīnano mai ka pali o Hōlei a ke ʻala līpoa o Kaimū, a hoʻokamaʻāina pū me ka ua Līhau o Pāhoa.

   She was a native with deep roots from this half of Puna, with its hala-scented walls, and the hīnano flowers from the cliff of Hōlei to the līpoa fragrance of Kaimū, and she became well-acquainted with the Līhau rain of Pāhoa.

   From an article of condolence for Mrs. Hannah Kaunukapu Kaloi Kamelamela. Hawaiian source: Ahia, Aki, Kaina. English trans. by author. Additional source: Pukui, *ʻŌlelo* 170. *Note: Pukui says that the Līhau is "the icy cold rain of Pāhoa, Puna, Hawaiʻi."*

Rain of Panaʻewa, Hawaiʻi

3. ʻOhuʻohu Panaʻewa i ka ua Līhau
   E hoʻopulu ana i ka liko lehua

   Panaʻewa is adorned with the Līhau rain
   Wetting the lehua buds

   From an article on a wedding in Panaʻewa. Hawaiian source: Kauanoano. English trans. by author.

Rain of Puʻuhue, Hawaiʻi

4. Oʻu hoa i ka uka o Puʻuhue
   Ua luhea i ka ua līhau
   A luhe lā i ka uka ʻiu
   I ka wiliʻōkaʻi a ka ua

My companion in the upland of Pu'uhue
Weighed down by the līhau rain
Drooping in the lofty uplands
In the whirling of the rain

From the song "Puuhue" by Mrs. Eliza Holt. Hawaiian source: Holstein 125. English trans. by author. *Note: The līhau rain is sometimes associated with the word "wili'ōka'i," which means "to whirl."*

Rain of Maui

5. Ua Līhau—Maui.

From a list of rain names and their descriptions. Source: "Names . . . rain."

Rain of Halemano, O'ahu

6. 'O ka noe mai a ka ua līhau i uka o Halemano ē
Me he kui ala e houhou mai ana i ka 'ili i ke anu ē

The mist of the līhau rain in the uplands of Halemano
Like a needle pricking the skin in the cold

From a kanikau, or lament, for Daniel Manuhoa Woodward. Hawaiian source: Woodward. English trans. by author.

Rain of Malamanui, O'ahu

7. Ku'u ipo i ke kai 'uwalu 'ili
Ua lupea i ka ua līhau
A luhe la'i i Malamanui
I ka wili'ōka'i a ka ua

My sweetheart in the sea that claws the skin
Attractive in the līhau rain
And limp with contentment at Malamanui
In the whirl of the rain

From the song "Lilii Mokihana" by Leleiōhoku. Hawaiian source: Holstein 17. English trans. by author. *Note: "Malamanui" is the name of an area near Kalena, O'ahu.*

Rain of Hā'upu, Kaua'i

8. 'O ka ua Līhau o Hā'upu
Me ka ua Noe anu o Kōloa

The Līhau rain of Hā'upu
With the cold Noe rain of Kōloa

From a kanikau, or lament, for Eda Kawaikauomaunahina Kalua. Hawaiian source: Kaapuwai. English trans. by author.

Rain of Mokihana, Kauaʻi

9. *See Noe (#12).*

Lucky rain

10. Gentle cool rain that was considered lucky for fishermen; moist and fresh, as plants in the dew or rain; cool, fresh, as dew-laden air. Onaona ke ʻala o ka maile i ka līhau ʻia e ka ua noe (*Sweet fragrance of the maile, kept fresh and moist by the misty rain*). Līhau mai nei ʻoe (*You are freshly adorned as the cool dew-laden plants*).

    Source: Pukui and Elbert, 1986, 205.

Līheʻelehua. Rain of Mānoa Valley, Oʻahu, that falls on lehua flowers.

From a list of rain names and their descriptions. Source: "Names . . . rain."

Lilikoʻi. Rain associated with Haleakalā, Maui. Also the name of a forested area in Haʻikū, Maui.

1. I ka hora ʻelima a ʻoi o kēia kakahiaka, ua pau loa mākou i ke ala, akā, ʻaʻole naʻe i haʻalele koke i kahi moe no ka mea e kawewe iho ana nō ka ua Lilikoʻi, ua hoʻopala ʻōhelo, o kuahiwi, Haleakalā. Ua kiʻei aku au a me koʻu wahi hoa i waho, ua paʻapū nā wahi a pau i ka noe, a me ka uhiwai, ua like me ka pūnohu uahi e kū ana i kahi hoʻokahi. . . .

    Ua uhi paʻa ʻia ka mauna i ka ʻohu a me ka noe, mai luna mai a hiki loa mai i lalo o ka honua paʻa, a ua ʻike ʻole ʻia aku kekahi wahi o ke kuahiwi. Eia nō hoʻi ke hoʻopulu kilihune nei ka ua Lilikoʻi o Haleakalā i ka ʻōhelo a me ka lau nahele o ka wao kele.

    This morning after five o'clock we had all awoken but did not immediately leave our resting place because the Lilikoʻi rain—rain that ripens the ʻōhelo berry—of the mountain Haleakalā was pouring down with a roar. My friend and I peered outside; everywhere was covered with mist and fog, like a stationary column of smoke. . . .

    The mountain was completely covered with clouds and mist from top to bottom, and not a single part of the mountain was visible. The Lilikoʻi rain of Haleakalā is still wetting with a fine drizzle the ʻōhelo and other plants of the forest.

    From an account by Thomas K. Maunupau of an expedition through Kaupō, Maui. Hawaiian source: Maunupau, *Huakai* 65, 82. English trans. by author.

Līlīlehua. Rain associated with Kāʻanapali, Maui, and with Pālolo, Oʻahu, and found in other areas. Also the name of a wind. "Līlī lehua" means "lehua blossom chill" or "tiny drops on the lehua blossom."

Rain of East Maui

1. *See Līlīpua (#1).*

Rain of Honokōwai, Maui

2. Hopu Kahana i ka ua Līlīlehua
   Palahe'a nō i ke pili o Honokōwai
   'Uku li'i ka pua onaona, nonoho i ka mau'u
   Nonohe nō, me he līlīlehua ala
   Ka 'apoa e ka Nahua o nā pali

   Kahana seizes the Līlīlehua rain
   Stained by the pili grass of Honokōwai
   The fragrant blossoms are tiny, dwelling upon the grass
   Attractive like the līlīlehua flower
   Caught by the Nahua of the cliffs

   From a mele ho'oilina, or legacy song, for 'Emalani Kaleleonālani. Hawaiian source: Nogelmeier 102. English trans. by author.

3. *For more examples of Līlīlehua of Honokōwai, see Nahua (#8) and Pa'ūpili (#13).*

Rain of Kā'anapali, Maui

4. Ka ua Līlīlehua o Kā'anapali. *The Līlīlehua rain of Kā'anapali.*

   A traditional saying. Hawaiian source: Pukui, *'Ōlelo* 170. English trans. by author.

5. Ku'u kāne mai ka lā haoa o Kahakuloa
   Ku'u kāne mai ka ua Līlīlehua o Kā'anapali

   My beloved husband from the scorching sun of Kahakuloa
   My dear husband from the Līlīlehua rain of Kā'anapali

   From a kanikau, or lament, for Kamakaokalani. Hawaiian source: Wiliama. English trans. by author.

Rain of Kahana, Maui

6. Mai ka ua Līlīlehua lā o Kahana
   Aloha ia wahi a kāua i hele ai
   Ho'omaha aku i ka malu o ka 'ulu o Lele

   From the Līlīlehua rain of Kahana
   Affection for this place where we traveled
   Resting in the shade of the 'ulu trees of Lele

   From a kanikau, or lament, for S. Lono. Hawaiian source: Wala. English trans. by author.

7. 'A'ahu 'o Kamaka'eha i ke kapa kūkai
   He kapa no ka po'e 'akuku o Lele
   . . . .

Pua ma'o ke kapa, he moe 'ao'ao kilohana
Kaula'i ana i ka ua Līlīlehua
I ka ua Pa'ūpili o Kahana

Kamaka'eha wears the kapa dipped frequently in the sea
A kapa of the kapa-beating people of Lele

. . . .

Kapa dyed with ma'o blossoms, the kilohana layer for a bed
Hanging in the Līlīlehua rain
In the Pa'ūpili rain of Kahana

From a mele kapa, or kapa chant, for Lili'uokalani, who also was known as Kamaka'eha. Hawaiian source: "He mele kapa no Liliu." English trans. by author.

8. *For another example of Līlīlehua of Kahana, see Pōpō'aha'ula (#1).*

Rain of Keka'a, Maui

9. E ho'i ka nani i Keka'a lā
I ka palena 'āina o Papa lā
I ka ua Līlīlehua lā

May the beauty return to Keka'a
To the land boundary of Papa
In the Līlīlehua rain

From a mele inoa, or name chant, for Manoanoa. Hawaiian source: Maele. English trans. by author.

10. *For more examples of Līlīlehua of Keka'a, see Nahua (#9, #10).*

Rain of Pi'iholo, Maui

11. Ua like nā koa o Kahekili me ka ua Līlīlehua o Pi'iholo i hapa 'ia mai e ka ua 'Ulalena, ka wena 'ula i ka lau o ke koa o Liliko'i. *Kahekili's warriors were like the Līlīlehua rain of Pi'iholo, fragmented by the 'Ulalena rain, the red glow upon the leaves of the koa of Liliko'i.*

From a description of Kahekili's warriors, who, with a divided front, overcame the forces of Kalani'ōpu'u. Hawaiian source: Kamakau, "Ka moolelo o Kamehameha I" 12/15/1866. English trans. by author. Additional sources: Kamakau, *Ke Kumu Aupuni* 23; Kamakau, *Ruling* 87.

Rain of Waiehu, Maui

12. E aloha a'e ana ho'i
I ka ua Līlīlehua
I ka lawe mālie i ka pili
Ko'iawe i ka wai o Waiehu

Greetings to
The Līlīlehua rain
Softly moving in
Lightly showering the waters of Waiehu

From a mele inoa, or name chant, for Kaʻiulani by Leleiōhoku. Hawaiian source: Holstein 6. English trans. by author.

Rain of Waiehu, Maui, and of Pālolo, Oʻahu

13. Name of a wind and rain, famous at Pālolo, Oʻahu, and Waiehu, Maui.

Source: Pukui and Elbert, 1986, 206.

Rain of Kawahapoko, Lānaʻi

14. Aloha ʻo Kawahapoko i ka ʻolu kohai
Kaumaha ʻū luʻuluʻu i kō aloha
Aloha ʻino, minamina wale ua ō
ʻIke aku i ka ua Līlīlehua
Ma [ka] lehua a ka wahine

Love for Kawahapoko that sways and bends
Bent over with grief and sadness for your love
What a misfortune, great sorrow, it continues
See the Līlīlehua rain
On the woman's lehua blossom

From a kanikau, or lament, for Kahaukomo. Hawaiian source: Kahiamoe. English trans. by author.

Rain of Kaʻau, Pālolo, Oʻahu

15. Kuʻu keiki mai ka hale kanaka nui
Kuʻu keiki mai ka ua Līlīlehua o Kaʻau

My beloved child from the home with many people
My dear child from the Līlīlehua rain of Kaʻau

From a kanikau, or lament, for Kapela. Hawaiian source: D. W. Pauahi. English trans. by author.

Rain of Lēʻahi, Oʻahu

16. Kuʻu hoa i ka ua Līlīlehua
Ka ua nihi mai ma Lēʻahi

My companion in the Līlīlehua rain
The rain that tiptoes upon Lēʻahi

From a kanikau, or lament, for Hina. Hawaiian source: Kaakopua. English trans. by author.

Rain of Mānoa, O'ahu

17. *See Kūkalahale (#11).*

Rain of Pālolo, O'ahu

18. 'O kēia Pōhakukīkēkē, he mo'o wahine ia. He wahine u'i kēia mo'o. 'A'ole na'e 'o Pōhakukīkēkē kona inoa mua, akā, 'o Kaualīlīlehuaopālolo kona inoa mai kona mau mākua mai.

'Oiai ua 'ono loa ko Pāhoa pu'u i ka u'i uwa'uwali a me ka maika'i 'une'unehe o ka u'i o ka "ua Līlīlehua o Pālolo," no laila, mī'ala mau loa ua Pāhoa nei ma kēlā āhua e ho'omomoni ai i ka 'ae o kona pu'u i kā ha'i mea i hānai ai a nui nepunepu a pu'ipu'i ho'i.

Pōhakukīkēkē was a mo'o woman, and she was quite beautiful. Pōhakukīkēkē was not her original name, for her parents had named her after the Līlīlehua rain of Pālolo, Kaualīlīlehuaopālolo.

Hungering for the soft loveliness and tender beauty of this young girl of Pālolo's Līlīlehua rains, Pāhoa would always hasten out to that hill, where he would salivate over this girl who had been raised to be so plump and succulent.

From the legend of Hi'iakaikapoliopele. Hawaiian source: Ho'oulumāhiehie, *Ka Mo'olelo* 315. English trans.: Ho'oulumāhiehie, *Epic* 293.

19. 'Ae, e ka mea heluhelu, he moa ho'okalakupua 'i'o nō kēia, e like nō ho'i me ko Kamapua'a 'ano kino 'e'epa, a 'o ka inoa o kēia moa 'o Ho'okahikaniakamoa, mai Kahiki mai.

A 'o Ka'uhalemoa, 'o ia kona muli mai, 'o ia ka moa i ka uka o Pālolo, nona ho'i ka ua Līlīlehua i 'ōlelo 'ia.

Indeed, dear reader, this chicken is truly an extraordinary creature, who, much like Kamapua'a, had a strange body, and the name of this chicken is Ho'okahikaniakamoa, from Kahiki.

And his younger brother is Ka'uhalemoa, the chicken of the uplands of Pālolo, where the Līlīlehua rain is said to be from.

From the legend of Kamapua'a. Hawaiian source: "He molelo . . . no Kamapuaa" 6/25/1891. English trans. by author. Additional source: Kame'eleihiwa 26. *Note: "Ka'uhalemoa" is a variant of "Ka'auhelemoa," the name of a supernatural fighting chicken.*

20. Ho'i ke aloha a i Pālolo
I ka ua Līlīlehua e kilihune nei
Ka ua nō ia 'olu ka mana'o
Hō'oni a'e nei i ku'u pu'uwai

. . . .

Ha'ina 'ia mai ana ka puana
I ka ua Līlīlehua e kilihune nei

Love returns to Pālolo
To the Līlīlehua rain misting down
It is the rain that soothes the mind
Stirring up feelings in my heart
. . . .
Tell the story through the refrain
Of the Līlīlehua rain misting down

From the song "Pālolo" by Charles E. King. Source: Wilcox et al. 213. *Note: Wilcox et al. say, "Līlīlehua is the name of a chilly rain and wind, famous in Pālolo, Oʻahu."*

21. Kuʻu kāne mai ka ua Līlīlehua o Pālolo
Mai ka makani Holoʻūhā o Kekaha

My darling husband from the Līlīlehua rain of Pālolo
From the Holoʻūhā wind of Kekaha

From a kanikau, or lament, for J. Henry. Hawaiian source: Kahinawe, "He kanikau." English trans. by author.

22. Kaimukī Shopping Center, which was once the King's Daughter's Home, sits on a mound that was the back of a moʻo, a legendary reptile. This moʻo loved Līlīlehua, who lived in Pālolo Valley. Līlīlehua fell in love with another person, and the moʻo became jealous. Līlīlehua was turned into a rain that does not pass Waiʻalae Avenue, thus avoiding the moʻo.

From an email sent to the author by Kimo Alama Keaulana on April 4, 2013. *Note: The King's Daughter's Home was established to care for elderly Caucasian women in Hawaiʻi.*

23. *For more examples of Līlīlehua of Pālolo, see Kuahine (#16) and Makaʻūpili (#3).*

Rain of Waʻahila Ridge, Oʻahu

24. *See Waʻahila (#26).*

Rain of Waialua, Oʻahu

25. Kuʻu kāne mai ka ua Līlīlehua o Waialua
Mai ka ua Pōʻaipuni o Kemoʻo
Nānā aku ʻo ka lā laʻilaʻi o Kaʻena
Auē kuʻu kāne ē

My dear husband from the Līlīlehua rain of Waialua
From the Pōʻaipuni rain of Kemoʻo
Behold, the still sun of Kaʻena
Pity for my dear husband!

From a kanikau, or lament, for J. Henry. Hawaiian source: Kahinawe, "Ka hope." English trans. by author.

Rain of Wai'ōma'o, O'ahu

26. He kanikau aloha kēia nou, e B. Mahune

. . . .

Nānā aku, 'o ka ua Līlīlehua
Aia i ka nahele i Wai'ōma'o

This is an affectionate chant of mourning for you, B. Mahune

. . . .

Look, it is the Līlīlehua rain
There in the forest at Wai'ōma'o

Hawaiian source: Wailea. English trans. by author.

27. Pipili ka ua i ka nahele
Ua kilikilihune kili hau
Kilikili hau Līlīlehua
No Wai'ōma'o i ka uka la'i

The rain clings close to the forest
Delicate, fine, chilly rain
Cold, misty Līlīlehua rain
There at Wai'ōma'o in the quiet upland

From the love song "Pipili ka ua i ka nahele." Source: Lili'uokalani 260. *Note: The source mentions that the setting for the song is "in the cool land area once known as Wai'ōma'o, deep in Honolulu's Pālolo Valley. A stream and a road, both named Wai'ōma'o, are reminders of the old name for the area, which means 'green water.' The rain and wind of the area were called 'Līlīlehua' (literally, 'lehua chill'), and in referring to it in this song, Lili'uokalani makes clear why she and other haku mele chose cold and rain as metaphors for love."*

Rain of Nounou, Kaua'i

28. Milohia ka wai o Nāmalokama
I 'ula Nounou i ka ua Līlīlehua
He hua na'u ke aloha e mālama nei
E hi'ipoi nei i Waiohuli lā ē
Huli mai
E huli mai nō 'oe, eia nō mākou
Nā hoa pili o ka ua me ka makani

The waters of Nāmalokama whirl
Nounou turns red in the Līlīlehua rain
Love is a fruit I am tending
Cherishing at Waiohuli
Face here

Turn, here we are
The close companions of the rain and the wind

From a farewell song for the newspaper *Ka Hae Hawaii*. Hawaiian source: Lohiau, "He mele no ka nupepa." English trans. by author.

Misty rain

29. He ʻala nei e moani mai nei
Na ka ua noe Līlīlehua
I lawe mai a kuʻu poli
Hoʻopumehana i kuʻu poli

There is a breath so soft and balmy
Brought by sweet zephyrs, Līlīlehua
And while wafted to my bosom
It brings a yearning for one I love

From the song "He ʻala nei e māpu mai nei," also known as "Ahe lau makani," by Liliʻuokalani and others. Source: Liliʻuokalani 31. Additional source: Wilcox et al. 3.

Rain of the forest

30. ʻAuhea wale ʻoe, e ka ua Līlīlehua
Kilihune i ka nahele
ʻEu hoʻi kāua i ka ua Kanilehua
E kui i lei nou

Are you listening, O Līlīlehua rain
Showering lightly in the forest
Let us move quickly in the Kanilehua rain
To string a lei for you

From the song "Pua magnolia." Hawaiian source: Holstein 95. English trans. by author.

Lilinoe. Fine mist associated with mountains and cliffs of Hawaiʻi and Maui. Also the name of a goddess of mists.

Rain of the Hāmākua Cliffs, Hawaiʻi

1. Hōʻea akula lākou nei i Hāmākua. A laila, huli aʻela ʻo Hiʻiaka a nānā aʻela i nā pali hulaʻana, a kānaenae aʻela ʻo ia i kēia kau.

A ua Lilinoe, ʻehu wale i nā pali
Pōpō ke kapa a paʻa ma ka lima

They arrived in Hāmākua, at which point Hiʻiaka turned, and looking out at the sheer sea cliffs, offered this chant.

O Lilinoe rain, shrouding the cliffs
Bundle the kapa and hold it in the hand

Hawaiian source: Hoʻoulumāhiehie, *Ka Moʻolelo* 357. English trans.: Hoʻoulumāhiehie, *Epic* 332.

Mist of Maunakea, Hawaiʻi

2. Then we have Lilinoe who is the kupua of the mists. And you can see Lilinoe as she comes down over the mountain sometimes. When you're driving from Saddle Road up to Hale Pōhaku, you can look over and see Lilinoe in some of the valleys. She flows up and over, very gently, very soft, like very fine kapa, white kapa. She fills in the valleys and you can see her hands just filtering out very thinly into the valleys.

   It's a remarkable sight and it really makes you very aware of the mist, of how mist flows around. And it's obvious that it's very, very beautiful.

   From a November 4, 2002, interview with Keawe Vredenburg. Source: Na Maka o ka Aina.

Rain of Haleakalā, Maui

3. Ua Lilinoe o Haleakalā—he ua kaulana no ia kuahiwi.

   Lilinoe rain of Haleakalā—a famous rain belonging to that mountain.

   From an account of East Maui rains and their descriptions, as told to Thomas K. Maunupau by Josua Ahuliʻi of Kaupō, Maui. Source: Maunupau, *Huakai* 27, 108.

Līlīpua. Rain associated with East Maui and Oʻahu.

Rain of East Maui

1. Kuʻu wahine mai ka ua Līlīpua
   Mai ka ua Līlīlehua heʻe koko
   Makaʻūpili heʻe koko ʻula i ke kula

   My darling wife from the Līlīpua rain
   From the fleeting rainbow-hued Līlīlehua rain
   And the Makaʻūpili rain that spreads a rainbow over the plains

   From a kanikau, or lament, for Batesepa Puhia, the wife of Davida Malo. Hawaiian source: Malo, "Eia." English trans. by author.

Rain of Oʻahu

2. *See Pōpōkapa (#13).*

Līpoa. Same as *Uaakalīpoa*. Rain associated with Heʻeia, Oʻahu. Also the name of a seaweed.

Rain of Heʻeia, Oʻahu

1. Aloha nō ʻoe, e ka Māluaaluakele
   E ka makani anu o kuʻu ʻāina

He ʻāina, he hoa aloha no ka pali
He kāne hoʻi, he ipo no ka ua Līpoa

Beloved are you, O Mālualuakele
The cold wind of my homeland
A home, a friend for the cliffs
A mate, a sweetheart for the Līpoa rain

From a mele in the legend of Kamaakamahiʻai, recited by Kahelekūlani of Heʻeia, Oʻahu, as she left her husband, Waikūmailani, who had found another woman. Hawaiian source: Kaualilinoe, "Ka moolelo" 11/5/1870. English trans.: Kaualilinoe, "Legend" 83. *Note: Limu līpoa is known to be found between Ahuolaka and Kapapa in Kāneʻohe Bay.*

Waterless rain

2. ʻIno ma kō hana ē
He ua ka Līpoa wai ʻole
Ale Mahiki i ke kēhau
I ka luna i Kanikū ē, aloha
I hoʻi mai nei lā
E kuʻu kaikuaʻana ē

Foul because of your deed
The waterless Līpoa rain
Mahiki is swallowed up in the mist
To the top to Kanikū; greetings
Come back here
My older brother

From a mele by Kamapuaʻa to the prophet Waikūmailani, who was chasing the spirit of Kamapuaʻa's brother. Source: Kahiolo, *He Moolelo* 84–85.

Loku. Probably related to *Lokuhala*. Downpour; associated with Hanalei, Kauaʻi. "Loku" can refer to the feeling of deep emotion, pain, or sorrow. It is both the name of a specific rain and a generally descriptive term; its various usages are determined by the context.

1. Ka ua Loku o Hanalei. *The pouring [Loku] rain of Hanalei.*

A traditional saying. Source: Pukui, *ʻŌlelo* 170. Additional source: Kahananui and Anthony 108.

2. ʻIke wale aku nō ʻoe
I ka ua Loku o Hanalei
Lipolipo wale maila
Ka nahele o Hoʻohie

So apparent to you
Is the Loku rain of Hanalei
Darkening
The forest of Hoʻohie

From the chant "Ula noweo." Hawaiian source: Barrère, Pukui, Kelly 87. English trans. by author. Additional source: "Ula no weo [*sic*]." *Note: Barrère, Pukui, Kelly say, "This Kauaʻi [chant] was used when the people were learning the letters on that island. Later, it was revised and made into a mele inoa or name song for Kapiʻolani, the granddaughter of Kaumualiʻi, the hereditary chief of Kauaʻi. Kapiʻolani was dearly loved, but more so by those of the island of her illustrious ancestors."*

3. Kau aku ka manaʻo no Hanalei
E ʻike i ka nani o ka ua Loku

My thoughts are on Hanalei
To see the beauty of the Loku rain

From the hula "Aloha Niihau" by Kalekahee. Hawaiian source: Tava and Keale 20. English trans. by author.

4. Kaulana wale e ka ua o Hanalei
E nihi aʻe nei i nā pali
E hoʻopili ʻia me ka lauaʻe
Me he ipo hoʻoheno nei i ka poli
Ka hoene mai nō a ke kai
Me he ala e ʻī mai ana
E hoʻi mai nō kāua lā e pili
Ka ua Loku kaulana aʻo Hanalei

Famous indeed is the rain of Hanalei
Sweeping along the cliffs
To be joined with the lauaʻe fern
Like a sweetheart cuddling in embrace
The sea murmurs softly
As though it were saying
You and I should be together again
Like the famous pouring [Loku] rains of Hanalei

From the song "Ka ua Loku" by Alfred ʻAlohikea. Source: Wilcox et al. 94. Additional source: Elbert and Mahoe 62.

5. E ka ua Loku o Hanalei
Kui ʻia mai nā mokihana
Na ka Lūpua e lawe mai
I wehi, i ʻohu no Kalani

O Loku rain of Hanalei
The mokihana fruits are to be strung
The Lūpua wind will bring them
A decoration, an adornment for the chief

From the song "Pela kapu o Kakae." Hawaiian source: Holstein 33. English trans. by author.

Lokuhala. Probably related to *Loku*. Rain associated with Hawai'i and Kaua'i. "Loku hala" means "to drench hala trees."

Rain of Puna, Hawai'i

1. Aia i laila ke aloha
   Aia i laila ka 'ano'i
   Aia i laila ka pilina
   Me ka ua Lokuhala o Puna

   There is the love
   There is the desire
   There is the union
   With the Lokuhala rain of Puna

   From the song "He inoa no Kaiulani" by Leleiōhoku. Hawaiian source: Holstein 6. English trans. by author.

Rain of Hanalei, Kaua'i

2. I ka hiki 'ana aku ho'i i ua ua Lokuhala nei o Hanalei, 'a'ole ua mau hoaloha nei, 'o ka hala 'ana aku nō ia, hala ke aloha, naue me ka 'ano'ai.

   When I reached Hanalei, land of the heavy [Lokuhala] rains that fall through the hala trees, these friends of mine were not at home, they had just left; gone was love and with it went affection.

   From a narrative by Daniel Kahā'ulelio about his visit to Kaua'i. Source: Kahā'ulelio 264–65.

Luahine. Rain associated with Mānoa, O'ahu. Also the name of a hill in Mānoa. "Luahine" means "old woman."

1. The Luahine rain starts at Paliluahine on the eastern corner of Mānoa. It moves very slowly, like an old lady, along the Woodlawn side of Mānoa. Then it tires and ends at Kauhi, or a rock known as "sleeping giant," along Wa'ahila Ridge.

   From an email sent to the author by Kimo Alama Keaulana on April 4, 2013.

2. When the girl was finally dead, her mother melted into the rain called Luahineomānoa.

From the story of Kahalaopuna by Mary Kawena Pukui. Source: Sterling and Summers 289.

Lūlaukō. Rain at Limunui, Kaua'i. *Lit.*, rain that scatters sugarcane leaves.

Source: Kent 379.

Lūlehua. Same as *Lūlehuahala*. Similar to *Mololani*. Rain associated with Pana'ewa, Hawai'i. "Lū lehua" means "to scatter lehua flowers." It is both the name of a specific rain and a generally descriptive term; its various usages are determined by the context.

1. Ka ua Lūlehua o Pana'ewa shakes down the lehua.

   From a list of rain names and their descriptions. Source: Kelsey, "Rains." Additional source: Pukui, *'Ōlelo* 172. *Note: Pukui says that the Lūlehua is "the heavy rain of the lehua forests of Pana'ewa in Hilo, Hawai'i." She also says that it was "famed in chants of old."*

2. 'O Hilo ia lā o ka ua kinakinai
   He ua Lūlehua ia no Pana'ewa
   I kinai i ka ua o ke kila
   He ua mao 'ole kaulana o Hilo

   This is Hilo of the endless rain
   A [Lūlehua] rain that pelts the lehua of Pana'ewa
   A beating, relentless rain
   The famous endless rain of Hilo

   From a mele hei, or string-figure chant. Source: Pukui, *Nā Mele Welo* 96–97. Additional source: Pukui and Korn 188–89.

3. *For another example of Lūlehua of Pana'ewa, see Kanilehua (#37).*

Lūlehuahala. Same as *Lūlehua*. Rain associated with Pana'ewa, Hawai'i.

1. *See Kanilehua (#18).*

M

Ma'akua. Rain associated with Hau'ula and Makaua, O'ahu.

Rain of Hau'ula, O'ahu

1. Aloha 'ino nō ku'u wahine

   . . . .

   Ku'u wahine mai ka ua Ma'akua o Hau'ula

   What great pity for my darling wife

   . . . .

   My beloved wife from the Ma'akua rain of Hau'ula

   From a kanikau, or lament, for Luakauwahine. Hawaiian source: Pawai. English trans. by author.

2. He hiwahiwa nui 'oe na ku'u kino
   'O ku'u kino kai mehana i kō poli
   Ua ka ua Ma'akua, e pehi ana i ka 'ilikai o ka moana
   Auē ku'u kāne, ku'u kāne ho'i ē

   You were a great adornment for my body
   My body was warmed in your arms
   The Ma'akua rain comes down, pelting the surface of the ocean
   Pity for my dear husband! Oh, my beloved husband!

From a kanikau, or lament, for John Kealoha. Hawaiian source: Kaaikaula. English trans. by author.

Rain of Makaua, Oʻahu

3. Iā kākou e hele aku ai e hōʻea aku ana kākou i Makaua, ka ʻāina o ka ua Maʻakua. Aia i laila kahi i noho ai ʻo ka ʻīlio hā a Kāne, nona ka inoa ʻo Kauhikeʻīmakaokalani, a he makua kāne nō naʻe ia no kākou....

   I ia wā, pane akula ʻo Hiʻiaka, "E noho nō ʻoe i ke kai o Makaua, e walea hoʻi i ka pehia e ka ua Maʻakua, e hoʻomanawanui ana hoʻi i ka hau anu o nā pali hāuliuli."

   As we go along we will reach Makaua, land of the Maʻakua rain. That is where the ʻīlio hā of Kāne dwells, named Kauhikeʻīmakaokalani, an uncle of ours....

   Hiʻiaka replied, "You must stay here at the shore of Makaua, enjoy the pelting Maʻakua rain, and patiently endure the cold dew of the verdant cliffs."

   In the first paragraph Hiʻiakaikapoliopele is speaking to Wahineʻōmaʻo. In the second paragraph she is speaking to her uncle Kauhikeʻīmakaokalani. Hawaiian source: Hoʻoulumāhiehie, *Ka Moʻolelo* 160, 162. English trans.: Hoʻoulumāhiehie, *Epic* 151, 153.

Mailani. Rain associated with Hilo, Hawaiʻi.

1. *See Kanilehua (#18).*

Makahekili. Rain name. Also the name of a plant.

1. *See Puʻukoa (#3).*

Makahuna. Rain associated with Oʻahu.

Rain of Pālolo, Oʻahu

1. ʻO ʻoe ia, e Pāhoa
   Wahine noho ua Makahuna o Pālolo
   Hoʻolono mai ana ʻo ka leo
   Leo ualo a kama hele

   It is you, O Pāhoa
   Woman who dwells in the Makahuna rain of Pālolo
   Listening to the voice
   The beckoning call of the traveler

   From a mele by Hiʻiakaikapoliopele calling out to the moʻo woman Pāhoa. Hawaiian source: Hoʻoulumāhiehie, *Ka Moʻolelo* 312. English trans.: Hoʻoulumāhiehie, *Epic* 291.

Rain of Waikīkī, Oʻahu

2. Kuʻu kāne i ka makani Hauālia
   ʻO ka Makahuna i Hāwāwā ē

Wā ihola, ke wā wale maila nō
Ka ua hilahila moe awakea

My husband of the Haualia wind
The Makahuna rain at Hāwāwā
Boisterous, making an uproar
The shy rain that settles down at midday

From a mele by Hiʻiakaikapoliopele on hearing the clamor of people in the house she has just left in Waikīkī. Hawaiian source: Hoʻoulumāhiehie, *Ka Moʻolelo* 311. English trans.: Hoʻoulumāhiehie, *Epic* 290. Additional source: Emerson, *Pele* 185.

Makaiwi. Rain associated with Waʻahila Ridge, Oʻahu.

1. *See Waʻahila (#26).*

Makakoʻi. Rain associated with Kauaʻi. "Maka koʻi" refers to the edge of an adze.

Rain of Haleleʻa, Kauaʻi

1. Ka ua Makakoʻi o Haleleʻa. *The adze-edged [Makakoʻi] rain of Haleleʻa.*

A traditional saying. Source: Pukui, *ʻŌlelo* 172. *Note: Pukui says that the Makakoʻi is "a rain so cold that it feels like the sharp edge of an adze on the skin."*

Rain of Hanalei, Kauaʻi

2. Ua Makakoʻi—Hanalei, Kauaʻi.

From a list of rain names and their descriptions. Source: "Names . . . rain."

Rain of Kawaikōī, Kauaʻi

3. A ka luna i Kahelekua
Mamake, māʻeʻele i ke anu
Kūʻouʻou i ke koʻekoʻe
Kua i ka ua Makakoʻi i Kawaikōī
Kōī, e ale ka luna o Koʻianana

Atop Kahelekua
Seized, stiffened by the cold
Bent over in the moist cold
Hewn in the Makakoʻi rain at Kawaikōī
Flowing with force, swallowing the top of Koʻianana

From an ancient chant. Hawaiian source: "Mele kahiko." English trans. by author. Additional sources: *Na Mele Aimoku* 249; "Eia ana."

Makalaukoa. Same as *Laukoa*. Rain associated with Nuʻuanu, Oʻahu.

Rain of Kahuailanawai, Oʻahu

1. Wahine hoʻokēwai o Kahaukomo
Kākele wai o Kahuailanawai
Ua anu wale i ka ua Makalaukoa

Woman of Kahaukomo who creates moisture
Gliding on the water of Kahuailanawai
Very cold in the Makalaukoa rain

From the chant "Kuʻu wahine i ka ua ʻUlalena." Hawaiian source: Kanahele 103. English trans. by author. Additional source: L. K.

Rain of Lanihuli, Oʻahu

2. Kukuhi kūlana i Lanihuli
Nānā iho i ke oho o ke kāwelu
Hiwahiwa ka lau liko o ka ʻawapuhi
I ka ua ē, ka ua Makalaukoa

Stepping into position on Lanihuli
Looking down at the leaves of the kāwelu grass
Dark and precious is the leaf bud of the ginger
In the rain, the Makalaukoa rain

From a mele hānau, or birth chant, for ʻEmalani Kaleleonālani. Source: Nogelmeier 239.

Rain of Nuʻuanu, Oʻahu

3. Ua Makalaukoa—Nuʻuanu; the rain that falls among the koa trees on top of the mountains.

From a list of rain names and their descriptions. Source: "Names . . . rain." Additional source: Pukui and Elbert, 1986, 509.

Makanoe. Rain associated with Puna, Kauaʻi. "Maka noe" means "misty face."

1. Noho i Puna ka ua Makanoe
He ua Makanoe anu ia no uka
He haili hoaka na ke Koʻolau
He welelau makani ia no ka Māluakele

. . . .

"Launa aku māua me kuʻu wahine
I ka ua Makanoe i Hauaʻiliki ē"

. . . .

Dwelling in Puna is the Makanoe rain
This is a cold Makanoe rain from the uplands
A memory cast by the Koʻolau wind
This is the tip of the wind from the Māluakele

. . . .

"My darling wife and I come together
In the Makanoe rain at Hauaʻiliki"

. . . .

From a kanikau, or lament, for David Kahalepouli Piikoi. Hawaiian source: Paukaikena. English trans. by author.

Makaʻūpili. Rain associated with Haleleʻa, Kauaʻi. Also found on Maui and Oʻahu. Also the name of a wind. "Maka ʻūpili" means "eyes shut tight."

Rain of ʻAuwaiawao, Maui

1. *See Nahua (#10).*

Rain of East Maui

2. *See Līlīpua (#1).*

Rain of Lēʻahi, Oʻahu

3. Nani Lēʻahi ke kuahiwi o Kaimanahila
I pili aloha lāua me ka ua Makaʻūpili
I hoʻi e pili me ka ua Līlīlehua
I hoʻomakua i ka ua Kuahine o Mānoa
I nā ʻauamo kiʻi a ka ua Nāulu i ke kula
I loku ʻia e ka ua ke Kulaokahuʻa
I nā kīkīpani a ka ua ʻAwaʻawa
Pūpū ka lāʻau a ka ua Hōliʻo
Ua maʻū i ka wai a ka ua Kiʻowao

Beautiful is Lēʻahi, the peak of Diamond Head
That lovingly joins the Makaʻūpili rain
Which returns to join the Līlīlehua rain
That fosters the Kuahine rain of Mānoa
With the carrying poles of the Nāulu rain of the plains
Kulaokahuʻa is poured upon by the rain
By every last drop of the ʻAwaʻawa rain
The trees of the Hōliʻo rain huddle together
Moistened by the water of the Kiʻowao rain

From a mele inoa, or name chant, for Heneri W. Auld. Hawaiian source: Hoopai. English trans. by author.

Rain of Māmalahoa, Kauaʻi

4. Anu Māmalahoa, lī e ke anu i ke koʻekoʻe
Ahu kupanaha i ka ua Makaʻūpili

Māmalahoa is cold, shivering in the damp chill
A mass of wonder in the Maka'ūpili rain

From the legend of 'A'ahoaka. Hawaiian source: "He moolelo no Aahoaka" 3/3/1877. English trans. by author.

Rain of Pe'ape'a, Kaua'i

5. 'Ōlelo 'o Kānaenae, he lepo waiwai 'ole, ua ihola ka ua Maka'ūpili o uka o Pe'ape'a, kahe maila kā ka 'ōmaka wai, kū maila ka 'ehu o ka lepo, 'a'ohe kaua, moe ia.

   Kānaenae said the soil was worthless. Then the Maka'ūpili rained in the uplands of Pe'ape'a, the stream flowed from its source, the dust from the ground rose up; there would be no war, so he slept.

   From the legend of Kapakohano. Hawaiian source: Haleole, "Ka mooolelo." English trans. by author.

6. *For another example of Maka'ūpili of Pe'ape'a, see Kapakapa'ahoa (#1).*

Makawiliwili. Rain name. Also the name of an 'ili in He'eia, O'ahu.

1. *See Pu'ukoa (#3).*

mākolekole. Rainbow-hued rain.

1. The mākolekole rain falls when the sun is shining.

   From a list of rain names and their descriptions. Source: Kelsey, "Rains."

2. Ua mākolekole—a rain with rainbow colors.

   Source: Pukui and Elbert, 1986, 230.

Mālana. Same as *Mālanalana*. Related to *'E'elekoa* and *Mālana'e'elekoa*. Rain associated with Waimea, Hawai'i. "Mālana" means "buoyant, to float, to move together."

Rain of Waimea, Hawai'i

1. Kū akula 'oe i ka Mālana a ke Kīpu'upu'u
   Holu ka maka o ka 'ōhā wai o Uli

   You are pierced by the Mālana and the Kīpu'upu'u
   The buds of the 'ōhā wai of Uli are bending

   From a mele inoa, or name chant, for Kamehameha I. Hawaiian source: *Nā Leo Hawai'i Kahiko,* Record I, Side A, Band 9. English trans. by author. Additional source: Moehonua and Lilikalani.

2. Ua Mālana—Waimea, Hawai'i.

   From a list of rain names and their descriptions. Source: "Names . . . rain."

3. ʻO ka inoa o ka pūʻali koa ma lalo o Kamehameha, he Mālana, a ua ʻōlelo ʻia, ua hiki aku ko lākou nui i ka ʻekolu lau, ua like ia me 1,200 kānaka. ʻO ka inoa o ka pūʻali ʻelua, ma lalo o Nānuʻuakaleiʻōpuʻu, he Kīpuʻupuʻu; a he ʻekolu nō hoʻi lau ko lākou nui.

   ʻO kēia poʻe koa, he poʻe kānaka ʻōpiopio lākou no Waimea; a ʻo ia ke kumu o ko lākou kapa ʻia ʻana ʻo "Kīpuʻupuʻu."

   The division of warriors under Kamehameha was named Mālana, and it was said that their numbers reached up to three lau, which is the same as 1,200 men. The name of the second division, under Nānuʻuakaleiʻōpuʻu, was Kīpuʻupuʻu, and they also numbered three lau.

   As for these warriors, they were young men from Waimea, and that is why they were called "Kīpuʻupuʻu."

   From a history of Kamehameha I. Hawaiian source: Hoʻoulumāhiehie, "He moolelo" 3/12/1906. English trans. by author. Additional sources: Desha, *He Moʻolelo* 1: 274; Desha, *Kamehameha* 191. *Note: These armies were named after the rains of Waimea—the Mālana and the Kīpuʻupuʻu.*

4. The name of a fine rain from the northeast at Waimea as it moves along before the trade wind.

   Source: Andrews 377.

5. *For another example of Mālana of Waimea, see Kīpuʻupuʻu (#25).*

Rain associated with steering paddles

6. Paʻa nā hale ʻohu makani o Kūkahauʻula
   I ka hoʻolale a ke Kēhau makani o lalo
   I nā hoe kāuli a ka ua Mālana

   The houses of wind and mist of Kūkahauʻula are complete
   Hastened by the Kēhau, wind from below
   By the steering blades of the Mālana rain

   From a chant for Kawaikini. Hawaiian source: Lohiau, "He mele no Kawaikini." English trans. by author.

Mālanaʻeʻelekoa. Related to *ʻEʻelekoa*, *Mālana*, and *Mālanalana*. Rain associated with Kohala, Hawaiʻi. Perhaps a combination of two rain names: *Mālana* and *ʻEʻelekoa.*

Rain of Mahiki, Hawaiʻi

1. *See Kīpuʻupuʻu (#5).*

Rain of ʻŌuli (or Uli), Hawaiʻi

2. *See Leikokoʻula (#1) and Paliloa (#5).*

Mālanalana. Same as *Mālana*. Related to *'E'elekoa* and *Mālana'e'elekoa*. Drifting rain associated with Waimea, Hawai'i. "Mālanalana" means "buoyant, to float, to move together."

1. Ku'u kāne mai ka ua Mālanalana
Lana ku'u aloha, lana me he wai lā
Ka waiwai nui nō ia e noho nei, 'o ke aloha iā 'oe

My dear husband of the drifting Mālanalana rain
Uplifted is my love, floating as though on water
It is the greatest treasure that now exists, the love of you

From a kanikau, or lament, for Kamehameha IV by his wife, 'Emalani Kaleleonālani. Source: Nogelmeier 338.

Mālua. Same as *Mālualua*. Rain associated with Hilo, Hawai'i, and with Kaunakahakai, Moloka'i. Also the name of a wind.

Rain of Hilo, Hawai'i

1. *See Ho'olua (#1, #2) and Uluau (#2).*

Rain of Kaunakahakai, Moloka'i

2. Hā'ule iho, he Mālua ka ua
He Pelu ka makani
Hauālialia Kaunakahakai

It falls, Mālua is the rain
Pelu is the wind
The Hauālialia wind is at Kaunakahakai

From the legend of Pāka'a and Kūapāka'a. Hawaiian source: Fornander, *Fornander* 5: 101. English trans. by author. Additional sources: Kuapuu, "He wahi moolelo" 5/8/1861; Nakuina, *Moolelo* 62; Nakuina, *Wind* 56. *Note: "Kaunakahakai" is the same as "Kaunakakai."*

Rain that causes loincloths to shred

3. Ke hō'ike'ike maila ka 'ama'u
I nā wai hae malo, wai a ka Mālua

The 'ama'u fern displays itself
In the malo-shredding waters, waters of the Mālua rains

From a mele recalled by Ho'oulumāhiehie. Hawaiian source: Ho'oulumāhiehie, *Ka Mo'olelo* 417–18. English trans.: Ho'oulumāhiehie, *Epic* 390.

Mālualua. Same as *Mālua*. Rain associated with the *Kea* rain of Hāna, Maui. Also the name of a wind.

1. *See Kea (#3, #4).*

Maluko'i. Name of a rain associated with Kahalu'u, O'ahu.

Source: Pukui and Elbert, 1986, 234.

Moani'ala. Rain associated with Hilo, Hawai'i. Also the name of a wind. "Moani 'ala" means "fragrant breeze."

1. 'O 'oe kā ia, e ku'u kāne Laupua makani o Ko'oko'olau
   Ku'u kāne i ka ua Moani'ala hala
   Lei lehua o Hilo i ke onaona

   It is truly you, my dear husband, Laupua wind of Ko'oko'olau
   My beloved husband in the Moani'ala rain, scented with hala
   Lei of lehua of fragrant Hilo

   From the legend of Nāmakaokapāo'o. Hawaiian source: Ilihia 5/26/1894. English trans. by author.

2. E 'olu'olu ana nō paha nā makamaka o ke *Kuokoa* e 'ike a e lohe ho'i i nā mea i hana 'ia ma Hilo, ka 'āina a ka ua Moani'ala i noho ai?

   Would the friends of the Kuokoa *enjoy seeing and hearing about the things happening in Hilo, the land where the Moani'ala rain resides?*

   From a letter to the newspaper *Kuokoa* about New Year's Day festivities in Hilo, Hawai'i. Hawaiian source: R. English trans. by author.

3. *For another example of Moani'ala of Hilo, see Kanilehua (#21).*

Moanianilehua. Same as *Moanilehua.* Rain associated with Puna, Hawai'i. Also the name of a wind. "Moaniani lehua" means "wafted lehua fragrance."

1. Ka ua Moanianilehua o Puna. *The [Moanianilehua] rain that brings the fragrance of the lehua of Puna.*

   A traditional saying. Source: Pukui, *'Ōlelo* 172. Additional source: Andrews 393.

Moanilehua. Same as *Moanianilehua.* Rain associated with Hawai'i. Also the name of a wind. "Moani lehua" means "wafted lehua fragrance."

Rain of Hilo, Hawai'i

1. 'A'ala Hilo i ka ua Moanilehua
   I ho'opē nō i ka hala me ka lehua
   Nā lehua hānau makahiapo a Pā'ie'ie

   Hilo is fragrant in the Moanilehua rain
   That dampens the hala and the lehua
   The firstborn lehua of Pā'ie'ie

   From a mele māka'ika'i, or travel chant, for 'Emalani Kaleleonālani, credited to Mrs. Mahie. Source: Nogelmeier 104.

Rain of Waiolono, Hawaiʻi

2. Kū akula Keaʻau, lele ka makani ma waho
   Ulumano, ma ke kaha o Waiolono
   Ua Moanilehua aʻela ma uka

   Keaʻau stands forth, the winds fly out
   Ulumano wind, on the barren land of Waiolono
   The Moanilehua is raining upland

   From a mele performed in the ʻai haʻa style. Hawaiian source: Emerson, *Unwritten* 77. English trans. by author.

Moelehua. Also the name of a wind.

1. Moelehua. *Sleeping lehua.*

   Source: Pukui and Elbert, 1986, 509.

Mokihana. Rain associated with Wailua, Kauaʻi. Also the name of a valley and stream in Waimea, Kauaʻi, and of a native tree of Kauaʻi and its fruits.

1. I Wailua koʻu hoa luhi e uē nei i Halehuki ē
   Pulu ka ʻuhane i ka ua Mokihana
   Ke wehe lā i ke oho o ke kāwelu ē

   At Wailua my weary companion cries, at Halehuki
   The spirit is drenched in the Mokihana rain
   Opening up the leaves of the kāwelu grass

   From a kanikau, or lament, for M. E. Manu. Hawaiian source: Hoapili. English trans. by author.

Mololani. Similar to *Lūlehua*. Rain associated with Koʻolaupoko, Oʻahu. Also found at Panaʻewa, Hawaiʻi. Also the name of a wind and of a crater in Kāneʻohe, Oʻahu.

Rain of Panaʻewa, Hawaiʻi

1. Ua Mololani
   No ke kākaʻikahi o kona mau paka ua, a no ke kīaweawe loloa hoʻi i ka wā e hāʻule mai ai i luna o ka lehua, no laila, ua kapa ʻia e ko nei kini, "He ua Mololani, he ua lū lehua no Panaʻewa." E hoʻomaka kēia ua e iho iho mai nā ʻōmaka ao ʻōpua o ka lewa, a e hoʻohehelo aʻe nō ʻo ia me he mau maʻawe lopi keʻokeʻo lā i hoʻokuʻuwelu ʻia, a e kilihune aʻe kona mau paka ua ma luna o nā pōhai lehua o Panaʻewa, a hoʻomau aku nō hoʻi i nā lehua o Haili ma kona wēlau komohana loa. A ma laila aku nalowale i loko o nā ʻohu wai o ka nahele.

Mololani Rain

Because its raindrops are so few, and because it streams long and gracefully when it falls upon the lehua flowers, it is called by the multitudes here, "The Mololani rain, a rain of Panaʻewa that scatters lehua blossoms." This rain begins to come down from the edges of the cloud billows in the upper heavens. And it shines brightly like white strands of thread that flutter in the wind. And its raindrops shower lightly upon Panaʻewa's lehua clusters and continue on to the lehua of Haili at its most western tip. And from there it disappears into the watery mists of the forest.

From a description of rains by Lilia H. Richards and company, who traveled to the uplands of Puna in 1876. Hawaiian source: Richards et al. English trans. by author.

Rain of Heʻeia, Oʻahu

2. *See Kiʻowao (#24).*

Rain of Kahaluʻu, Oʻahu

3. Ua Mololani—Kahaluʻu, Koʻolaupoko, Oʻahu.

From a list of rain names and their descriptions. Source: "Names . . . rain." Additional source: Pukui and Elbert, 1986, 253.

Rain of Kekele, Oʻahu

4. Ke ʻalawa iho ʻoe i ka ulu hala o Kekele
ʻAʻole āu mea nani ʻole o laila
He maopopo ka pua o ka hala ke hopu aku me ke ʻala
Noke hala ʻole ka wai a ka hīnalo i ka lāʻau
Kupu ka maka o ka ʻawapuhi i ka ua ʻia e ka ua
Maʻū ke oho o ka palai ke hui pū iho
Ua ulu ʻia mai e ka ua Mololani
Pau ʻole ka nani, ka maikaʻi o ia wahi

When you glance at the hala groves of Kekele
You will find nothing without beauty there
The hala fruit is recognized by its sweet scent
Persisting unceasingly is the water of the hīnalo on the tree
The bud of the ʻawapuhi springs up when watered by the rain
The fronds of the palai fern are dampened when joined together
Grown by the Mololani rain
Endless is the beauty and splendor of this place

From a mele inoa, or name chant, for Kaumakaokane. Hawaiian source: Kaiehu. English trans. by author.

Rain of Mololani, Kāneʻohe, Oʻahu

5. Rain over a crater of that [same] name on Mōkapu Peninsula, Oʻahu.

   Source: Kent 378.

Rain of Olomana and Ahiki, Oʻahu

6. Kuʻu kāne mai ka ua Mololani
   E hehi mai ana iā Olomana me Ahiki

   My dear husband from the Mololani rain
   Treading on Olomana and Ahiki

   From a kanikau, or lament, for Kamakahoohie. Hawaiian source: Hiilei. English trans. by author.

N

Nahae. Rain associated with Alaka'i, Kaua'i. "Nahae" means "to shred."

1. ʻOiai ʻo ka nanā ʻo Kauaikananā
   ʻO ka mana o ka ua Nahae i Alakaʻi

   While the surly one is Kauaikananā
   The power is in the shredding [Nahae] rain at Alakaʻi

   From a mele mākaʻikaʻi, or travel chant, for ʻEmalani Kaleleonālani by Kapapa. Source: Nogelmeier 132. *Note: Kuapuu, "He wahi moolelo" 4/10/1861, says that "Kauaikananā" is the name of a valley in Waimea, Kauaʻi. It is also the name of a stream there.*

Nahua. Rain and wind associated with Kāʻanapali, Maui; also found in other areas. "Nahua" means "to pelt, sting."

Rain of Hoeuli, Maui

1. Lei Hoeuli i ka ua o ka Nahua
   Ua nanahu pōʻaha ʻula i ke pili
   Ka ua a ka ua ʻUla nū i ka nahele
   ʻO ka wai ke kahe ala i kai o Lele ē, i laila

   Hoeuli wears a lei of rain from the Nahua
   That struck and encircled the pili grass with red
   The ʻUla rain roars in the forest
   The water is flowing seaward of Lele, there

From a mele lei, or song presenting a lei, originally composed for Lunalilo and inherited by Kalākaua. This portion of the mele was composed by H. K. Kapakuhaili. Hawaiian source: Kapakuhaili, "He lei." English trans. by author. Additional sources: Kapakuhaili, "Lei pua maikai ouli"; *Na Mele Aimoku* 158.

Rain of Honokōwai, Maui

2. Me he lole ʻula i pili ala ka Nahua
   Ka pili ʻaʻai i ke kula o Honokōwai

   Like a red cloth clinging there is the Nahua
   Clinging—bright and brilliant—to the plains of Honokōwai

   From a mele for Lāhainā and Kāʻanapali on Maui. Hawaiian source: Samuela. English trans. by author. Additional sources: "Ka moolelo o He'ma [*sic*]" 10/1/1920; "Story of Hema" 20.

3. *For another example of Nahua of Honokōwai, see Līlīlehua (#2).*

Rain of Kāʻanapali, Maui

4. Kuʻu wahine mai ka ua Nahua o Kāʻanapali
   He pali nui naʻu ke aloha e mālama nei
   ʻĪnaʻi pū nō me ka waimaka

   My darling wife from the Nahua rain of Kāʻanapali
   Love is a great cliff I care for
   Accompanied by tears

   From a kanikau, or lament, for Luakauwahine. Hawaiian source: Pawai. English trans. by author.

5. Auē! Kuʻu moʻopuna ē!
   Kuʻu moʻopuna mai ka ua Nahua o Kāʻanapali
   Mai ka ua ʻāwili mai i ke pili
   I pili ʻia nō e kāua ka pō a me ke ao

   Alas! My dear grandchild!
   My beloved grandchild from the Nahua rain of Kāʻanapali
   From the rain that entwines the pili grass
   Night and day were joined by us

   From a kanikau, or lament, for Nalohaelua. Hawaiian source: Kaloma. English trans. by author.

Rain of Kahana, Maui

6. ʻO au kahi, e ka Nahua
   E ua nei i ka uka o Kahana

   Include me too, O Nahua rain
   Showering upon the upland of Kahana

From a mele by C. Kana'ina. Hawaiian source: *Na Mele Aimoku* 157. English trans. by author. Additional source: Kana'ina.

Rain of Keka'a, Maui

7. E momole aku ana i Keka'a ka ua Nahua. *The pelting [Nahua] rains wear down the rock at Keka'a.*

   A saying in the legend of Pāka'a and Kūapāka'a referring to Keawenuia'umi's long wait in having a tree hewn for a canoe. Hawaiian source: Nakuina, *Moolelo* 36. English trans.: Nakuina, *Wind* 31.

8. Kuleana i Keka'a ka ua Nahua
   Ka Līlīlehua anu i ka uka o Honowai

   The Nahua rain belongs in Keka'a
   The cold Līlīlehua rain in the upland of Honowai

   From a mele kālai'āina, or political chant, for 'Emalani Kaleleonālani, credited to Keaweaheulu. Hawaiian source: Nogelmeier 208. English trans. by author. *Note: "Honowai" is probably "Honokōwai," the name of an ahupua'a in Kā'anapali, Maui.*

9. Ku'u kāne mai ka ua Nahua
   Mai ka ua Līlīlehua i ke kula
   I ke kula nō ia ua o Keka'a

   My beloved husband from the Nahua rain
   From the Līlīlehua rain on the plains
   On the lowlands is this rain of Keka'a

   From a kanikau, or lament, for Kamehameha IV by his wife, 'Emalani Kaleleonālani. Hawaiian source: Emalani. English trans. by author. Additional source: Nogelmeier 327.

10. Kāhiko i Keka'a ka ua Nahua
    He ua Nahua, ua Līlīlehua
    Ua Maka'ūpili, ua Kaua'ula
    Ua noho i uka o 'Auwaiawao ē

    Keka'a is adorned with the Nahua rain
    The Nahua and Līlīlehua rains
    The Maka'ūpili rain, the Kaua'ula rain
    Occupying the uplands of 'Auwaiawao

    From a mele ho'āla, or chant to awaken, for Kalākaua. Hawaiian source: "Ka hoala." English trans. by author. Additional sources: Kaaua; Kanahele 10–11.

Rain of Lāhaināluna, Maui

11. No Lāhaināluna ke aloha lā
    . . . .
    'O 'oe ku'u lei mae 'ole lā

A ka manaʻo e hana nei lā
Ka ua Nahua i ke pili lā
Ke hehi maila i ka nahele lā

Love for Lāhaināluna

. . . .

You are my lei that never fades
That my mind crafts
The Nahua rain over the pili grass
Trampling through the forest

From a mele inoa, or name chant, for Manoanoa. Hawaiian source: Kekaumehameha. English trans. by author. *Note: "Lāhaināluna" may refer to Lāhaināluna Seminary.*

Rain of north Maui

12. The fine rain with the northeast trade winds on the northern part of Maui.

Source: Andrews 409.

Rain of Lānaʻi and Molokaʻi

13. Niniu Molokaʻi, poahi Lānaʻi
I ka ua Nahua

Molokaʻi revolves, Lānaʻi spins
In the Nahua rain

From a mele ʻāina, or song of the land, for Maui. Hawaiian source: "Nani Lahaina." English trans. by author. Additional sources: Kanahele 39–40, 54, 57; Kualu. *Note: "Nani Lahaina" mentions that Charles Kauluhinano Hoopii of Lāhainā, Maui, was the singer of this mele and that he learned it from Pamano, the sister of the composer. This source also says that the Nahua is "a rain from Nāpili, district of Lāhainā, Maui." Kanahele 54, 57 and Kualu use "nahunahu" instead of "Nahua."*

Cold, continuous rain

14. He ua Nahua, no ka ua mau i ke kakahiaka me ke ahiahi, ʻaʻole puka ma waho o ka hale ke kanaka, ʻo ka mea i hele i kahi e hele ai a loaʻa i ka ua, lōlō laʻa, ʻōpili, haʻukeke, nanahu hewa i ke alelo a moku, me ka lehelehe a moku, haʻalulu ke kino a puni, no laila mai kēia inoa ua Nahua.

A Nahua rain—Because it always rains in the morning and the evening, people do not leave their homes. Those who do go out and get caught in the rain become numb and cramped; they shiver, accidentally biting (nanahu) and cutting their tongues and lips; the entire body shakes; hence the name ua Nahua.

From a letter to the newspaper *Kuokoa* regarding famous rains. Hawaiian source: Kalaaukumuole, "Manao wehewehe." English trans. by author. Additional source: Kalaaukumuole, "Thought explanations." *Note: "Nahua" and "nanahu" are forms of the word "nahu," which means "to bite."*

Nahunahu. Pelting rain associated with Hawai'i. "Nahunahu" means "to bite."

Rain of Auahi, Hawai'i

1. Ku'u kāne mai ka lā wela o Auahi ē
   Momomoea kāua i ka ua Nahunahu

   My dear man from the burning sun of Auahi
   Where we two made our way through the biting Nahunahu rain

   From a kanikau, or lament, for Lohi'au by Hi'iakaikapoliopele. Hawaiian source: Ho'oulumāhiehie, *Ka Mo'olelo* 382. English trans.: Ho'oulumāhiehie, *Epic* 357.

Rain of Halaaniani, Hawai'i

2. 'Ike 'ole au i nā hala o Halaaniani
   I ke ālai 'ia mai e ka ua Nahunahu

   I do not see the hala trees of Halaaniani
   That are blocked by the Nahunahu rain

   From the song "Nani Haili pō i ka lehua" by Konia for her hānai daughter, Lili'uokalani. Hawaiian source: Wilcox et al. 197. English trans. by author. Additional sources: Holstein 145; Lili'uokalani 298.

Rain of Honoli'i, Hawai'i

3. 'Ino nā māpuna wai o Honoli'i ē
   Ke 'imi mai nei i ka hoa paio, 'o au ē
   'O au kā ka ua Nahunahu i huli maila

   The bubbling springs of Honoli'i are cruel
   Searching for an opponent, me
   I am the one sought by the Nahunahu rain

   From a mele by Hi'iakaikapoliopele to disperse mo'o who had come to the door of a house. Hawaiian source: Ho'oulumāhiehie, *Ka Mo'olelo* 105. English trans. by author. Additional source: Ho'oulumāhiehie, *Epic* 99.

Rain of Kaho'okū, Hawai'i

4. Make mā'e'ele i ke anu a Kaho'okū
   Ua neki wale i ka 'ino o ka ua Nahunahu
   I ka pilikia hāiki a Ka'auwea
   I ahona 'Ōma'olala i ka ua 'Awa

   Freezing to death in the cold of Kaho'okū
   Crowded together in the fury of the Nahunahu rain
   In the strained confinement of Ka'auwea
   'Ōma'olala fares better in the 'Awa rain

   From a chant for Kawaikini. Hawaiian source: Lohiau, "He mele no Kawaikini." English trans. by author.

Rain of Kaho'okū and Kapili, Hawai'i

5. Ka pōnaha lehua ma uka o Kaho'okū
Punia i ke 'Awa ka ua Nahunahu
Ua hohoa ka papa, ua moku kahawai
. . . .
'O Pua'akanu, oheohe me he kanaka 'oā lā i ka lā
Kēia ka maha lehua i kai o Kapili nei
I pili aku ho'i māua, o haele
E pi'i i ka uka e kui, e lei i ka lei
Ka lehua a ka ua Nahunahu

The cluster of lehua trees above Kaho'okū
The Nahunahu rain is surrounded by the 'Awa rain
The foundation is repeatedly struck, broken are the streams
. . . .
Pua'akanu is tall and straight like a man forsaken in the sun
This is the grove of lehua trees below Kapili
We were together, now we must depart
Ascend the uplands to string and to wear the lei
The lehua blossoms of the Nahunahu rain

From a mele by Lohi'au as cinders were being tossed upon him by Pele's sisters. Hawaiian source: Ho'oulumāhiehie, *Ka Mo'olelo* 368–69. English trans. by author. Additional sources: Ho'oulumāhiehie, *Epic* 343–44; Emerson, *Pele* 197–200.

Rain of Mali'o, Hawai'i

6. E Hawai'i o Keawe.
Aloha 'oukou a pau, nā hoa launa ma ka na'auao. 'O 'oukou nō ka po'e o kākou i pōmaika'i, no ka mea, aia nō 'oukou ke noho lā ma Kumukahi, ma kahi a ka lā i puka mai ai, ma ka hikina i Ha'eha'e. Ma laila nō nā lehua o Mali'o, me Kaho'okū i ka ua 'o Nahunahu, e nānā mai ana i nā kī o Wahinekapu, me ka makani he moani Pu'ulena, e lū ana i nā lehua o Hiulani.

O Hawai'i of Keawe.
Aloha to all of you, close friends in enlightenment. You truly are the blessed among us for you live at Kumukahi, where the sun rises in the east at Ha'eha'e. There are the lehua blossoms of Mali'o, and Kaho'okū in the Nahunahu rain, attending to the kī of Wahinekapu, with the Pu'ulena breeze that scatters the lehua of Hiulani.

From a letter encouraging readers to subscribe to *Ka Nupepa Kuokoa*. Hawaiian source: Pualewa. English trans. by author.

Rain of Maʻolala (or ʻŌmaʻolala), Hawaiʻi

7. I Haili i ke kula lehua ē, lei
Lei kāhiko i ka lāʻau
Ka hulu mamo weo i pōlena i ka ua
I uauahi lena i ka uahi lehua
I ʻōhinu, i ʻōmaʻo, i Maʻolala
Lala aku i ka ua Nahunahu ke anu
A he anu kahi a ka ua Noelehua

At Haili in the lehua fields are lei
Lei that bedeck the trees
With glowing mamo feathers, yellow in the rain
Yellowish, smoky in the lehua vapors
Shiny, green at Maʻolala
Shining in the Nahunahu rain, the cold
Cold is the place of the Noelehua rain

From a mele by Kamehameha Kauokoa. Hawaiian sources: Kauokoa; *Na Mele Aimoku* 269–70. English trans. by author.

Rain of Panaʻewa, Hawaiʻi

8. ʻO ka leo o ka ʻōʻō kani kuamauna
ʻŌʻō hulu mamo koʻokū
I pala i ka ua, ʻo ka ua Nahunahu
Manuheʻu ka maha o ke poʻo o ka lāʻau o Panaʻewa
I ke ʻakiʻakia e ka ua nihoniho

The voice of the ʻōʻō bird sounds atop the mountain
ʻŌʻō with puffy, yellow feathers
Yellowed in the Nahunahu rain
Bruised are the treetops of Panaʻewa
Nibbled upon by the toothed rain

From the chant "O ka leo o ka oo kani kuamauna." Hawaiian source: "O ka leo." English trans. by author. Additional source: *Na Mele Aimoku* 244.

Rain of Wahinekapu, Hawaiʻi

9. Aloha ʻino ke kupa Kaniahiku
Kū mai ka ua Nahunahu kiʻekiʻe i luna
Hoʻokakano lua i ka laʻi o Wahinekapu

Woe betide the native son of Kaniahiku
When the stormy [Nahunahu] rains gather high overhead
Threatening the peace of Wahinekapu

From a mele hoʻāeae, or love chant. Source: Pukui, *Nā Mele Welo* 82–83.

Rain that travels far

10. ʻAʻole e pio ke ahi iā ʻoe
He ahi na ke akua wahine
Na Pele nui ʻai apuapu
Hele akula he loa ka ua Nahunahu
Nahua e ke ahi, pau ka hulu puaʻa
Kū ka hohono i nā moku

You cannot extinguish the fire
The fire of the goddess
Of great Pele who ravages
The biting [Nahunahu] rain went a long distance
Stung by the fire, the hog's bristles were consumed
The stench engulfed the islands

From a mele to Makaliʻi by Kamapuaʻa telling of his own bravery as a warrior. Source: Kahiolo, *He Moolelo* 142–43.

Nāulu. Sudden shower; associated with Kawaihae, Hawaiʻi, and with Niʻihau, and found in other areas. Also the name of a shower cloud and a wind.

Rain of Hilo Bay, Hawaiʻi

1. A he nani Hilo Bay na ke onaona
Nohea i ka wai a ka Nāulu

And Hilo Bay is beautified by the sweet fragrance
Lovely in the waters of the sudden [Nāulu] rains

From a kānaenae, or chant of eulogy, for Joseph Nāwahīokalaniʻōpuʻu by JNH of Waiʻanae, Oʻahu. Hawaiian source: Sheldon, *Ka Puke* 370. English trans.: Sheldon, *Biography* 467.

Rain of Hilo Pali Kū, Hawaiʻi

2. Hāneʻe mai Līloa me he Uluaunui lā
Me he kuāua Nāulu nui i pano Hilo Pali Kū

Līloa sags like an Uluaunui wind
Like a heavy Nāulu shower that obscures Hilo Pali Kū

From "No ke lii wahine," a mele kūʻauhau, or genealogical chant, for ʻEmalani Kalele-onālani. Hawaiian source: Nogelmeier 285. English trans. by author.

Rain of Kahelo, Hawaiʻi

3. ʻO ʻoe a ʻo wau, pili kāua
Laʻi ai ke kaunu lā i Kahelo
ʻElo ʻoe ʻelo au i ka ua noe
Pulu pē i ka wai a ka Nāulu

You and I, we should be together
That lovemaking at Kahelo be tranquil
You are wet, I am wet from the misty noe rain
Soaked by the waters of the Nāulu

From the song "Kuu ipo hana maewaewa." Hawaiian source: Kahaanapilo. English trans. by author.

Rain of Kawaihae, Hawaiʻi

4. Ka ua Nāulu o Kawaihae. *The cloudless [Nāulu] rain of Kawaihae.*

   A traditional saying. Source: Pukui, *ʻŌlelo* 172. Additional source: Pukui and Elbert, 1986, 263. *Note: Pukui,* ʻŌlelo *172, says, "The rain of Kawaihae often surprises visitors because it seems to come out of a cloudless sky. A native knows by observing the winds and other signs of nature just what to expect."*

5. Ahuwale nā kualono
   ʻIke ʻia ka pae ʻōpua
   E kukū ana i ke kai
   I ke kai hāwanawana
   ʻŌlelo ʻo Kawaihae
   Hae ana e ka Nāulu

   The ridges are in plain view
   The cloud banks are seen
   Rising over the sea
   Over the whispering sea
   Kawaihae calls out
   Pour down, O Nāulu

   From the chant "A Kona Hema ʻo Kalani." Hawaiian source: Elbert and Mahoe 31. English trans. by author.

6. E wiki ʻoe, ʻapa nei, eia aʻe ʻo Kawaihae
   Hoʻohaehae Nāulu i ka makani Kuehuʻale

   Make haste, slowpoke, for here is Kawaihae
   Where the Nāulu showers stir up wave-gusting [Kuehuʻale] winds

   From the song "He aloha nō ʻo Honolulu" by Lot Kauwē. Source: Wilcox et al. 50. *Note: In this rendering, "Kuehuʻale" is used instead of "kuʻehu ʻale," which was printed in the source.*

Rain of Kona, Hawaiʻi

7. ʻO Kona ia o ke kai malino a ʻEhu ē!
   Ke ala a ʻEhu, ke ala a kāua i hele ai
   I ke ao, i ka pō, pōwehiwehi i ka ua Nāulu a weli

He weliweli, he maluhia i ke aloha iā ʻoe
Iā ʻoe, iā ʻoe, e Kalopelekei i ka lā ē!

Oh, Kona it is, of the tranquil sea of ʻEhu, oh!
The pathway of ʻEhu, the path we two traversed
By day, by night, made dim and threatening by the stormy Nāulu rains
Frightful, but peaceful because of love for you
For you, for you, O Kalopelekei of the day, oh!

From a kanikau, or lament, for Kamehameha IV by his wife, ʻEmalani Kaleleonālani. Source: Nogelmeier 337.

Rain of Mauliola, Hawaiʻi

8. Kuʻu hoa mai ka moku ʻo Halelani
. . . .
Mai kēlā ʻāina ʻai ʻole ʻo Mauliola
Nānā aku i nā keiki ke ola
Mai kēlā ʻāina wai ʻole e noho ʻia ai
A hāʻule ka ua Nāulu, loaʻa ka wai o ia kaha

My companion from the ship Halelani
. . . .
From that land Mauliola that has no food
Look to the children for life
From that waterless land that is lived upon
Then the Nāulu showers fall; the water of this place is gotten

From a kanikau, or lament, for J. Henry. Hawaiian source: Kahinawe, "He kanikau." English trans. by author.

Rain of Maunakea, Hawaiʻi

9. Uhi poʻipū ke ao Nāulu hākumakuma i ka piko o Maunakea
ʻŌlapa uila ke ahikao lele ʻoāʻoaka mālamalama nā kūkulu poepoe honua
Nakulu paʻapaʻaʻina, nākolokolo ʻuʻina akula nā pōhaku a ka hekili, wawalo nā paia lani
He mau hailona, he mau kāhoaka ia na ke kilo hōlona o ka liʻulā
E hoʻomāʻike mai ana e ʻike kūmaka i kona welo, i kona iwikuamoʻo
He ʻōiwi kā hoʻi na ke kilohana pio ʻole i ka pehia e nā kuāua Nāulu a ka loko ʻino ē
ʻO ia, he ʻoiaʻiʻo

The thick, dark Nāulu clouds engulf the summit of Maunakea
Fireworks flare up like the flash of lightning, illuminating the horizons of the earth

Rumbling, snapping, roaring, crackling rocks of thunder, echoing on the walls of heaven
These are signs, heavenly portents, to even the novice seer of mystical images
Revealing, that all may see, his ancestry, his lineage
A native born to excellence, inextinguishable in the pelting of the merciless Nāulu showers
It is so; it is true

From a biography of Joseph Nāwahīokalaniʻōpuʻu. Hawaiian source: Sheldon, *Ka Puke* 154. English trans. by author. Additional source: Sheldon, *Biography* 228.

10. *For another example of Nāulu of Maunakea, see Kēhau (#2).*

Rain of Panaʻewa, Hawaiʻi

11. Pāuli ka lani, hoʻoilo ka malama
Ka malama o ka ua, ʻo Hilinehu
Ua hoʻi ka wai hālalo o ka ʻākuli
Ka pī ʻia e ka lā i ka makani
Ola akula i ka wai a ka Nāulu
E kū ē, e kū lā!
Kua me he koʻi lā ka ua i ka lāʻau
Ua ʻoʻoki ka lehua o Panaʻewa

The heavens darken, for the wintry month here
The rainy month of Hilinehu
The water that gathered in the water holes
Has been dried out by sun and wind
So life is found again from the [Nāulu] rain clouds
Striking! Striking!
The rain strikes as an adze against the trees
Cutting away at the lehua of Panaʻewa

From a hula lāʻau, or stick dance, taught to Samuela Akoni Mika of Waiākea Homesteads, Hawaiʻi. Source: Pukui, *Nā Mele Welo* 174–75.

Rain of Waipiʻo, Hawaiʻi

12. Ua luhea, ua lupeʻa ia lipo i ke ʻala
Luhiehu i ka wai kau a ka ʻōpua
Pē ūpē i ka wai a ka Nāulu
Kulukulu wai lehua

Drooping, intoxicated by the fragrance is the deep forest
Verdant from the floating water of the billowy clouds
Crushed flat by the waters of the Nāulu
And the dripping of the lehua nectar

From the song "Nani Waipio" by Leleiōhoku. Hawaiian source: Holstein 26. English trans. by author.

Rain of ʻAlae, Maui

13. Hoʻopuʻupuʻua nā kahawai o ʻAlae i ka wai a ka Nāulu. *The streams of ʻAlae choke in the water from the Nāulu rain.*

From a mele that mentions Maui place names. Hawaiian source: Kualu. English trans. by author. Additional source: Kanahele 54, 57.

Rain of East Maui

14. Ua Nāulu—he ua i ka wā mālie hele i ke kuahiwi; he ao Nāulu me ka ua.

Nāulu rain—a rain that falls when it is calm and goes to the mountains; a Nāulu cloud with rain.

From an account of East Maui rains and their descriptions, as told to Thomas K. Maunupau by Josua Ahulii of Kaupō, Maui. Hawaiian source: Maunupau, *Huakai* 27. English trans. by author.

15. Ke koloau o ka pua māmane i ka lā
E ake nō a maʻū i ka wai a ka Nāulu
ʻO ia wai hoʻi a ke kupa e hahai nei
E alualu nei lā i ka maka o ka ʻōpua

The māmane blossom reaches out in the sunlight
Eager to be moistened by the water from the [Nāulu] clouds
It is the only thing that the native son follows after
And pursues to the face of the horizon clouds

From the chant "Maikaʻi Kaʻuiki." Source: Pukui, *Nā Mele Welo* 168–69.

Shower clouds of Haleakalā, Maui

16. *See ʻŪkiu (#3).*

Rain of Kamaʻomaʻo, Maui

17. *See ʻŪkiukiu (#4).*

Rain of Kula, Maui

18. Ke hoʻi nei ka ʻuhane i ka malu niu o Lele
I ka malu kuawa o Wailuku
I ka ua Nāulu noe anu o Kula
I ka ua noe uahi moe i ke pili
I pili ʻia ka ua me ka lā
Ke anu hoʻi me ke koʻekoʻe

The spirit is returning to the shade of the niu trees of Lele
To the shelter of Wailuku Valley

To the cold, misty Nāulu rain of Kula
To the smoky, misty rain that rests upon the pili grass
Joined are the rain and the sun
The cold and the chill

From a kanikau, or lament, for L. L. Ua by Ua's students at the Lāhaināluna Seminary. Hawaiian source: Fornander, *Fornander* 6: 525. English trans. by author.

19. Ō mai ʻo ʻEmalani, ke aliʻi nona ia inoa
Ia hoʻonuʻanuʻa i uka, ka uahi o Kula
Ua hāliʻi ke kūmoena wai a ka Nāulu
ʻAʻohe hana a ka ua ʻŪkiu e laʻi nei
E hoʻokō ʻia ana e ka ua ʻUlalena
E ka ua kāhiko kukui o Lilikoʻi ē, eia ʻoe ē

Emmalani responds, the chiefess for whom is the name
That conflagration in the uplands, the smoke of Kula
The watery mat of the Nāulu rain has been spread out
There's no action by the ʻŪkiu rain, now peaceful
It shall be carried out by the golden-red rain ʻUlalena
O rain that adorns the kukui trees of Lilikoʻi, here you are, oh

From a mele kālaiʻāina, or political chant, for ʻEmalani Kaleleonālani. Source: Nogelmeier 179–80.

Rain of Lauʻulu, Maui

20. Kāhiko Kokomo i ka ʻohu lā
ʻO ka hone aʻe a ka Nāulu lā
I ka luna wale o Lauʻulu lā

Kokomo is adorned with the vaporous mists
And the teasing of the Nāulu showers
There on high at Lauʻulu

From a mele inoa, or name chant, for Lele. Hawaiian source: Ku. English trans.: "Pā ka leo" 21.

Rain of Olinda, Maui

21. Hanohano Olinda, kuʻu home kuahiwi
E hoʻopumehana i ke ahi kapuahi
ʻO ka neʻe paʻa mai a ka ua Nāulu
Kohu maka kui kele ʻīnikiniki i ka ʻili
Haʻina ʻia mai ana ka puana
Kēhau kakahiaka ʻīnikiniki pāpālina

Majestic is Olinda, my mountain home
Get warmed by the blaze of the hearth

The steady approach of the sudden Nāulu rains
Is like needle points nipping the skin
Tell the story in the refrain
Morning [kēhau] frost that tingles the cheeks

From the song "Hanohano Olinda" by Alice Johnson. Source: Wilcox et al. 43.

Rain of Piʻiholo, Maui

22. A he Kiu ka makani
Hoʻohae ana i ka Nāulu
A he wai ʻolu ia
No ka luna lā o Piʻiholo

And Kiu is the wind
Tearing through the Nāulu rain
And what refreshing water it is
For Piʻiholo's upper reaches

From the song "Ke hone ae nei." Hawaiian source: Waialeale. English trans. by author. Additional source: Johnson 406–7.

23. *For another example of Nāulu of Piʻiholo, see ʻUlalena (#18).*

Rain from Kahoʻolawe to ʻUlupalakua, Maui

24. There is much verbal testimony that in former days, say 25 years ago and before, there were many light drifting showers at the south end of East Maui, at ʻUlupalakua, which originated over Kahoʻolawe and drifted across the channel with the Nāulu breeze.

From a 1910 report by Ralph S. Hosmer describing the degraded condition of Kahoʻolawe. Source: King 129.

Rain of Lānaʻi

25. The rainfall on the top-lands is thirty-five inches a year and on the lower slopes, twelve inches, but the porous volcanic rocks hold the precipitation poorly. There are a number of mountain springs and in Maunalei gulch a small stream. This water may partly or wholly disappear in periods of drought under which Lānaʻi suffers every three or four years. From September to November the rains are likely to appear in sudden local showers called Nāulu. In the winter months the island is visited by southwest or Kona rainstorms.

From a chapter on Lānaʻi's geography. Source: Emory 5.

Rain of Kāʻana, Molokaʻi

26. Nā lehua o Kāʻana
I ka noua e ka ua Nāulu
Kamumu akula i ke kaha o Pālāʻau

The lehua of Kāʻana
Pelted by the Nāulu rain
Rumbling across the expanse of Pālāʻau

From a kanikau, or lament, for Nuholani. Hawaiian source: Kaalehu. English trans. by author. Additional source: "Pā ka leo" 25. *Note: "Kāʻana" is the name of a hill in Kaluakoʻi.*

Rain of Kaiolohia, Molokaʻi

27. Waiho akāka ke kula o Kaiolohia, ka lele maʻopu a ka wai a ka Nāulu. *The Kaiolohia plain stays clear as the water of the Nāulu shower leaps downward.*

From a chant. Source: Pukui and Elbert, 1986, 241.

Rain of Kaluakoʻi, Molokaʻi

28. Ke uē nei
E uē aʻe ana i ka hoa o ke anu
ʻO ka Nāulu lā i Kaluakoʻi
Ke ua aʻela i kai o Kalāʻau ē

Crying
Crying to the companion of the cold
The Nāulu at Kaluakoʻi
Raining seaward of Kalāʻau

From a mele by Hiʻiakaikapoliopele asking for entry into a hall of hula. Hawaiian source: Hoʻoulumāhiehie, *Ka Moʻolelo* 128, 322. English trans. by author. Additional source: Hoʻoulumāhiehie, *Epic* 121, 300.

Rain of Maunaloa, Molokaʻi

29. Ua loa Maunaloa, loa i ka lā
Ua loa ke ala a ka hewahewa
Ke ukukuhia maila e ka lā
Ua piha ke kula ka hū a ka Moaʻe
Ua piha i ka wai hū a ka Nāulu
Nāna nō i hū a lana i laila

Far is Maunaloa, far to go in the sun
Long and wearisome the trail
For the sun dips up the moisture
The plain is filled to overflowing with the Moaʻe wind
Filled too with the water that overflows from the [Nāulu] rain clouds
It overflowed by itself and flowed there

From a mele komo, or chant of welcome. Source: "Ua loa."

Rain of Honolulu, Oʻahu

30. *See Makaʻūpili (#3).*

Rain of Kaʻala, Oʻahu

31. Lei Kaʻala i ka ua a ka Nāulu
Hoʻoluʻe ihola i lalo o Haleʻauʻau
Ka ua Leikōkōʻula i ke pili
I pilia ka mauʻu nēnē me ke kupukupu

Kaʻala is wreathed by the rain of the Nāulu
That then pours down upon Haleʻauʻau
The rainbow-wreath [Leikōkōʻula] rain on the pili grass
That draws the nēnē grass next to the kupukupu fern

From "Makalapua," a mele inoa, or name chant, for Liliʻuokalani. Source: Liliʻuokalani 294. Additional sources: Elbert and Mahoe 76; Wilcox et al. 164; Holstein 117.

32. ʻOhu iho nei Kaʻala i ka wai a ka Nāulu. *Kaʻala is adorned with the waters of the Nāulu.*

From a kanikau, or lament, for Miss L. Kahalepouli. Hawaiian source: Kealii-kanakaole. English trans. by author.

Rain of Kaupeʻa, Oʻahu

33. ʻAʻole au e hele i ke kaha o Kaupeʻa
Kēlā kaha kūpā koili a ka lā i ke kula
Ua kūpono aʻela ka lā i ka piko o Wākea
Ola i ke ahe a ka makani Māunuunu
I ka hapahapai mai a ka makani ʻAoʻaoa
Ke koi lā i ke ao o ka Nāulu e hanini i ka wai
Ola ihola nā kupa kamaʻāina i ka wai a ka ʻōpua
Ke halihali aʻela nā ʻōpua i ke awa lau

I shall not tread Kaupeʻa's expanse
That stretch where the sun beats down on the plain
The sun is right overhead, at the navel of Wākea
I am spared by the Māunuunu wind
By the uplifting ʻAoʻaoa breeze
Urging the Nāulu storm clouds to pour down their waters
The natives here survive on water from the clouds
Which billowing clouds carry along to the branching lochs

From a mele by Hiʻiakaikapoliopele as she traveled over the hot stretch of land near Puʻuloa, Oʻahu. Hawaiian source: Hoʻoulumāhiehie, *Ka Moʻolelo* 294–95. English trans.: Hoʻoulumāhiehie, *Epic* 275.

Rain of Kemoʻo, Oʻahu

34. Ana ʻole ka ua o Kamaoha ē
I ka hone aʻe a Kaʻōpua lā
A ka ua Nāulu i ke kula ē
I ke kula aku o Kemoʻo lā
Kohu ʻole ka ua o Halemano ē

The rain of Kamaoha is immeasurable
Murmuring sweetly at Kaʻōpua
The Nāulu showers on the plains
The plains of Kemoʻo
The rain of Halemano resembles no other

From a chant for Keaopolohiwa. Hawaiian source: Kanoho. English trans. by author. Additional source: "Pā ka leo" 30–31.

Rain from Kualoa to Kahana, Oʻahu

35. *See kualau (#8).*

Rain of Malamanui, Oʻahu

36. Ua laʻi luna o Malamanui
Ua ʻolu i ka wai a ka Nāulu
I ka wai ua lana mālie

It is peaceful atop Malamanui
Pleasant in the water of the Nāulu rain
In the calm, drifting water

From the song "Lia i ka wai mapuna." Hawaiian source: Holstein 137. English trans. by author.

Rain of Mānoa, Oʻahu

37. Nuʻuanu ē, Nuʻuanu hoʻi
Anu hewa i ka uka aʻo Mānoa
Ua anu ē, ua anu hoʻi
Pulu ʻelo i ka wai a ka Nāulu

Nuʻuanu, Nuʻuanu indeed
Menacingly cold in the uplands of Mānoa
So cold, so very cold
Soaked by the waters of the Nāulu

From the song "Leahi i Daimana Hila" for Liliʻuokalani by Ani Peahi. Hawaiian source: *Buke Mele* 76. English trans. by author.

Rain of Mikilua, Oʻahu

38. Uluhua, uluhua wale au
Uluhua i ka wai a ka Nāulu
Ia wai nihi aʻe i nā pali
Kālewa ma uka o Mikilua

I am discouraged and annoyed
Vexed at the Nāulu rains
That water that sweeps across the cliffs
Drifting inland of Mikilua

From the song "Uluhua wale au" by Theresa Cartwright. Source: Wilcox et al. 264.

Rain of Pūhāmaloʻo, Oʻahu

39. E waha ana ka ʻīlio i ke kula o Pūhāmaloʻo
Ke hoʻohaehae aʻela i ka Nāulu
Ua mokuhia kahawai, ua nihoa ka pali

The dog furrows the plain of Pūhāmaloʻo
Tearing at the Nāulu
The stream channels are ruptured, the cliffs are sharpened

From a mele recited by Hiʻiakaikapoliopele as she gazed upon Hawaiʻi from atop Pōhākea, Oʻahu, and saw that her friend Hōpoe had been killed in the fires of her sister Pele. Hawaiian source: Hoʻoulumāhiehie, *Ka Moʻolelo* 280. English trans. by author. Additional source: Hoʻoulumāhiehie, *Epic* 261.

Rain of Puʻuokona, Oʻahu

40. ʻO Puʻuokona kai kupu kele i ka makani
Kupu kela ke poʻo wai a ka Nāulu

Puʻuokona rises high in the wind
The water source of the Nāulu grows profusely

From a mele inoa, or name chant, for Kaumakaokane. Hawaiian source: Kaiehu. English trans. by author.

Rain of Waialua, Oʻahu

41. ʻAʻole wahi hoʻohalahala ʻana
A ka ua Nāulu i ke kula
He like wale nō ā mai Kaʻena a Waialua
Pili pono akula ka lā

Having nothing to criticize
The Nāulu rain on the plains
From Kaʻena to Waialua it is the same
The sun clings tightly

From a mele ho'āeae, or love chant. Hawaiian source: Pukui, *Nā Mele Welo* 80. English trans. by author. *Note: The line "He like wale nō ā mai Ka'ena a Waialua" is similar to the traditional saying, "Like nō Ka'ena me Waialua," meaning "Ka'ena and Waialua are one" (Pukui,* 'Ōlelo *215). Ka'ena Point is in the district of Waialua.*

Rain of Wai'anae, O'ahu

42. Kāhiko i ka pua ko'olau
He 'ohu kapu no ka wahine
I kui 'ia mai e Li'a
Ka wahine noho i ka Ulumano
No Kaiona noho i ka la'i
O ka ua Nāulu i ke kula

Bedecked with the ko'olau bloom
A sacred adornment for the woman
Strung into a lei by Li'a
The woman who resides in the Ulumano wind
For Kaiona who dwells in the calm
Of the Nāulu rain over the plains

From a mele inoa hula, or name-chant dance, for Kaiona. Hawaiian source: Handy and Pukui 137. English trans. by author.

Rain of Kaua'i

43. 'A'ohe pahuna hala 'ole a ka Māluakele
Tū nō, mino 'olu ka 'eli o ka lā'au
Ka 'ō'ō 'ia i ka ua a wali ka nahele
Pu'upu'ua i luna ke 'awa a ka Nāulu
Huahua'i nā huawai a ka ua i ka lani
Piha pono nā ki'o wai o Kūlanihāko'i
'A'ohe pakī wai a ke Tēhau

The Māluakele wind never misses in its thrusting
Stabbed indeed, the digging of the stick creates a pleasant indentation
Pierced in the rain, mashed and made supple is the forest
Swollen above is the cold rain of the Nāulu
The water gourds with rain from the heavens gush forth
The pools of Kūlanihāko'i are full to the brim
Not a sprinkling of the Tēhau rain

From a mele about Kaua'i, said to be composed by Kapa'akea. The chant is a mele inoa, or name chant, a mele 'āina, or chant for the land, and a mele hula, or hula chant. Hawaiian source: "Maikai Kauai." English trans. by author.

44. *For another example of Nāulu of Kaua'i, see Noelehua (#2).*

Rain of Kalehuawehe, Kauaʻi

45. ʻO Niolopua ʻoe, ʻo kuʻu kaikunāne
Aloha a ka ua ʻino no Hawaiʻi
I ka pali o ʻŪʻūlua
I ka nahā pali o Konoleʻa
I kūhewa i ka ua Nāulu
Ke lana lā i kai o Kalehuawehe

You are Niolopua, my beloved brother
How terrible are these rains from Hawaiʻi
Upon the cliff of ʻŪʻūlua
Upon the cracks of the cliff of Konoleʻa
Suddenly attacked by the Nāulu rain
Floating seaward of Kalehuawehe

From a mele from Wailua, Kauaʻi. Hawaiian source: "O Niolopua." English trans. by author.

Rain of Kaunalewa, Kauaʻi

46. Paʻa ka hale a ka Nāulu i Kaunalewa
Ke kaupaku ʻia lā e Kalanamaihiki

The house of the Nāulu at Kaunalewa is complete
Thatched by Kalanamaihiki

From the song "Ka lani Poloula" by Kalai o Kona. Hawaiian source: Fornander, *Fornander* 6: 536. English trans. by author.

Rain of Kona, Kauaʻi

47. Hana ua wai Nāulu ʻo Kona
Hana ua wai Nāulu ʻo Mānā
I hoʻonani ʻia e piha Keālia wai
Wai Kahelu, ua piha Kalanamaihiki
Na ka wai ua Kaunalewa
Maikaʻi iho i ka wai Lolomauna

Kona produces the Nāulu rainwater
Mānā produces the Nāulu rainwater
That enhances and fills the spring of Keālia
The waters of Kahelu, Kalanamaihiki is filled
By the rainwater of Kaunalewa
Beautified by the water of Lolomauna

From a chant originally composed for Lunalilo and inherited by Kalākaua. This portion of the mele was composed by Nāmāhana. Hawaiian source: *Na Mele Aimoku* 151. English trans. by author. Additional sources: Namahana; Kanahele 19–20.

Rain of Koʻolau, Kauaʻi

48. Nani Kilohana i ka wai o Kemamo
He liko lehua ia no Puʻukapele
He wai no ka ua Nāulu
I hoʻāu iho i ke Koʻolau

Beautiful is Kilohana in the water of Kemamo
It is like a new leaf of the lehua from Puʻukapele
Waters of the sudden Nāulu rains
That wash down over the Koʻolau district

From a mele mākaʻikaʻi, or travel chant, for ʻEmalani Kaleleonālani. Source: Nogelmeier 73.

Rain of Mānā, Kauaʻi

49. A ua wai Nāulu ka uka o Mānā
Ke hahai lā i ka liʻulā o Kaunalewa

The waters of the sudden Nāulu showers cover Mānā
Following the mirage of Kaunalewa

From a mele by Hiʻiakaikapoliopele upon her arrival at Hāʻena, Kauaʻi. Hawaiian source: Hoʻoulumāhiehie, *Ka Moʻolelo* 180. English trans.: Hoʻoulumāhiehie, *Epic* 168.

Rain of Pihanakalani, Kauaʻi

50. ʻO wau wale nō i ka uka lāʻau
I ke ano o Pihanakalani
Hoʻolale mai a ka Nāulu
Ulu nō ka manaʻo, ka liʻa i loko
I ka wai pua hau o Maluaka

I am alone in the woody uplands
In the sanctity of Pihanakalani
Urged by the Nāulu
Thoughts, desires grow within
For the hau flower waters of Maluaka

From the song "ʻO ka wai hālau i Wailua lā." Hawaiian source: Pukui, *Nā Mele Welo* 200. English trans. by author.

Rain of Niʻihau

51. ʻO Niʻihau nō e ka ʻoi a puni mai Hawaiʻi moku o Keawe, ʻo Kauaʻi moku o Manokalanipō, loku ʻia e ka ua Nāulu. *Niʻihau is the best of all from Hawaiʻi, island of Keawe, to Kauaʻi of Manokalanipō, drenched in the downpour of the Nāulu.*

A traditional saying. Source: Tava and Keale 29.

52. ʻO ʻoe kā kaʻi ka welo ʻana a ka lā i Lehua
I kūpuni iho ai ʻoe i ka Nāulu o Niʻihau
A me nā ʻale aʻe hoʻi o Kaulakahi

So it is you marching with the setting sun to Lehua
Where you are wrapped in the Nāulu of Niʻihau
And the waves of Kaulakahi

From a kanikau, or lament, for the newspaper *Ka Hae Hawaii*. Hawaiian source: Palaile. English trans. by author. *Note: "Kaulakahi" is the name of the channel between Kauaʻi and Niʻihau.*

Rain of Puʻuwai, Niʻihau

53. Kauanāulu. This was the original name of the town now called Puʻuwai; also the name of a man well known in the art of lua (Hawaiian art of self-defense). There is a saying for this area: "Kau ke ao Nāulu, ola ka ʻāina. Aloft is the Nāulu cloud, bringing life to the land."

From a book on Niʻihau traditions. Source: Tava and Keale 85. *Note: "Kauanāulu" is "ka ua Nāulu," "the Nāulu rain."*

54. Puʻuwai kau mai i luna, hoʻoheno ʻia nei e ka ua Nāulu. *Puʻuwai rises above, caressed by the Nāulu rain.*

From a book on Niʻihau traditions. Hawaiian source: Tava and Keale 1. English trans. by author.

Rain of Lehua and Niʻihau

55. He loko Niʻihau na ka Mikioi
Ua piha i ka wai ua a ka Nāulu
I hoʻokulukulu ma ka maha o Lehua

Niʻihau is a pond formed by the Mikioi wind
Filled with Nāulu rainwater
Dripping on the sides of Lehua

From a koʻi ʻeli honua, or origin chant, for ʻEmalani Kaleleonālani. Hawaiian source: Nogelmeier 252. English trans. by author. *Note: In this rendering, "Mikioi" is used instead of "mikioi," which was printed in the Hawaiian source.*

Rain of Lehua

56. A Lehua i ka wai a ka Nāulu
Ka hōʻeuʻeu a Kalanamaihiki
Hiki ke ao, ka Nāulu makani mai lalo

At Lehua in the water of the Nāulu rain
The stirring of Kalanamaihiki
The cloud arrives, the Nāulu, wind from below

From a mele inoa, or name chant, for Leleiōhoku. This portion of the mele was composed by Nahalepelapela. Hawaiian source: *Na Mele Aimoku* 311. English trans. by author.

57. He Moaʻe ka makani o Lehua
Hoʻoheno a ka ua Nāulu i ka wai huna a ka pāoʻo

A Moaʻe is the wind of Lehua
Glorified by the sudden Nāulu rains upon the secret water of the goby fish

From a wind chant recited by Pele. Hawaiian source: Hoʻoulumāhiehie, *Ka Moʻolelo* 14. English trans.: Hoʻoulumāhiehie, *Epic* 14.

58. Aloha Lehua ke ala iho a ka lā
I hoʻāhiahi aku i ka laʻalāʻau
Hele lilo hewa ka malu i ka Nāulu
I lima iki ʻia paha e Waiolono?

Beloved is Lehua on the descending path of the sun
That draws eventide over the shrubbery
The calm dissipates into sudden, stormy Nāulu rains
Touched and affected perhaps by Waiolono?

From a kanikau, or lament, for Albert Ka Haku O Hawaiʻi by his mother, ʻEmalani Kaleleonālani. Source: Nogelmeier 317.

59. Hoʻokahi ʻo Lehua e ʻalo nei
I kolu i ke ao a ka Nāulu

Lehua is alone to endure
Three with the Nāulu rain clouds

From the song "Makalapua" by E. C. Hawaiian source: *Buke Mele* 85. English trans. by author.

Beloved rain cloud

60. Aloha ke ao Nāulu o ka ʻāina
Aloha ka ua koko ʻula neneʻe i ka moana

Love for the Nāulu rain cloud of the land
Love for the rainbow-hued koko ʻula rain that marches over the ocean

From a kanikau, or lament, for Rode Koleiki. Hawaiian source: Kahale. English trans. by author.

Girl turned into rain

61. Ua hāpai aʻela ʻo Hina. . . . ʻO ka ʻehā o nā keiki i hānau mai ai, he kaikamahine, ʻo ia hoʻi ʻo Leialoha.

A i ka manawa i nui aʻe ai ʻo ia kaikamahine, ua lele akula ʻo ia i ka lani, a lilo kona kino he kino ua, a ua kapa ʻia ʻo Kauanāulu.

Hina was pregnant. . . . The fourth child born was a daughter, Leialoha.

And when this girl grew up, she flew into the sky, and her body took the form of rain, and it was called Kauanāulu.

From the legend of Kamapuaʻa. Hawaiian source: "He molelo . . . no Kamapuaa" 6/23/1891. English trans. by author. Additional source: Kameʻeleihiwa 15. *Note: "Kauanāulu" is "ka ua Nāulu," "the Nāulu rain."*

Rain mentioned in a prayer for rain

62. I hāʻule ʻole mai ke kuāua, a laila, ʻimi ke ola o ka mahi ʻai i ke akua, ʻo ke kāhea i ke akua i ka ua, a he poʻe pulapula i kūpono no ia ʻaoʻao, ua hāʻule koke ka ua. Penei ke kāhea ʻana:

E mōlia, e ʻālana iā ʻoe, e Kūkeaoloa
E Kūkeaopoko
E Kūkeaohoʻomihamiha i ka lani
I kēlā ao nui ʻeleʻele i ka maka o ka ʻōpua lā
E hāʻule mai, e ka ua Nāulu
E ka ua kūlokuloku
E ka ua lewalewa
E mokupawa ai ka ua mahina ʻai
E pulu mai kēlā iwi a kēia iwi
Mai kēlā kaikā a kēia kaikā
I pulu ka lepo
I momona ko kākou kīhāpai
I ulu kā kākou mau mea kanu

Should kuāua showers not fall, then life for the farmer would be sought from the gods. The gods would be called upon for rain, and for the rightful descendants of these family gods, the rain would soon fall. The call would go as follows:

Make a sacrifice, make an offering to you, O Kūkeaoloa
O Kūkeaopoko
O Kūkeaohoʻomihamiha in the sky
To that big black cloud in the face of the cloud bank
Fall, O Nāulu rain
O flowing kūlokuloku rain
O swinging lewalewa rain
So that the farming rains be intermittent
Soak that border to this
From that cultivated patch to this
So that the earth is saturated
So that our garden is fruitful
So that our plants grow

From a description of sweet potato cultivation. Hawaiian source: Kamakau, "Ka moolelo Hawaii" 11/18/1869. English trans. by author. Additional sources: Kamakau, *Works* 27; Solis 20–25. *Note: Solis 22 says that the Nāulu is a sudden shower, the kūlokuloku is a heavy shower, and the lewalewa is probably a light, steady rain that would be desirable following the planting of ʻuala. He also says that these rains "are kinolau of the deities that are being addressed by the farmer."*

Rain mentioned in references to types of people

63. Ka ʻiwa hoʻohaehae Nāulu. *An ʻiwa, frigate bird, teases the [Nāulu] showers.*

    Source: Pukui and Elbert, 1986, 46. *Note: Pukui and Elbert say that this phrase refers to an attractive person who provokes envy.*

64. Koʻūkoʻū i ka wai a ka Nāulu. *Damp from the water of the [Nāulu] showers.*

    Source: Pukui and Elbert, 1986, 167. *Note: Pukui and Elbert say that this refers to someone who is drunk.*

Rain over the ocean

65. Ua lilo nā ihe, nā pololū me nā kuʻia laumeki i wai ʻauʻau, ua lilo nā mea make me he mau paka ua lā o ka ua Nāulu e haluku ana i ka moana, me he ua kualau lā ke koʻiaweawe i ka ʻili o ke kai, me he waipuʻilani e kulukulu ana i ka lipo ʻaʻau kele.

    The javelins, long spears, and barbed spears poured down like bathwater. The lethal weapons were like drops of the Nāulu rain beating down upon the ocean, showering like the kualau rain over the surface of the sea, like a waterspout pouring over the rippling deep ocean.

    From a passage in the history of Kamehameha I in which weapons from Kahekili's warriors poured down like rain over the forces of Kalaniʻōpuʻu. Hawaiian sources: Kamakau, "Ka moolelo o Kamehameha I" 12/15/1866; Kamakau, *Ke Kumu Aupuni* 23. English trans. by author. Additional source: Kamakau, *Ruling* 87.

Rain that can occur anywhere

66. Ua Nāulu—hard rain but not lasting.

    From a list of rain names and their descriptions. Source: "Names . . . rain."

Rain that carries the kēhau mist

67. ʻAʻole i like i ka Nāulu
    Ia ua hoʻohali kēhau
    Me he ipu wai i ninia lā
    Nā hau o Kūmomoku

    Unlike the Nāulu shower
    The rain that bears the kēhau mist
    Like a water gourd that was poured
    Upon the hau trees of Kūmomoku

From a mele inoa, or name chant, for Kūaliʻi. Hawaiian source: *Na Mele Aimoku* 94. English trans. by author. Additional sources: Fornander, *Account* 382, 396; Fornander, *Fornander* 6: 458; Fornander, *Fornander* 4: 390–91.

Rain that makes the land glisten

68. He ʻā, he haʻihaʻi pua, he Nāulu kaupua liko moku. *A booby bird, a flower breaker, a Nāulu rain cloud that gathers, making the land glisten.*

From the chant "Haui Kalani." Hawaiian source: Fornander, *Fornander* 6: 381. English trans. by author. *Note: Fornander says that the Nāulu is "a shower with wind."*

Rain that softens the forest

69. He ipo aloha naʻu ka nani o Kūkiʻi e waiho nei
I lohia ʻia mai e nā lehua o Hōpoe, ʻau i ke kai
Ke kaʻikaʻikū ʻia maila a Kalanamaihiki
Hikiwawe ka hana a ka ua i ka nahele
Ke hoʻowali ʻia maila e ka ua Nāulu
Kū helahela ke kula o Kamāʻoa, nopu i ka lā
Kukuni wela i ka ʻili o ka malihini

The beauty of Kūkiʻi, my intimate lover, lies there
Sparkling from the lehua of Hōpoe, reaching out into the sea
Carried about until arriving at Kalanamaihiki
The rain works hastily upon the forest
Softened by the Nāulu rain
The plain of Kamāʻoa is extensive, heated by the sun
That scorches the skin of newcomers

From the mele hoʻāeae, or love chant, "He inoa no kuʻu aloha," also known as "Aloha ka uka ʻōpua holu i ka makani." Hawaiian source: Pukui, *Nā Mele Welo* 76. English trans. by author.

Rain that will string an adornment

70. He leo no ka ipo kaʻu i lohe iho
Naʻu e kākele a mau ia pua
Ua hoʻoholo like ʻia e ka Nāulu
E kui i wehi no ka liko

A voice from the sweetheart is what I heard
I shall cast and hook this blossom
A similar resolution is made by the Nāulu
To string an adornment for the leaf bud

From the song "Lei ʻawapuhi" by Mekia Kealakaʻi. Hawaiian source: Elbert and Mahoe 74. English trans. by author.

Rainwater associated with a beloved's return

71. Aia [a] ho'i mai ka lei aloha lā
'Olu iho ka lani, nanahe ka wai
Pū'ili i ka pua kamaha'o a loko
Hi'ilani i ka wai o ka Nāulu lā

Oh, when the beloved one returns
The heavens should be so pleasant, the waters so sweetly appealing to hear
Embrace the wonderful bloom within
Cradle the showery waters of the Nāulu rain

From the song "E 'ae paha wau i ka ualo." Source: Lili'uokalani 104.

Shower of fine rain

72. Heavy mists; a shower of fine rain, apparently without clouds, or a single cloud; he ua kūha'o.

Source: Andrews 407.

Strong rain of short duration

73. He ua Nāulu he ua 'uku ia, he ikaika na'e. *A Nāulu rain, a brief rain, though it is intense.*

Hawaiian source: Malo, *Ka Moolelo* 20. English trans. by author. Additional source: Malo, *Hawaiian* 14.

Sudden shower

74. Kāua i laila, e ku'u hoa
He ua kūhewa ia na ka Nāulu

We are there, my friend
A sudden shower from the Nāulu

From the chant "Aia i Pokahi ka liko uahi manu." Hawaiian source: *Na Mele Aimoku* 290. English trans. by author.

75. He ua Nāulu ka ua e loa'a i ka Makali'i, he ua kūhewa ia, aia ma kona wahi e hā'ule ai, ua ulu ka laulele, ka pōpolo, a ua ola ia wahi i ka lau nāhelehele.

Sudden showers (ua Nāulu) fall during the Makali'i season. These are rains accompanied by wind gusts, and where they fall, laulele and pōpolo plants spring up, and that place comes to life with wild growth.

From a description of sweet potato cultivation. Hawaiian source: Kamakau, "Ka moolelo Hawaii" 11/18/1869. English trans.: Kamakau, *Works* 31.

nihi lau kukui. Rain associated with Mā'ili, O'ahu. "Nihi lau kukui" means "to creep softly upon kukui leaves."

1. Ho'olono a'e au 'o ka nehe
   O ka ua nihi lau kukui
   Ke loku maila i Mā'ili
   I ka lau o ka pilipili

   I listen to the rustling
   Of the nihi lau kukui rain
   Pouring down upon Mā'ili
   Over the leaves of the pilipili

   From the song "Lau ahihi o Lihue." Hawaiian source: Holstein 121. English trans. by author.

Nihipali. Rain associated with Ka'ū, Hawai'i, and with Moelana, O'ahu. Also the name of a wind. "Nihi pali" means "to creep along the cliffs." It is both the name of a specific rain and a generally descriptive term; its various usages are determined by the context.

Rain of 'Au'aulele, Hawai'i

1. *See Hā'ao (#6).*

Rain of Hā'ao, Hawai'i

2. Ku'u kaikamahine i ka ua Nihipali o Hā'ao
   Aloha ia wahi a kāua e pili ai

   My dear daughter in the Nihipali rain of Hā'ao
   Beloved is this place where we would be together

   From a kanikau, or lament, for Miss Mary Binning by Pa'ahana. Hawaiian source: Handy and Pukui 155–56. English trans. by author.

Rain of Moelana, O'ahu

3. Ioseba Nawai aloha
   Ku'u kāne mai ka ua Nihipali ma Moelana ē
   Lana mai ka mana'o i pā 'ia e ke aloha ē
   Ku'u aloha pau 'ole lā

   Beloved Ioseba Nawai
   My dear husband from the Nihipali rain at Moelana
   Buoyant thoughts are struck by love
   My endless love

   From a kanikau, or lament, for Ioseba Nawai. Hawaiian source: Noholoa. English trans. by author.

4. Ka ua Nihipali o Moelana. *The [Nihipali] rain that sneaks along the cliffs at Moelana.*

   A traditional saying. Source: Pukui, *ʻŌlelo* 172. *Note: Pukui says that the Nihipali is "the rain at Moelana, below the Nuʻuanu Pali."*

Noe. Same as *Noenoe.* Fog; mist that is lighter than the uhiwai, but heavier than the ʻohu, ʻehu, and ʻehuehu. Associated with Hawaiʻi, Maui, and Kauaʻi and found on other islands. Also the name of a wind. "Noe" is both the name of a specific rain and a generally descriptive term; its various usages are determined by the context.

Rain of Kamāʻoa, Hawaiʻi

1. Kuʻu kāne mai ka ua Noe o ke pili o Kamāʻoa. *My dear husband from the Noe mist over the pili grass of Kamāʻoa.*

   From a kanikau, or lament, for Kaoo. Hawaiian source: Kamaka, "He kanikau." English trans. by author.

Rain of ʻŌʻōkuauli, Hawaiʻi

2. *See ʻAwa (#9).*

Rain of Pāʻieʻie, Hawaiʻi

3. Ke noua aʻela e ka ua noe
   He ua hana nāwele i ka pua
   Pulu ihola ka maka o ka lehua
   Nā lehua i ka nahele i Pāʻieʻie

   Being buffeted by the misty [noe] rain
   A rain working ever so softly upon the flowers
   Dampening the delicate points of the lehua
   The lehua in the forest at Pāʻieʻie

   From the song "Ka makani Līhau Pua." Source: Liliʻuokalani 180.

Rain of Poliʻahu, Hawaiʻi

4. Līlī luna Poliʻahu i ka ua ʻo Noe. *The top of Poliʻahu shivers in the cold of the Noe rain.*

   From a chant for Kawaikini. Hawaiian source: Lohiau, "He mele no Kawaikini." English trans. by author.

Rain of Wahinekapu, Hawaiʻi

5. O hele ana ʻoe, e ka Noe, e ka ʻAwa
   E nā kī a Wahinekapu
   E ka ua lele aʻe ma luna
   O Kalāhikiola lā
   O hele ana ē!

May you journey swiftly, O Noe, O 'Awa rain
O kī of Wahinekapu
O rain that flies up
Above Kalāhikiola
Travel fast!

A call to Hiʻiakaikapoliopele by Pāʻūopalaʻe as the former sails from Honokāne, Kohala, for Maui. Hawaiian source: Emerson, *Pele* 65. English trans. by author.

Rain of East Maui

6. Ua Noe—he ua liʻiliʻi me ka noe pili mau i ke kula.

   Noe rain—a light rain and mist that always cling to the plains.

   From an account of East Maui rains and their descriptions, as told to Thomas K. Maunupau by Josua Ahulii of Kaupō, Maui. Hawaiian source: Maunupau, "Huaka'i." English trans. by author. Additional source: Maunupau, *Huakai* 108.

Mist of Kalena, Maui

7. 'Auhea wale ana hoʻi ʻoe
   E ka Noe lipo i ka nahele
   Lipolipo i ke oho o ka palai
   Hoa pili o ka 'Ulalena

   Where are you
   O dark Noe mist of the forest
   Dark green from the palai fronds
   Close friend of the 'Ulalena

   From the song "Wai hui o Kalena." Hawaiian source: Holstein 167. English trans. by author.

Rain of Lilikoʻi, Maui

8. E Lāhaināluna, e aloha ʻoe
   Ka ua Noe kā ʻoe o Lilikoʻi
   E hākoʻi lua nei ka puʻuwai

   O Lāhaināluna, love to you
   You are the Noe rain of Lilikoʻi
   That causes the heart to flutter

   From a kanikau, or lament, for Lāhaināluna Seminary by its students upon hearing that the building was decaying. Hawaiian source: Fornander, *Fornander* 6: 528. English trans. by author.

Rain of Alakaʻi, Kauaʻi

9. *See Noelehua (#2) and 'Ulalena (#26).*

Rain of Kōloa, Kaua'i

10. E ka ua Noe o Kōloa, ua pau kou ho'opulu 'ana i ka nui kino o ku'u aloha.
    O Noe rain of Kōloa, your drenching of my love's entire body has ended.

    From an obituary for Eda Kawaikauomaunahina Kalua. Hawaiian source: Kalua. English trans. by author.

11. *For another example of Noe of Kōloa, see Līhau (#8).*

Rain of Wai'ahulu, Kaua'i

12. 'O ka Noe kai Wai'ahulu
    'O ka Līhau ko Mokihana
    'O ka ua hoa Kapa'ahoa
    'O ia kai kilikā i ka pua

    The Noe rain is at Wai'ahulu
    The Līhau rain is of Mokihana
    The companion rain is the Kapa'ahoa
    The one that showers lightly upon flowers

    From a mele for Kamehameha III by Naihe. Hawaiian source: Naihe. English trans. by author.

Rain of Kā'eo, Ni'ihau

13. Kauanoelaniokaeo. 'O ia ta inoa o ta'u keiki 'elua. Mo'olelo mai to'u mau mātua i ta mana'o nui o tēlā inoa, 'o ia ho'i, e hiti mai ana ta ua. Tēlā pali 'o Kā'eo, he hō'ailona tēlā. 'O ta po'e noho ma Ni'ihau, noi lākou i ke Atua i ta wai. Ta hō-'ailona tēlā. Te uhi hou mai tēlā ua noe, e ua ana no 'elua, 'ekolu lā. Ketahi manawa, mao aku. Nui ta wai e lo'a a hiti iā lākou te mālama i ta wai. Tā ke Atua aloha tēlā i tāna po'e haipule.

    Kā'eo is a peak on Ni'ihau. I named my second son after that peak. His name is Kauanoelaniokā'eo. My parents told me about the significance of that name. It indicates that the rain is coming. People who live on Ni'ihau ask God for water and that is the sign. When the misty [noe] rain covers Kā'eo, we know there will be rain for two or three days—sometimes longer. Then there is lots of water. That is a sign of God's love for those who worship him.

    From a section on Ni'ihau place names. Source: Kanahele, Licayan, Nizo 41, 77.

Rain associated with beauty

14. *See Koko (#6).*

Rain of the plains

15. 'Auhea wale 'oe, e ka ua Noe
    Ka ua ko'iaweawe i ke kula
    Ku'u hoa 'o ka ua Leikoko'ula
    Ke pili lau holu o Luanui

Where are you, O misty [Noe] rain
The rain that streams down on the plain
My mate is the rainbow-hued [Leikokoʻula] rain
On the swaying pili leaves of Luanui

From a mele hula, or hula chant, contributed by Solomono Huihui. Source: Pukui, *Nā Mele Welo* 190–91. Additional source: "Auhea . . . ka ua noe." *Note: "Luanui" is the name of several areas on Oʻahu, including an ʻili in Kaʻalaea, Koʻolaupoko.*

Noelehua. Rain associated with Waiʻaleʻale, Kauaʻi. Also found on Hawaiʻi. "Noe lehua" means "lehua mist."

Rain of Maʻolala (or ʻŌmaʻolala), Hawaiʻi

1. *See Nahunahu (#7).*

Rain of Waiʻaleʻale, Kauaʻi

2. Kapu ka luna o Waiʻaleʻale i ka ua Noelehua
Lehua, ʻo ka lehua maka noe
Ua nonohe wale i Hauaʻiliki
I ʻiliki ʻia e ka Noe luna o Alakaʻi
Kaʻi ʻāuna lākou a Kawaikini a Kawaikōī
I Kahelekua, i hele hiō i ke ala kīpapa a Ola
E ola i ka wai ua a ka Nāulu

The upland of Waiʻaleʻale is sacred with the Noelehua rain
Lehua, the lehua maka noe shrub
Very attractive at Hauaʻiliki
Poured down upon by the Noe atop Alakaʻi
They move like a flock to Kawaikini, to Kawaikōī
At Kahelekua, going along the sloped, paved path of Ola
Live by the waters of the Nāulu rain

From a mele inoa, or name chant, for Albert Ka Haku O Hawaiʻi by his uncle, Kamehameha V. Hawaiian source: Alii Iolani. English trans. by author. Additional source: Pukui and Elbert, 1986, 271. *Note: "Ke ala kīpapa a Ola" is a trail above Hanalei, Kauaʻi, that was built under the direction of Ola, a Kauaʻi chief.*

3. Ka ua Noelehua o Waiʻaleʻale. *The Noelehua rain of Waiʻaleʻale.*

A traditional saying. Hawaiian source: Pukui, *ʻŌlelo* 172. English trans. by author. *Note: Pukui says that the Noelehua is "the rain of Waiʻaleʻale that moistens the lehua blossoms there."*

Noenoe. Same as *Noe.* Misty or foggy rain; associated with Puna, Hawaiʻi, and found in other areas. "Noenoe" can refer to feeling foggy from intoxication. It is both the name of a specific rain and a generally descriptive term; its various usages are determined by the context.

1. Lī mala anu ka mauna i ka noenoe
   Naenae anu i ka ua pūʻukiʻuki
   Ka nonole iwi haʻihaʻi o ka pua
   Ua ʻeha nō, malakia i ka Puʻulena
   Ka malahia lau ahe a ka moani

   Cold, chilled, aching is the mountain in the noenoe mist
   Panting, cold in the pūʻukiʻuki rain
   The pliable bones of the flower are broken
   Pained, bruised by the Puʻulena wind
   The aching caused by the many soft breaths of the breeze

   From the mele "O kaaukuku kukuku ku a ua [*sic*]." This portion was composed by Ululani. Hawaiian source: *Na Mele Aimoku* 217. English trans. by author. *Note: The mountain referred to above is probably Maunaloa, Hawaiʻi.*

2. *See ʻAwaʻawa (#7).*

Nounouʻili. Rain of Waimea, Kauaʻi. "Nounou ʻili" means "to pelt the skin."

1. Ka ua Nounouʻili o Waimea. *The skin-pelting [Nounouʻili] rain of Waimea.*

   A traditional saying. Source: Pukui, *ʻŌlelo* 172.

ʻohu. Fog, smoke, light cloud on a mountain; mist that is lighter than the uhiwai and noe, but heavier than the ʻehu and ʻehuehu.

Mist of Maunakea, Hawaiʻi

1. A pau lākou i ka hiamoe, pā maila ka makani maikaʻi ma ka ʻāina mai, hoʻololi aʻela ʻo Kūapākaʻa i nā waʻa a me ka peʻa, a holo pono i Hawaiʻi. I ia lā a pō, a ao aʻe, ʻike akula lākou i ka piko o Maunakea i loko o ka ʻohu, e māʻalo ana me he ʻōpua lā.

   When they had all gone to sleep, a favorable wind from the land blew; Kūapākaʻa altered the course of the canoes and the sails and headed straight for Hawaiʻi. It took them all that day and night until the following day before they saw the top of Maunakea rising out of the ʻohu, passing by like a cloud.

   From the legend of Pākaʻa and Kūapākaʻa. Hawaiian source: Fornander, *Fornander* 5: 125. English trans. by author.

Fog of Kuʻia, Maui

2. *See Kanikanilehua (#2).*

Crawling mist

3. E ka ʻohu kolo mai i uka
   E ka ʻohu kolo mai i kai

   . . . .

E ala, e ka ua, e ka makani
E ala! E ala!

O ʻohu that crawls toward the uplands
O ʻohu that crawls toward the sea

. . . .

Rise, O rain and wind
Rise! Rise!

From the legend of Kaulanapōkiʻi. Hawaiian source: Fornander, *Fornander* 4: 569. English trans. by author.

Mist at sea

4. I ia holo ʻana, kau ʻo Kāneʻāpua i luna o nā waʻa a holo akula. ʻElua hoʻokele o luna, ʻo Hoʻokeleihilo, ʻo Hoʻokeleikaʻū; ma hope o nā hoʻokele ʻo Kāneʻāpua. ʻO ka manawa ia, paʻa ʻo luna i ka ʻohu me ka pōhina, ʻaʻohe ʻike o mua iā hope o nā waʻa; pēlā ko lākou holo ʻana a hiki i ka ʻāina ʻo Kānehunamoku, he kino ʻīlio ke ʻano o ka ʻāina ma kā lākou ʻike aku.

   During this voyage, Kāneʻāpua boarded the canoes and they set sail. There were two navigators aboard, Hoʻokeleihilo and Hoʻokeleikaʻū; behind them was Kāne-ʻāpua. At this time, ʻohu and fog engulfed them; those in the front could not see those in the back of the canoes. That is how they sailed until reaching the land of Kānehunamoku, which appeared to them in the form of a dog.

   From the legend of Wahanui. Hawaiian source: Fornander, *Fornander* 4: 519. English trans. by author.

Rain associated with great beauty

5. *See Koko (#6).*
6. *For more examples of ʻohu, see Hāʻao (#6), Lanipaʻina (#2), Mālana (#6), ʻŪkiu (#3), and ʻUlalena (#14).*

ʻŌʻiliʻili maka akua. Rain that falls irregularly and is associated with the presence of a god.

1. Ka ua ʻōʻiliʻili maka akua. *The rain that appears here and there to denote the presence of a god.*

   A traditional saying. Source: Pukui, *ʻŌlelo* 172. *Note: Pukui says that this was "said of the rain that falls with a drop here and a drop there instead of falling in a shower."*

Olopawa. Same as *Paiolopawa.* Rain associated with Nuʻalele, Maui. Also the name of a wind. "Olo pawa" means "the sounds of dawn."

1. Noho mai Nuʻalele i ke kai
   Kāhiko i ka ua Olopawa

Nuʻalele dwells in the sea
Adorned by the Olopawa rain

From a mele inoa, or name chant, for Keikeiloselani. Hawaiian source: Luukia. English trans.: "Pā ka leo" 23.

ʻŌnini. Same as *ʻŌninipua*, *ʻŌninipuaʻiʻa*, and *ʻŌninipuaʻiʻo*. Sea rain associated with Punahoa, Maui.

1. Alau koʻu uluhua
I ka pōʻai a ka ua ʻŌnini
Ke hoʻohehelo aʻela i Punahoa
ʻŌmaka nalu o Keʻanini

My anger is divided
By the circling of the ʻŌnini rain
Haughty at Punahoa
The swelling surf of Keʻanini

From the song "Wai o Kanewai" by J. A., H. A., and M. K. Hawaiian source: Holstein 106. English trans. by author. *Note: Andrews 46 says that the term "alau" refers to a location "where a wind is parted, as the east wind at Hāna, Maui." "Alau" is also the name of an island off the coast of Hāna. "Keʻanini" is the name of a Hāna surf spot.*

ʻŌninipua. Same as *ʻŌnini*, *ʻŌninipuaʻiʻa*, and *ʻŌninipuaʻiʻo*. Sea rain associated with Hāna, Maui.

1. Kuʻu wahine mai ka Iʻaiki o Hāna
Mai ka ua ʻŌninipua i ke kai

My beloved wife from the Iʻaiki wind of Hāna
From the ʻŌninipua rain over the sea

From a kanikau, or lament, for Luakauwahine. Hawaiian source: Pawai. English trans. by author.

ʻŌninipuaʻiʻa. Same as *ʻŌnini*, *ʻŌninipua*, and *ʻŌninipuaʻiʻo*. Sea rain associated with Kahananui, Maui. Also found on Oʻahu.

Rain over the sea near Kahananui, Maui

1. Kuʻu moʻopuna mai ka ua ʻŌninipuaʻiʻa i ke kai
Ke kiola aʻela i uka o Kahananui

My dear grandchild from the ʻŌninipuaʻiʻa rain over the sea
Tossed upon the shore of Kahananui

From a kanikau, or lament, for Nalohaelua. Hawaiian source: Kaloma. English trans. by author.

Rain of O‘ahu

2. Ku‘u hoa mai ka ua ‘Ōninipua‘i‘a
   Hoa hele no‘u i ka ua me ka makani
   Hoa hele no‘u i nā pali o Ko‘olau

   My dear companion from the ‘Ōninipua‘i‘a rain
   Traveling companion of mine in the rain and wind
   Traveling companion of mine over the Ko‘olau Mountains

   From a kanikau, or lament, for H. Haumea. Hawaiian source: David. English trans. by author.

‘Ōninipua‘i‘o. Same as *‘Ōnini*, *‘Ōninipua*, and *‘Ōninipua‘i‘a*. Sea rain associated with Hāna, Maui.

1. Ua pau, ke ho‘i nei māua me ka ua ‘Ōninipua‘i‘o i ka moana. *The end; we are returning with the ‘Ōninipua‘i‘o rain to the ocean.*

   From the closing to a list of winds of Hāna, Maui. Hawaiian source: Kailiehu. English trans. by author. Additional source: Johnson 272–73.

P

Paiolopawa. Same as *Olopawa*. Rain and wind associated with Hāna, Maui. Also the name of a place in Hāna, Maui. "Pai olo pawa" means "to amplify the sounds of dawn."

Rain of Hāna, Maui

1. *See Lau'awa (#1).*

Rain of Ka'uiki, Maui

2. Kū Ka'uiki, pakohana i ka makani
Ho'ohalulu i ke kai o Kaihalulu
Ke 'oi a lau i ka makani
E kali ana i ka ua Paiolopawa
I ka hili kua 'ia e ke Koholālele

Ka'uiki stands, struck barren by the wind
Roaring at the sea of Kaihalulu
Made sharp and pointed by the wind
Awaiting the Paiolopawa rain
Hit on the back by the Koholālele wind

From a mele for Ke'elikōlani by Pipi. Hawaiian source: Fornander, *Fornander* 6: 486. English trans. by author.

3. Kani kō paha ka ua, pa'a ka ua i ka lani
Holoi e Kaihuakala i nā kī'aha a ka lani wai
A ka ua Paiolopawa

The rain resounds, a rain stuck in the sky
Kaihuakala washes the receptacles of the watery sky
Of the Paiolopawa rain

From the chant "Maika'i Ka'uiki." Hawaiian source: Pukui, *Nā Mele Welo* 164. English trans. by author. *Note: In this rendering, "Kaihuakala" is used instead of "ka ihu a ka lā," and "Paiolopawa" instead of "pai olo pawa," which were printed in the Hawaiian source. "Kaihuakala" is the name of a hill inland of Ka'uiki.*

Palahīnano. Rain associated with Haili, Hawai'i. "Pala hīnano" means "yellow pandanus blossom."

1. 'Alawa a'e 'oe 'o ka ua Palahīnano
Ua makahehi wale i ke kula o Haili

Take a glance to see the Palahīnano rain
Delightful indeed is the plain of Haili

From a mele māka'ika'i, or travel chant, for 'Emalani Kaleleonālani, credited to Mrs. Mahie. Source: Nogelmeier 104.

palahī pua'a. Intermittent rain. "Palahī pua'a" means "pig diarrhea."

Rain of Kona, Hawai'i

1. "Mai hea mai 'oe?"
"Mai Kona mai."
"Pehea ka ua o Kona?"
"Palahī pua'a ka ua o Kona."
"A pehea ke aku?"
"Hī ka pā, hī ka malau."

"Where are you from?"
"From Kona."
"How is the rain of Kona?"
"The rain of Kona pours like the watery excreta of a hog [palahī pua'a]."
"How are the aku fish?"
"They run loose from the hook and the bait carrier."

A traditional saying. Source: Pukui, *'Ōlelo* 222. *Note: Pukui says that this was "said in fun of one suffering from loose bowels."*

Intermittent rain

2. Ka ua palahī puaʻa—falls, stops, falls, stops, etc.

   From a list of rain names and their descriptions. Source: Kelsey, "Rains."

Palihala. Rain associated with Waʻahila Ridge, Oʻahu. "Pali hala" means "cliff of hala trees."

1. *See Waʻahila (#26).*

Paliloa. Rain associated with Waimea, Hawaiʻi, and found in other areas. Similar to "Kapaliloa," the name of a place in Waimea, Hawaiʻi. "Pali loa" means "tall cliffs."

Rain of Kohala, Hawaiʻi

1. Aniani mai ka makani
   Ma Kohala mai ke ʻala

   . . . .

   Kuaʻāina i ka laʻi
   Pulu i ka ua Paliloa

   The wind blows softly
   The fragrance comes from Kohala

   . . . .

   Countryside in the calm
   Soaked by the Paliloa rain

   From a mele inoa, or name chant, for Kamehameha V. Hawaiian source: Hopoiku. English trans. by author.

Rain of Mahiki, Hawaiʻi

2. Aloha ʻino nō kuʻu kaikamahine, kuʻu lei ʻāʻī, kuʻu hoa hele o ke alanui ē

   . . . .

   Aloha ka ua Paliloa, wai huna a ka manu i ka lāʻau, mai ka lepo kulaʻi kanaka o Mahiki

   Pity for my daughter, the lei that adorns my neck, my traveling companion on the road

   . . . .

   Love to the Paliloa rain, hidden water of the birds in the trees, from the dirt of Mahiki that knocks people down

   From a mele ʻuhane, or lament, for Kealohapauole, presumably by Kaikamahineole. Hawaiian source: "He uhane." English trans. by author.

3. *For another example of Paliloa of Mahiki, see Leikokoʻula (#1).*

Rain of Maluleia, Hawaiʻi

4. Hoʻopākī lua ka pua o ke koʻolau
Hoʻopala pahē ana i Maluleia
He poluea na ka ua Paliloa

Crushing the koʻolau flower
Softening, ripening, and almost spoiling at Maluleia
Dizzied by the Paliloa rain

From a kanikau, or lament, for Kamehameha IV. Hawaiian source: Moehonua and Lilikalani. English trans. by author.

5. Ka pua i ka nahele o Maluleia
I anana ʻia e ka Paliloa
E ka ua Mālanaʻeʻelekoa i ʻŌuli
He ua aloha nō ia no ko laila kupa
E walea ai i ke hau pā kaiaulu

The flower in the forest of Maluleia
Measured by the Paliloa rain
By the Mālanaʻeʻelekoa rain at ʻŌuli
A rain loved by the locals of that place
Who relax in the cool breeze that touches the community

From a mele mākaʻikaʻi, or travel chant, for ʻEmalani Kaleleonālani. Hawaiian source: Nogelmeier 121. English trans. by author. *Note: In this rendering, "Mālanaʻeʻelekoa" is used instead of "malana ʻEʻelekoa," which was used in the Hawaiian source.*

Rain of Uli (or ʻŌuli), Hawaiʻi

6. *See Kīpuʻupuʻu (#8).*

Rain of Waimea, Hawaiʻi

7. Ka ua Paliloa o Waimea. *The Paliloa rain of Waimea.*

A traditional saying. Hawaiian source: Pukui, *ʻŌlelo* 172. English trans. by author. *Note: Pukui says that the Paliloa is "the rain of Waimea, Hawaiʻi, that sweeps down the cliffs."*

Rain of Waiʻōpua, Hawaiʻi

8. Kau pono i Waiʻōpua ka Nāulu
He hōʻailona makani ia na ka ua Paliloa

The Nāulu rests directly atop Waiʻōpua
It is a sign of wind for the Paliloa rain

From a mele kālaiʻāina, or political chant, for ʻEmalani Kaleleonālani. Hawaiian source: Nogelmeier 180. English trans. by author.

Rain of Halemano, Oʻahu
9. *See Kiʻowao (#6).*

Rain of Hāʻena, Kauaʻi
10. ʻAʻohe hana a ka ua Paliloa
Ke nihi aʻela ma ke kua o Hāʻena

The Paliloa rain seems inactive
It creeps slowly at the back of Hāʻena

From the legend of ʻAʻahoaka. Hawaiian source: "He moolelo no Aahoaka" 2/24/1877. English trans. by author.

Panehala. Possibly related to *Panehewa*. Rain associated with Maluaka, Kauaʻi.

1. He hoa i ka ua o ʻEleao
Ka ua Panehala o Maluaka
Ka nahele kī wai o ka ua
Ka ʻāhui hala i ka makani

A friend in the rain of ʻEleao
The Panehala rain of Maluaka
The forest upon which the rain shoots its water
The hala fruit in the wind

From a mele for ʻEmalani Kaleleonālani. The chant is a mele inoa, or name chant, and a mele mākaʻikaʻi, or travel chant. Hawaiian source: Nogelmeier 74. English trans. by author. Additional source: Holstein 140.

Panehewa. Possibly related to *Panehala*. Rain associated with Manuʻakepa, Kauaʻi.

1. Hikilele au me kuʻu hoa
Hoa noʻu i ka ua o ʻEleao
Ka ua Panehewa o Manuʻakepa

My companion and I are startled
The companion in the rain of ʻEleao
The uncertain [Panehewa] rain of Manuʻakepa

From a mele hōʻihi, or honorific chant, for ʻEmalani Kaleleonālani. Source: Nogelmeier 14.

Papawai. Rain name associated with Olowalu, West Maui.

Source: Pukui and Elbert, 1986, 319.

Paʻūpili. Rain associated with Lāhainā, Maui. Also found at Kāʻanapali, Maui, and on Hawaiʻi, Oʻahu, and Kauaʻi. "Paʻū pili" means "to soak pili grass."

Rain of Lilinoe, Hawaiʻi

1. He ua noe Paʻūpili hau no Lilinoe. *An icy Paʻūpili mist of Lilinoe.*

   From a chant for Kawaikini. Hawaiian source: Lohiau, "He mele no Kawaikini." English trans. by author.

Rain of Hālona, Maui

2. E aloha aʻe ana nō wau
   I ka ua Paʻūpili
   Ka ninihi koʻiawe
   I nā pali ma uka o Hālona

   I am truly loving
   The Paʻūpili rain
   That moves lightly
   Along the cliffs above Hālona

   From the song "Hālona" by J. Elia. Hawaiian source: Elbert and Mahoe 40. English trans. by author.

Rain of Kahana, Maui

3. *See Līlīlehua (#7).*

Rain of Kanahā, Maui

4. I ka holo ʻana i Maui e hoʻāo me Piʻilani, ua kāhiko ʻia hoʻi ʻo luna o nā waʻa i nā mea ʻula, a me ka lā ʻula, a me nā kānaka ʻula. I ka hiki ʻana hoʻi i Lāhainā, ua hoʻihoʻi ʻia aku ka ua Paʻūpili i ka uka o Kanahā, a ʻo ka ua Koko hoʻi o Kāʻanapali ke kāhiko no Lāʻieloheloheikawai.

   On [Lāʻielohelohe's] sail to Maui to marry Piʻilani, the decks of her canoes were decorated with red things, and a red sail, and attendants in red. On her arrival in Lāhainā, the Paʻūpili rain was returned to the uplands of Kanahā, and the Koko rain of Kāʻanapali was the adornment of Lāʻieloheloheikawai.

   From the legend of Lāʻielohelohe. Hawaiian source: Kamakau, "Ka moolelo o Hawaii nei" 8/26/1865. English trans. by author.

Rain of Kekaʻa, Maui

5. Pohu laʻi maka[ni] ma loko o Kekaʻa—kāua
   He hohonu kā ia na ka ʻUmihau—kāua
   Ke kākua maila lā i ka ua Paʻūpili—kāua
   He ane kōkua paha ke Kauaʻula—kāua

   Calm and quiet is the wind within Kekaʻa—the two of us
   Made deep by the ʻUmihau wind—the two of us
   Fastened in the Paʻūpili rain—the two of us
   A helpful breeze is the Kauaʻula—the two of us

From a mele inoa, or name chant, for Albert Ka Haku O Hawaiʻi. Hawaiian source: Kamaiopili. English trans. by author. *Note: The Paʻūpili rain is sometimes associated with the word "kākua," meaning "to bind or fasten on," suggesting another possible interpretation of the rain name: "pāʻū pili," or "skirt that clings."*

6. Lohia lau ʻohe ia ka laʻi o Lele
E unuhia a oki me he waʻa kioloa lā
Ka ʻōʻili o ka pua i ka mālie
ʻUmea e ka lā i kāwalawala hiolo
Kākua iho ka ua Paʻūpili ē
ʻAe
Pili iho i ke kula o Kekaʻa
Pili nānā i ka ua Leikokoʻula

The calm of Lele shines like ʻohe leaves
Long and tapered like a kioloa canoe
Billowy clouds appear in the calm
Attracted by the sunlight that falls and scatters
The Paʻūpili rain envelops
Yes
Clinging to the plains of Kekaʻa
Clinging, resting in the Leikokoʻula rain

From a mele for Lāhainā and Kāʻanapali on Maui. Hawaiian source: Samuela. English trans. by author. Additional sources: "Ka moolelo o He'ma [*sic*]" 10/1/1920; "Story of Hema" 20.

Rain of Lāhainā (or Lele), Maui

7. Kuʻu hoa mai ka malu ʻulu o Lele
Kuʻu hoa mai ka laʻi a ke Kiowea
Kuʻu hoa mai ka ua Paʻūpili o Lāhainā

My companion from the shade of the ʻulu trees of Lele
My companion from the calm of the Kiowea wind
My companion from the Paʻūpili rain of Lāhainā

From a mele aloha, or farewell song, for Nāhiʻenaʻena by Kini. Hawaiian source: Fornander, *Fornander* 6: 441. English trans. by author.

8. Keikei Lāhainā i ka ua Paʻūpili. *Majestic Lāhainā in the Paʻūpili rain.*

A traditional saying. Source: Pukui, *ʻŌlelo* 183.

9. E ʻai ana kā ʻoe, e Puailahaina, ka ua Paʻūpili
Pili iho ka laʻi, pili i ka pohu
Ke kōhi ala i ka hono o Hauola

You are indeed going to eat, O Puailahaina, of the Paʻūpili rain
The calm clings, clings to the stillness
Restraining the bay of Hauola

From a chant for ʻEmalani Kaleleonālani for the moment she sat down to eat. Hawaiian source: Nogelmeier 192. English trans. by author.

10. Nani Lāhainā i ka ua Paʻūpili
I pili aloha ʻia me ka makani he Maʻaʻa
. . . .
Keikei Lāhainā i ka ua Paʻūpili
I pili aloha ʻia
E Maunahoʻomaha me Kekaʻa

Beautiful is Lāhainā in the Paʻūpili rain
Joined lovingly by the Maʻaʻa wind
. . . .
Lāhainā is glorious in the Paʻūpili rain
Held in the bonds of love
By Maunahoʻomaha and Kekaʻa

From a mele olioli chanted by Charles Kauluhinano Hoopii of Lāhainā, Maui, who learned it from Pamano, the younger sister of the composer. Hawaiian source: "Nani Lahaina." English trans. by author. Additional sources: Kanahele 39–40, 54, and 57; Kualu.

11. Kau pono i Lāhainā ka ua Paʻūpili
Ua lī ʻē i nā ukana a ka Maʻaʻa

There in Lāhainā is the Paʻūpili rain
Already chilled by the baggage of the Maʻaʻa wind

From a mele kālaiʻāina, or political chant, for ʻEmalani Kaleleonālani. Source: Nogelmeier 178.

12. Ka ua Paʻūpili o Lele. *The pili-soaking [Paʻūpili] rain of Lele.*

A traditional saying. Source: Pukui, *ʻŌlelo* 172. *Note: Pukui says, "The plains of Lāhainā, Maui, were covered with pili grass in ancient days. When the rain poured the grass was well soaked."*

13. Ke oho ua Paʻūpili o Lele
Līlīlehua i uka o Honokawai

Tresses that sweep like the Paʻūpili rain of Lele
Like the Līlīlehua wind and rain above Honokawai

From a mele hōʻihi, or honorific chant, praising the hair of ʻEmalani Kaleleonālani. Source: Nogelmeier 60. *Note: "Honokawai" is a variant of "Honokōwai," the name of an ahupuaʻa in Kāʻanapali, Maui.*

14. Aloha kuʻu hoa o ka ua Paʻūpili
Kuʻu hoa o ka malu o ka ʻulu
O ke ahe a ka makani he ʻAʻa
Kuʻu hoa o ka piʻina o Paʻupaʻu a me nā lehua o Līhau

Love for my dear companion of the Paʻūpili rain
My companion of the shade of the ʻulu trees
Of the gentle blowing of the wind, an ʻAʻa
My companion of the rise of Paʻupaʻu and the lehua blossoms of Līhau

From a kanikau, or lament, for L. Kahuakai. Hawaiian source: Kaauwaepaa. English trans. by author.

15. *For another example of Paʻūpili of Lāhainā, see Kanilehua (#5).*

Rain of Lāhaināluna, Maui

16. Huli aku ke alo nānā iā uka
ʻO ka hāliʻi mai a ka ua Paʻūpili
I pili ʻia e ka ua ka lā wela o Lāhaināluna

The face turns to look at the uplands
The spreading of the Paʻūpili rain
The rain joins the hot sun of Lāhaināluna

From a kānaenae, or chant of eulogy, for Paulo Ku. Hawaiian source: Paulo. English trans. by author.

Rain of Waʻahila, Oʻahu

17. Kuʻu keiki mai ka wai o Kewalo
Auē kuʻu keiki
Kuʻu keiki mai ka ua Paʻūpili o Waʻahila

My darling child from the waters of Kewalo
Alas, my beloved child
My dear child from the Paʻūpili rain of Waʻahila

From a kanikau, or lament, for Kapela. Hawaiian source: D. W. Pauahi. English trans. by author.

Rain of Hanalei, Kauaʻi

18. Auē! Ka mea aloha ʻo ka wahine i naue aku nei
Kuʻu wahine mai ka ua Paʻūpili o Hanalei
Ke pōhai maila i luna o Hīhīmanu

Alas! The beloved one, the woman who has moved on
My dear wife from the Paʻūpili rain of Hanalei
That circles above Hīhīmanu

From a kanikau, or lament, for Esetera Wahahee. Hawaiian source: Kalionui. English trans. by author.

Rain of Kawaikini, Kauaʻi

19. ʻO kuʻu lehua kēnā i Mālamalamaiki
    Naʻu nō i kui i ka uka o Kawaikini
    He lei, he kāhiko no ka ua Paʻūpili
    E pili aku ana māua i ke koʻekoʻe
    A ʻalo aku māua i ka ua me ka makani

    That is my beloved lehua at Mālamalamaiki
    I am the one who strung it in the uplands of Kawaikini
    A lei, an adornment for the Paʻūpili rain
    We will cling together in the cold
    And we will endure the rain and the wind

    From the legend of Kamaakamahiʻai. Hawaiian source: Kaualilinoe, "Ka moolelo" 10/15/1870. English trans. by author. Additional source: Kaualilinoe, "Legend of Kamaakamahiai" 77.

Rain of Līhuʻe, Kauaʻi

20. He ua Paʻūpili (pili soaking) ko Līhuʻe, Kauaʻi. *Līhuʻe, Kauaʻi, has a Paʻūpili rain.*

    From a textbook on the Hawaiian language. Hawaiian source: Kahananui and Anthony 108. English trans. by author.

21. Nani wale Līhuʻe i ka laʻi
    I ka noe a ka ua Paʻūpili

    So beautiful is Līhuʻe in the calm
    In the mist of the Paʻūpili rain

    From the song "Wailua alo lahilahi," also known as "Nani wale Līhuʻe," credited by some to Liliʻuokalani and Kapoli and by others to Leleiōhoku and Mrs. Kamakau. Source: Liliʻuokalani 275. Additional sources: Holstein 22; Wilcox et al. 201. *Note: Wilcox et al. say, "In this mele . . . winds and weather suggest a range of experiences enjoyed by these lovers, from the soft Paʻūpili rain that moistens the pili grass to the Mālualua, a gusty north wind."*

22. Ua nani wale ʻo Līhuʻe
    I ka ua Paʻūpili hale
    I ka wai huʻihuʻi anu
    Kahi wai aʻo Kemamo

    So very beautiful is Līhuʻe
    In the drenching [Paʻūpili] rain that clings to the house
    With the cold, refreshing waters
    From the springs of Kemamo

    From the song "Maikaʻi Kauaʻi." Source: Wilcox et al. 160.

Rain of Peʻapeʻa, Kauaʻi

23. He ua Paʻūpili ko Peʻapeʻa
    Ke kāhiko maila i ka lāʻau
    ʻO ke koa, pulu ka lau o ka ʻamaʻu
    Lau palai o Mahaikona

    Peʻapeʻa has a Paʻūpili rain
    It is adorning the trees
    The koa, the ʻamaʻu fronds are soaked
    Palai fronds of Mahaikona

    From a makena, or lament, for the passing of Kamehameha IV. Hawaiian source: Nogelmeier 356. English trans. by author.

Other examples

24. Ua ʻike ʻia ʻoe e Rain Paʻūpili. *You are seen by Paʻūpili rain.*

    From a song. Source: Pukui and Elbert, 1986, 321.

25. I ka lani nō ka ua, hāneʻe ke pulu
    E hoʻomaikaʻi ana i ka ua Paʻūpili
    I hoa hui no Uwilakūlani
    No ke kama nona ia inoa, ke ō ke haʻi lā

    While the rain is in the sky, the mulch slides down
    Bettering the Paʻūpili rain
    An associate of Uwilakūlani
    Of this child to whom belongs this name chant, to answer, to tell

    From a mele inoa, or name chant, for Heneri W. Auld. Hawaiian source: Hoopai. English trans. by author.

Paupua. Rain associated with Kalanipuʻu, Kauaʻi. Also the name of a wind.

1. Aloha kuʻu hoa o ka ua he Paupua
   Ke hiʻi ʻia aʻela e Kalanipuʻu
   E ka noe anu a ka Waiolohia ʻeā

   Love for my dear companion of the rain, a Paupua
   Held by Kalanipuʻu
   By the cold mist of the Waiolohia rain

   From a kanikau, or lament, for L. Kahuakai. Hawaiian source: Kaauwaepaa. English trans. by author.

Peʻehala. Same as *Peʻepeʻepūhala*, *Peʻepūhala*, and *Peʻepūhalahīnano*. Rain associated with Hunahunalaukī, Maui. "Peʻe hala" means "to hide under hala trees."

1. Lilo akula ua i ka moana
   He maka o Hāwini ia ua

He ua ʻalo ma ka lae
Nihi pali, nihi lae
Nihi i ka lae o Mokupapa
Hele wale ka ua a kipa wale
Ka ua Peʻehala huna kai o—
O Hunahunalaukī

The rain is consumed by the ocean
A rain favored by Hāwini
The rain accompanies the headland
Creeping along the cliffs, creeping along the capes
Softly moving to the point of Mokupapa
Casually it goes and visits, uninvited
The Peʻehala rain, like sea spray of—
Of Hunahunalaukī

From a mele by Keaulumoku expressing his love for the Hāmākua districts of Maui, where he had lived with Kamehamehanuiʻailūʻau and Kahekili. Hawaiian source: Kamakau, "Ka moolelo o Kamehameha I" 2/23/1867. English trans. by author. Additional sources: Kamakau, *Ke Kumu Aupuni* 66; Kamakau, *Ruling* 113–14. *Note: "Hāwini" is the name of a bay and "Mokupapa" the name of an ahupuaʻa in Hāmākualoa, Maui.*

Peʻepākaiaulu. Rain associated with Kohala, Hawaiʻi.

Rain of Kalāhikiola, Hawaiʻi

1. Aloha ʻino nō kuʻu kaikamahine, kuʻu lei ʻāʻī, kuʻu hoa hele o ke alanui ē

. . . .

Oʻu hoa hele i kai o Kohala, i ka ua Peʻepākaiaulu o Kalāhikiola ē

Pity for my dear daughter, the lei that adorns my neck, my traveling companion on the road

. . . .

My traveling companion by the sea of Kohala, in the Peʻepākaiaulu rain of Kalāhikiola

From a mele ʻuhane, or lament, for Kealohapauole, presumably by Kaikamahineole. Hawaiian source: "He uhane." English trans. by author.

Rain of Waimea, Hawaiʻi

2. Hoʻolāʻau mai ʻo Kawelowelo
Ua Peʻepākaiaulu o Waimea

Kawelowelo stiffens
In the Peʻepākaiaulu rain of Waimea

From a mele inoa, or name chant, for Kamehameha V. Hawaiian source: Emerson, *Unwritten* 162. English trans. by author. *Note: Emerson says that it is "a fierce rain squall such as arises suddenly in the uplands of Waimea, Hawaiʻi. The traveler, to protect himself, crouches (peʻe) behind a hummock of grass, or builds up in all haste a barricade (pā) of light stuff as a partial shelter against the oncoming storm."*

3. *For another example of Peʻepākaiaulu of Waimea, see Kīpuʻupuʻu (#35).*

Peʻepāpōhaku. Same as *Peʻepeʻepōhaku* and *Peʻepōhaku*. Rain associated with East Maui. "Peʻe pā pōhaku" means "to hide behind a rock wall."

Rain of East Maui

1. Ua Peʻepāpōhaku—He ua hoʻonāukiuki, ʻano like me ka ua Hāleuʻole i ka ʻiliki iho a mao aʻe nō a holo akula ka poʻe a peʻe i ka pā pōhaku, a ʻo ia ke kumu i kapa ʻia ai kēia inoa. ʻO ka inoa o kēia ua, he inoa na kamaliʻi paha i kapa no ka wā e pāʻani ai lākou i loko o ka ua, ʻo ia ka manaʻo o ka mea kākau, ʻaʻole na kahiko.

   Peʻepāpōhaku rain—A rain that causes annoyance, somewhat like the Hāleuʻole rain in pouring down suddenly and then clearing up; people run away and hide behind rock walls, and that is the reason it is called by this name. The name of this rain may have been given to it by children while playing in the rain; that is the belief of this author; it is not an ancient name.

   From an account of East Maui rains and their descriptions, as told to Thomas K. Maunupau by Josua Ahulii of Kaupō, Maui. Hawaiian source: Maunupau, *Huakai* 27. English trans. by author.

Rain of Kaupō, Maui

2. E uē ana au i kuʻu makua kāne
   Iā Paulo, kanaka i ka wēkiu
   Pua makani kolokolonahe o ke kula
   Kuʻu makua kāne mai ka ua Peʻepāpōhaku o Kaupō
   I pili ʻia e ka ua ke anu me ke koʻekoʻe

   I weep for my father
   For Paulo, man of the highest station
   A gentle wind blows over the plains
   My dear father from the Peʻepāpōhaku rain of Kaupō
   The rain accompanies the cold and the dampness

   From a kānaenae, or chant of eulogy, for Paulo Ku. Hawaiian source: Paulo. English trans. by author.

Peʻepeʻepōhaku. Same as *Peʻepāpōhaku* and *Peʻepōhaku*. Rain associated with Hāmākua, Maui. "Peʻepeʻe pōhaku" means "to hide behind rocks."

1. A Hāmākua au
Noho i ka ulu hala
Malihini au i ka hiki ʻana
I ka ua Peʻepeʻepōhaku

I was at Hāmākua
Sitting in a hala grove
I arrived a stranger
To the Peʻepeʻepōhaku rain

From a mele hoʻoipoipo, or love song. Hawaiian source: Emerson, *Unwritten* 122. English trans. by author.

Peʻepeʻepūhala. Same as *Peʻehala*, *Peʻepūhala*, and *Peʻepūhalahīnano.* Rain associated with Hāmākualoa, Maui. "Peʻepeʻe pūhala" means "to hide under hala trees."

1. Pehea lā auaneʻi ka pono?
Ma kahi nō a ʻolua e hele ai
Mai Kula a Hāmākua
Peʻepeʻepūhala ka ua o Hāmākua
Ka ua o Hunahunalaukī
E hūnā aku ʻoe i kuʻu mea i laila

How can we make things right?
Where you two go
From Kula to Hāmākua
Peʻepeʻepūhala is the rain of Hāmākua
The rain of Hunahunalaukī
You hide my beloved one there

From a prayer by Kawaunuiaola, a clever woman of Kula, Maui, who got her husband to return to her. Hawaiian source: Fornander, *Fornander* 4: 555. English trans. by author.

Peʻepōhaku. Same as *Peʻepāpōhaku* and *Peʻepeʻepōhaku.* Rain associated with Kaupō, Maui. "Peʻe pōhaku" means "to hide behind rocks."

1. Ka ua Peʻepōhaku o Kaupō. *The [Peʻepōhaku] rain of Kaupō that makes one hide behind a rock.*

A traditional saying. Source: Pukui, *ʻŌlelo* 172. *Note: Pukui says, "It falls so suddenly that one flees behind rocks for shelter."*

Peʻepūhala. Same as *Peʻehala*, *Peʻepeʻepūhala*, and *Peʻepūhalahīnano.* Rain associated with Hāmākualoa, Maui. "Peʻe pūhala" means "to hide under hala trees."

Rain of Hāmākualoa, Maui

1. Ka ua Peʻepūhala o Hāmākua. *The [Peʻepūhala] rain of Hāmākua that makes one run to the hala tree to hide.*

   Source: Sterling 97.

Rain of Huelo, Maui

2. Ka ua Peʻepūhala o Huelo. *The [Peʻepūhala] rain of Huelo that makes one hide in a hala grove.*

   A traditional saying. Source: Pukui, *ʻŌlelo* 172.

Peʻepūhalahīnano. Same as *Peʻehala*, *Peʻepeʻepūhala*, and *Peʻepūhala*. "Peʻe pūhala hīnano" means "to hide under the hīnano-bearing hala trees."

1. Rain name.

   Source: Pukui and Elbert, 1986, 322.

Pehihala. Rain associated with Panaʻewa, Hawaiʻi, and with Hāmākua, Maui. "Pehi hala" means "to pelt the hala fruit."

Rain of Panaʻewa, Hawaiʻi

1. E akahele iho ʻoe i nā lehua o Hiʻilani, hehene auē
   Inā i Puna ke ʻala me ke onaona, hehene auē
   He kāhiko ia no Hōpoe i ke kai, hehene auē
   Ke hopo lā i ka ua Pehihala o Panaʻewa, hehene auē
   I ke ʻāluka a nā manu i ka ʻōhelo papa, hehene auē

   Be careful around the lehua blossoms of Hiʻilani, tralala
   If the fragrance and aroma are in Puna, tralala
   These are adornments for Hōpoe in the sea, tralala
   Perturbed by the Pehihala rain of Panaʻewa, tralala
   And by the birds flocking over the strawberries, tralala

   From a name chant for Kawailahaole. Hawaiian source: Kahoohulimoku. English trans. by author. *Note: "Hiʻilani" may be the same as "Hiulani," a place in Puna, Hawaiʻi.*

Rain of Hāmākua, Maui

2. Ka ua Pehihala o Hāmākua. *The [Pehihala] rain of Hāmākua that pelts the pandanus fruit clusters.*

   A traditional saying. Source: Pukui, *ʻŌlelo* 172–73.

3. *For another example of Pehihala of Hāmākua, see ʻŪkiu (#8).*

Pehilehua. Rain associated with Nuʻuanu, Oʻahu. "Pehi lehua" means "to pelt the lehua flowers."

1. Nani Nuʻuanu i ka ua Pehilehua a ka Puʻulena
I hoʻonaue i ke oho o ke kāwelu o Lanihuli

Nuʻuanu is glorious in the Pehilehua rain and the Puʻulena wind
That set to motion the leaves of the kāwelu grass of Lanihuli

From a mele credited to Penimana. Hawaiian source: Pukui, *Nā Mele Welo* 20. English trans. by author.

Pilinahe. Rain name.

Source: Pukui and Elbert, 1986, 330.

Pōʻaihala. Same as *Pōʻaihale.* Rain associated with Kahaluʻu, Oʻahu. "Pōʻai hala" means "to surround hala trees."

Rain of Kahaluʻu, Oʻahu

1. Ua Pōʻaihala—Kahaluʻu, Koʻolaupoko, Oʻahu.

From a list of rain names and their descriptions. Source: "Names . . . rain."

2. He kanikau aloha no Mary Haliilua

. . . .

Kuʻu wahine mai ka Aʻe makani o Koʻolau
Mai ka ua Pōʻaihala o Kahaluʻu
Luʻuluʻu au i kō aloha e noho nei

A loving lament for Mary Haliilua

. . . .

My darling wife from the Aʻe wind of the Koʻolau side
From the Pōʻaihala rain of Kahaluʻu
I am overwhelmed by your love, which remains

From a kanikau, or lament, for Mary Kaawaloa i Haliilua. Hawaiian source: Keliihaleole. English trans. by author.

3. ʻIno Koʻolau ē, ʻino Koʻolau
ʻAikena i ka ua o Koʻolau
Ke ua maila i Māʻeliʻeli
Ke hoʻowaʻawaʻa maila i Heʻeia
Ke kūpā lā ka ua i ke kai
Haʻa hula leʻa ka ua
I ʻĀhuimanu, ka ua hōʻoni
Hoʻonaue i ka puʻu koʻa
Ka ua Pōʻaihala o Kahaluʻu
Luʻuluʻu ē, luʻuluʻu iho nei au

I ka pūʻolo waimaka o ka ʻōnohi
Ke kulu iho nei ē
E uē kāua ē

Vile is Koʻolau, stormy is Koʻolau
Worn and eroded by the Koʻolau rain
Raining down upon Māʻeliʻeli
Carving grooves into Heʻeia
Digging at the ocean
The rain merrily dances
At ʻĀhuimanu, making things move
Shaking the coral piles
The Pōʻaihala rain of Kahaluʻu
So oppressive, I am overladen
With a bundle of tears in my eyes
Flowing down
Let us weep together

From a mele pule kau by Hiʻiaka, a prayer chant addressed to nature by Hiʻiaka-ikapoliopele. Hawaiian source: "Ino Koolau." English trans. by author. Additional sources: *Nā Leo Hawaiʻi Kahiko,* Record I, Side A, Band 5; Hoʻoulumāhiehie, *Ka Moʻolelo* 156–57, 436; Hoʻoulumāhiehie, *Epic* 147–48, 407; Emerson, *Pele* 90–91.

4. Kaʻa hewa ka ua Pōʻaihala o Kahaluʻu, ʻeā
Lūlū lima ke aloha i ʻĀhulimanu, ʻeā

The Pōʻaihala rain of Kahaluʻu twirls awry
Love shakes hands in greeting at ʻĀhulimanu

From a mele mākaʻikaʻi, or travel chant, for ʻEmalani Kaleleonālani by Moehonua. Source: Nogelmeier 153.

5. E ʻike aku ana i ka ua Pōʻaihala o Kahaluʻu
Ua luʻuluʻu au i kō aloha e noho nei
Ua pēpē a wali koʻu nui kino
I ka hana a ka ua koʻiawe i nā pali
Pulu ʻelo au i ka lihilihi o kō aloha

Looking at the Pōʻaihala rain of Kahaluʻu
I am overcome by your love, which remains
My entire body is crushed and made supple
By the work of the koʻiawe rain moving lightly on the cliffs
I am soaked by the smallest amount of your affection

From a kanikau, or lament, for John Kealoha. Hawaiian source: Kaaikaula. English trans. by author.

Skin-numbing rain

6. Honi ana i ke anu i ka mea huʻihuʻi
   Huʻi hewa e ka ʻili i ka ua Pōʻaihala
   Lei ana i ka mokihana, i ka wewehi o Kaiona
   Līhau pue i ke anu hauʻoki o Kaleponi

   Taking in the cold and the chill
   Numbingly cold is the skin in the Pōʻaihala rain
   Bedecked in mokihana, in the finery of Kaiona
   Huddling, chilled to the bone in the frosty cold of California

   From the song "Wahine hele la o Kaiona," also known as "He inoa no Pauahi," for Bernice Pauahi Bishop by Leleiōhoku. Hawaiian source: Holstein 7. English trans. by author. Additional source: *I Mua e nā Pōkiʻi.*

Pōʻaihale. Same as *Pōʻaihala*. Rain associated with Kahaluʻu and northern Koʻolaupoko, Oʻahu. Also found on Hawaiʻi. "Pōʻai hale" means "to surround houses."

Rain of Māʻeliʻeli, Hawaiʻi

1. Aia i ka luna o Māʻeliʻeli
   Ka neneʻe a ka ua Pōʻaihale

   There atop Māʻeliʻeli
   Is the creeping of the Pōʻaihale rain

   From the chant "Piano ahiahi" by Pālea of Kaʻū, Hawaiʻi. Hawaiian source: Pukui and Korn 107. English trans. by author. *Note: Pukui and Korn describe the Pōʻaihale rain as a mist that creeps in and out between old thatched houses. Māʻeliʻeli, situated above Waiʻōhinu in Kaʻū, Hawaiʻi, is also the name of a hill along the Kahaluʻu/Heʻeia border in Koʻolaupoko, Oʻahu.*

Rain from ʻĀhulimanu to Kualoa Point, Oʻahu

2. Hiki ʻoe i ka ʻolu o Heʻeia Kea
   ʻIke ʻoe i ka ua Pōʻaihale o ʻĀhulimanu
   Ke nihi maila i ka lae o Kualoa

   You arrive in the coolness of Heʻeia Kea
   You see the Pōʻaihale rain of ʻĀhulimanu
   Creeping to the point of Kualoa

   From a mele inoa, or name chant, for Kalawaia. Hawaiian source: Paakai. English trans. by author.

Rain of Kahaluʻu, Oʻahu

3. Ka ua Pōʻaihale o Kahaluʻu. Rain that whisked about on all sides of a house at Kahaluʻu. Hiʻiaka sang the song while walking along Kāneʻohe Bay. The

"whisking rain" was also described as "the house-circling rain of Kahaluʻu." This phenomenon of course was caused by the wind.

From an entry in the winds section of Kent. Source: Kent 439. Additional source: Pukui, *ʻŌlelo* 173.

4. He Pōʻaihale ko Kahaluʻu. *Kahaluʻu has the Pōʻaihale.*

   From a chant in the Kamapuaʻa legend that mentions winds and rains around Oʻahu. Source: Akana 16, 13.

5. Aia kuʻu lei i Mailehuna
   Lei mamo ʻo Hilo i ka lehua
   Makemake au e inu i ka wai o lalo
   Lei Kahaluʻu i ka ua Pōʻaihale

   My cherished lei is at Mailehuna
   Hilo wears a yellow lei of lehua blossoms
   I desire to drink the waters from below
   Kahaluʻu is garlanded with the Pōʻaihale rain

   From the song "Nai aupuni." Hawaiian source: Holstein 166. English trans. by author.

6. Aloha ʻino nō kuʻu wahine
   . . . .
   Kuʻu wahine mai ka ua Pōʻaihale o Kahaluʻu

   What great pity for my darling wife
   . . . .
   My beloved wife from the Pōʻaihale rain of Kahaluʻu

   From a kanikau, or lament, for Luakauwahine. Hawaiian source: Pawai. English trans. by author.

7. He kanikau aloha kēia nou, e B. Mahune
   . . . .
   Luhi wale hoʻi ka ʻuhane, e Mahune, i ka loa ē
   He loa ka ua Pōʻaihale ma uka o Kahaluʻu

   This is an affectionate chant of mourning for you, B. Mahune
   . . . .
   The spirit of Mahune is very tired of the distance
   Distant is the Pōʻaihale rain in the uplands of Kahaluʻu

   Hawaiian source: Wailea. English trans. by author.

Rain of Puʻuniki, Oʻahu

8. E ʻeli au i kō aloha, e ka ua Pōʻaihale
   ʻO ia ua nihi aʻe ma Puʻuniki
   Kilihēhē i ka lau ʻākōlea o Palikea
   Ke pōʻai puni ala iā Kahaluʻu

I shall delve for your love, O Pōʻaihale rain
That rain that creeps along at Puʻuniki
Drenching the ʻākōlea leaves of Palikea
Encircling all of Kahaluʻu

From a mele by Lohiʻau as he was hiding from Hiʻiakaikapoliopele. Hawaiian source: Hoʻoulumāhiehie, *Ka Moʻolelo* 437. English trans.: Hoʻoulumāhiehie, *Epic* 407.

Pōʻailauʻawa. Rain associated with Moelana, Oʻahu. "Pōʻai lau ʻawa" means "to surround the ʻawa leaves."

1. *See Kanikoʻo (#1).*

Pōʻaipuni. Rain associated with Kemoʻo, Oʻahu, and with Kūmaka, Kauaʻi. "Pōʻai puni" means "to encircle."

Rain of Kemoʻo, Oʻahu

1. *See Līlīlehua (#25).*

Rain of Kūmaka, Kauaʻi

2. Ka ua Pōʻaipuni o Kūmaka. *The [Pōʻaipuni] rain of Kūmaka that completely surrounds.*

A traditional saying. Source: Pukui, *ʻŌlelo* 173. *Note: Pukui says, "The rain and mists of Kūmaka, Kauaʻi, completely screen homes, trees, and so forth from view."*

Pohākōʻeleʻele. Rain said to occur during the month of ʻIkuā. "Pohā kōʻeleʻele" means "cracking explosion," as of a storm.

1. I loko o ka malama ʻo ʻIkuā, ʻo Pohākōʻeleʻele ka ua. *During the month of ʻIkuā occurs the rain called Pohākōʻeleʻele.*

From "Na la a me na malama," an article on the days and months in the Hawaiian calendar, which specifies ʻIkuā as being in September and October. Source: Fornander, *Fornander* 5: 662–63.

Polohīnano. Same as *Polopuahīnano.* Rain associated with Puna, Hawaiʻi. "Polohīnano" means "white hīnano blossom with its stem."

Rain of Maukele, Hawaiʻi

1. Lohi ʻo Maukele, ʻanapa i ka lā o ʻOkiʻokiaho
He aho ia no ka ua Polohīnano

Maukele sparkles, shining under the sun of ʻOkiʻokiaho
It is a cord for the Polohīnano rain

From a kanikau, or lament, for Kamehameha IV. Hawaiian source: Moehonua and Lilikalani. English trans. by author.

Rain of Puna, Hawai'i

2. 'Ae; no Puna paia 'ala i ka ua Polohīnano ku'u pōki'i i hānau ai, akā na'e ho'i, na ka ua Kanilehua o Hilo 'o ia i hānai a nui. He milimili a he lei 'ā'ī 'o ia na Laika Nawailau i ka ua Pōpōlehua—ua lū lau lehua o Hilo ē!

   Yes, my younger sister was born in Puna of the fragrant walls in the Polohīnano rain. However, it was the Kanilehua rain of Hilo that raised her. She was a favorite, a beloved child of Laika Nawailau in the Pōpōlehua rain—rain that scatters the leaves of the 'ōhi'a of Hilo!

   From an obituary for Mary K. Pe'a. Hawaiian source: Spencer. English trans. by author.

Polopuahīnano. Same as *Polohīnano.* Rain associated with Hilo, Hawai'i. "Polo pua hīnano" means "white hīnano blossom with its stem."

1. *See Ho'olua (#3).*

Po'okole. Rain said to occur during the month of Welo.

1. I ka pau 'ana o ka mahi 'ai 'ana, waiho aku ka mea mahi 'ai 'o ka hā'ule mai o ka ua; i loko o ka malama 'o Welo 'o Po'okole ka ua, 'o ia ka ua, he ao kū iā kākou; a laila, hele aku ke kanaka mahi 'ai e 'ako i ka lau.

   After he has cut down the weeds he waits for the rain to fall; during the month of Welo, Po'okole (short head) is the rain, that is the rain which we style a cloudburst; then the farmer proceeds to crop off stalks.

   From "Na la a me na malama," an article on the days and months in the Hawaiian calendar, which specifies Welo as being in March and April. Source: Fornander, *Fornander* 5: 664–65. *Note: In the article, the Po'okole rain is called an "ao kū," which Fornander translates as "cloudburst." Pukui and Elbert, 1986, 27 describe "aokū" as a rain cloud or mist.*

Po'olipilipi. Same as *Ko'ilipilipi.* Rain associated with Hilo, Hawai'i, and with Kalihi, O'ahu. "Po'o lipilipi" means "adze-like head."

Rain of Hilo, Hawai'i

1. He ua nō ka ua Po'olipilipi i ka Unulau
   Ua mohala ka pua lehua i ke ke'eke'ehi 'ia e ka ua
   He ua ho'opumehana o ka 'āina o Hilo

   Indeed, the Po'olipilipi rain falls in the Unulau wind
   The lehua flower has blossomed in the treading of the rain
   A warming rain of the land of Hilo

   From a call to the chiefs and people of Hilo by Kūapāka'a, who was trying to awaken them to set sail. Hawaiian source: Nakuina, *Moolelo* 91. English trans. by author. Additional sources: Nakuina, *Moolelo* 89; Nakuina, *Wind* 86, 88.

2. 'O Hilo ua Po'olipilipi i ka Unulau
Ua mae ka pua o ka lehua
I ka hehihehi a ke kuāua

Hilo of the Po'olipilipi rain in the Unulau wind
The lehua blossom has withered
In the trampling of the shower

From a call to Keawenuia'umi and his Hawai'i Island chiefs by Kūapāka'a, who was trying to awaken them to set sail. Hawaiian source: Fornander, *Fornander* 5: 119. English trans. by author.

Rain of Kalihi, O'ahu

3. Ka Ua Po'olipilipi o Kalihi

I ke au kahiko, aia he kāne a me kekahi wahine, a ma muli o nā 'ao'ao mau o nā keiki a kānaka, ua hāpai mau maila ke kuko i loko o lāua a 'elua, a i ka hānau 'ana mai, he hewa; a i mea e kō ai ko lāua mau 'i'ini, ho'okahi wale nō hana pono, 'o ka holo i ka nahele e pe'e ai, 'oiai ua 'aihue malū lāua iā lāua iho. He 'oia'i'o nō, i kekahi lā, me ka 'ike 'ole mai o nā mākua o kēia kaikamahine, ua 'āha'i akula ka pupuhi iā lāua nei a noho i ka nahele o ke awāwa o Kalihi Uka, a ma loko o kekahi ana, ua pe'e ihola lāua ma laila no ka ho'opau 'ana i ke koena o nā wahi i koe o ke kīhāpai 'a'ole i kanu 'ia. I kēia manawa ho'i a lāua nei e nononoke nei, aia nā kulu paka ua 'elo'elo o ua uka lā ke holina ma'ū ihola iā lāua nei, akā, 'a'ohe a lāua nei nānā a'e 'o ia mea, noke nō.

I ka mao 'ana a'e o kēia kuakoko mua, 'ō'ili akula kekahi o lāua e nānā i ka ua inā paha ua mālie, akā, 'a'ohe wahi mea a mao a'e, e noke ana nō. No ka mao 'ole a'e o ka loku 'ana a ka ua, ua hā'ule akula lāua nei hiamoe. Hala he mau hola a i ko lāua ala 'ana a'e, e noke ana nō ka ua, hā'ule akula nō lāua nei hiamoe. Hala he mau lā, hala he mau pō o ka noke 'ana a ka ua, 'a'ohe wahi mea a mao iki 'o ka ua, a 'o kā lāua nei hana nō ka moe.

'O ka hele ia o ka noke 'ana o ka ua a li'u wale, a 'o kā lāua lā hana nō ka hiamoe. 'O ka hopena i loa'a iā lāua ma kēia moe 'ana ma muli o ia ho'okina a ka ua, i ko lāua ala 'ana a'e, aia kā ua 'ōpaha a lipilipi ko lāua mau po'o i ka ua mea 'o ka hiamoe i ka noke mau a ka ua i ka pō a me ke ao.

He ua ai nō ko Hilo, 'a'ohe na'e he hana 'ino i ke po'o o kānaka e like a'ela me ko Kalihi ka ho'olipilipi i ke po'o o ke kalohe, 'e'ole na'e ho'i lāua, ō 'ia ai ia ua akamai i ke kālai po'o—ka ua Po'olipilipi o Kalihi.

The Po'olipilipi Rain of Kalihi

In earlier times, there was a man and a woman, and because of the natural tendencies of offspring of the human race, lustful desires were conceived within the two of them, and out of those desires was born sin. In order to fulfill their yearning, there was only one thing to do—go to the forest and hide, so they secretly stole themselves away. It is true—one day, without the knowledge of the girl's parents, they ran away and disappeared and stayed in the forest of Kalihi Uka Valley. And they hid in a cave there until they could finish off the remaining areas of the garden that had not been planted. While they were carrying on, drenching raindrops of these uplands flowed down, wetly seeking them out, but they took no notice of this; they continued in their work.

When the first laborious exertion ended, one of them emerged to look at the rain, to check if it was calm. But it had not cleared up one bit; the rain was still going. Since the downpour had not ceased, they fell asleep. Several hours passed, and when they awoke, it was still raining, so they fell asleep again. Days and nights passed and the rain remained; it had not let up, not a single bit, so they just slept.

The rain continued to persist for such a long time, and all they could do was sleep. What resulted from their being forced by the rain to sleep was that when they awoke, their heads were flattened and tapered from sleeping while the rain persisted all night and all day.

Hilo also has a rain for lovemaking; however, it does not do harm to people's heads like Kalihi's rain, which sharpens the heads of naughty people. Because of these two people, the rain became known for its skill in shaping heads—the Po'olipilipi rain of Kalihi.

Hawaiian source: Poepoe, "Na mea kaulana." English trans. by author. Additional sources: Poepoe, "Ka-elemu-wai o Kalihi"; Sterling and Summers 326.

4. Ua Po'olipilipi—Kalihi, Honolulu, O'ahu; because it is always raining that the people get up to see if it is raining and lay down so often that one side of their heads gets flat and the other side round.

   From a list of rain names and their descriptions. Source: "Names . . . rain."

Po'onui. Numbingly cold rain said to occur during the month of Welehu. "Po'o nui" means "big head" and may refer to being troublesome, top-heavy, or a nuisance. Po'onui is both the name of a specific rain and a generally descriptive term; its various usages are determined by the context.

Rain of Welehu

1. Ma ia malama, ua anu, a ua uhi mai ka hau o nā kuahiwi. He manawa kēwai ia o ka mauna a me ka 'uhē. A 'o ka ua o ka ho'oilo ia wā ma ke 'ano pinepine; he ua puehu, 'o ia ka ua i kapa 'ia he ua Po'onui, no ke 'ōkakala o ke po'o a kūnāhihi ke kino i ke anu o ka po'e noho uka.

During this month, it is cold, and the mountain tops are covered in snow. It is a time when the mountains are wet and cold. The rain most frequent during this part of winter is a light, scattered rain. It is the rain that is called a Poʻonui rain for the chicken skin on the heads and the numbness of the bodies of the upland dwellers in the cold.

From a description of the month of Welehu, which Kepelino specifies as being December. Hawaiian source: Kepelino 97. English trans. by author. Additional source: Pukui and Elbert, 1986, 364.

Pōpōʻahaʻula. Rain associated with Kahana, Maui.

1. Anu Kahana i ka ua Pōpōʻahaʻula
Wena ʻula i ka ua Līlīlehua
Līlīlehua ʻoē nā pali
A lehua-lehua-lehua ʻula i ke kula
Kanu pōʻaha ʻula i ka nahele
I kuʻikuʻi a ʻeha ka lāʻau
Ke koko o ka ua ke kahe lā i ke pili

Cold is Kahana in the Pōpōʻahaʻula rain
Glowing red in the Līlīlehua rain
Līlīlehua, the cliffs murmur
Blossoms of lehua, lehua, lehua, red on the plains
Planting red clusters of flowers in the forest
The trees are hit and beaten until pained
The blood of the rain flowing onto the pili grass

From a mele inoa hoʻoilina, an inherited name chant, for the Kamehameha line of rulers. Hawaiian source: Haakulou. English trans. by author.

Pōpōkapa. Same as *Pōpōua*. Rain associated with Nuʻuanu, Oʻahu. Also found on other parts of Oʻahu and on Hawaiʻi. "Pōpō kapa" means "to bundle up kapa."

Rain of Hilo, Hawaiʻi

1. Kuʻu keiki mai ka ua Pōpōkapa o Hilo
E ʻāhaʻi ana e hūnā ma ka poli
ʻŌpili maʻule i ka ua a ka Waʻahia

My beloved child from the Pōpōkapa rain of Hilo
Fleeing, to hide in the bosom
Numb and weak in the rain of the Waʻahia

From a mele ʻuhane, or lament, for Keliikuewa. Hawaiian source: Namuo. English trans. by author.

Rain of ʻĪpuʻu, Hawaiʻi

2. He ʻuhane, he aloha no kuʻu keiki
ʻO ʻoe nō ia e Maikau e waiho maila ē
. . . .
Kuʻu keiki i ka ua Pōpōkapa o ʻĪpuʻu ē

A dirge, a loving remembrance for my dear child
It is you, Maikau, who lies there
. . . .
My beloved child in the Pōpōkapa rain of ʻĪpuʻu

From a kanikau, or lament, for M. E. Manu. Hawaiian source: Hoapili. English trans. by author.

Rain of Kalihi, Oʻahu

3. Kuʻu wahine mai ka ua Pōpōkapa o Kalihi
Ke ʻāhaʻi lā ma ke kua ka malu
Ke nounou maila e ka Waʻahila
Ka ua kāʻili wahine o kuʻu ʻāina
Huli aʻela Kaʻala kau i luna
Waiho wale kai o Pōkaʻī

My darling wife from the Pōpōkapa rain of Kalihi
Carried off on the back, where there is shelter
Pelted by the Waʻahila
The woman-snatching rain of my beloved land
Kaʻala, set on high, turns
The seaward land of Pōkaʻī is left plainly visible

From a mele by Halemano for his wife as he played the game of kilu. Hawaiian source: Fornander, *Fornander* 5: 253. English trans. by author.

Rain of Kekele, Oʻahu

4. *See Waʻahila (#22).*

Rain of Nuʻuanu, Oʻahu

5. Ka ua Pōpōkapa o Nuʻuanu. *The tapa-bundling [Pōpōkapa] rain of Nuʻuanu.*

A traditional saying. Source: Pukui, *ʻŌlelo* 173. *Note: Pukui says, "The Pōpōkapa rain is so called because anyone who came up Nuʻuanu Pali from the windward side had to bundle his garments and hold his arms against his chest to keep from getting wet."*

6. Ka ua Pōpōkapa of Nuʻuanu is a rain that blows through the Nuʻuanu Gap (Nuku o Nuʻuanu) down as far as Pauoa. It is called the "kapa bundling" rain because those facing it might keep their kapa dry by rolling it into a bundle and putting it behind their backs.

From a list of rain names and their descriptions. Source: Kelsey, "Rains."

7. Ua Pōpōkapa—rain that caused the natives coming from the town to bundle up their clothes.

   From a list of rain names and their descriptions. Source: "Names . . . rain."

8. Kuʻu kāne, e kuʻu kāne hoʻi
   Kuʻu kāne mai ka wā heu ʻole o māua
   Mai ka ua Pōpōkapa o Nuʻuanu

   My dear husband, oh, my dear husband
   My beloved husband from the time of our youth
   From the Pōpōkapa rain of Nuʻuanu

   From a kanikau, or lament, for Kamehameha IV by his wife, ʻEmalani Kaleleonālani. Hawaiian source: Nogelmeier 336. English trans. by author. *Note: Nogelmeier says that the Pōpōkapa is "a sudden rain of Nuʻuanu, Kapālama, and Kalihi, requiring drying kapa to be balled up quickly to save it from the rain."*

9. Nuʻuanu wai puna a ke aloha
   Kū kilakila a ka ua Pōpōkapa
   Pūkuʻi kāua lā i laila
   Kuʻu lei maʻo, ʻo ʻoe ia

   Nuʻuanu, spring water of love
   Majestic in the Pōpōkapa rain
   We nestle together there
   My lei of maʻo blossoms, it is you

   From the song "Nuuanu waipuna" by Major Kealakai. Hawaiian source: Holstein 64. English trans. by author. *Note: In this rendering, "wai puna" is used instead of "Waipuna," which was printed in the Hawaiian source.*

10. ʻO ka holu māewa mai a ke kāwelu o Lanihuli
    Hali māʻeʻele i ka ua Pōpōkapa o Nuʻuanu
    I ka neʻe mai a ka ua Waʻahila i ke pili
    I ka pili nō kā me ke kāne a hele
    Auē ka mea minamina ʻo ke kāne ē

    The fluttering sway of the kāwelu grass of Lanihuli
    Carried numb through the Pōpōkapa rain of Nuʻuanu
    As the Waʻahila rain moves over the pili grass
    In the bond with my husband who has gone
    How tragic, my husband

    From a kanikau, or lament, for Nehemia Hepa. Hawaiian source: Kapaalua. English trans. by author.

11. Kanikau, he aloha no T. Mapuhoe
Uē helu mai nei ʻo Kauinui
Kuʻu keiki mai ka ihu o nā moku
Hoʻolono ka ʻuhane i ke kani a ke oeoe o Kīlauea
Nānā ka ʻuhane i ka ua Pōpōkapa o Nuʻuanu
Honi hoʻomau i ke ʻala o ka hala o Kekele
Mai ke kāwelu holu i ka makani o Lanihuli
Aloha ia wahi a kāua i pili ai
Auē! Kuʻu lei aloha, he keiki ē

A loving lament for T. Mapuhoe
Kauinui is weeping and recounting your deeds
My beloved child from the prow of the ships
The spirit hears the sound of the whistle of the ship Kīlauea
The spirit looks upon the Pōpōkapa rain of Nuʻuanu
Constantly smelling the fragrance of the hala trees of Kekele
From the kāwelu grass that sways in the wind of Lanihuli
Love for this place where we clung together
Alas! My dear lei of affection, a child

From a kanikau, or lament, for Tonetakine Mapuhoe. Hawaiian source: Kauinui. English trans. by author.

12. Aloha ʻino nō kuʻu wahine
. . . .
Kuʻu wahine mai ka ihu o nā moku
Mai ka ua Pōpōkapa o Nuʻuanu

What great pity for my darling wife
. . . .
My beloved wife from the prow of the ships
From the Pōpōkapa rain of Nuʻuanu

From a kanikau, or lament, for Luakauwahine. Hawaiian source: Pawai. English trans. by author.

13. Kuʻu makuahine hoʻi
Kuʻu makuahine o ka ua Pōpōkapa o Nuʻuanu
E hahai ana ma ke kua o Waipuhia
Kuʻu hoa o ka ua me ka lā wela
O ke anu hoʻi me ke koʻekoʻe
Kuʻu makuahine i ka ua Līlīpua heʻe koko ʻula i ke kula ā
ʻO ʻoe maila ia; aloha ʻoe; e uē nō au ā

Oh, my beloved mother
My dear mother of the Pōpōkapa rain of Nuʻuanu
Following over the ridge of Waipuhia
My beloved companion of the rain and the hot sun
Of the cold and the damp
My dear mother in the fleeting rainbow-hued Līlīpua rain over the plains, oh
It is you; love to you; I cry indeed, oh

From a kānaenae, or chant of eulogy, for H. Kamahiai. Hawaiian source: Makuaole. English trans. by author.

14. *For another example of Pōpōkapa of Nuʻuanu, see Waʻahila (#24).*

Other examples

15. He kanikau aloha nui kēia nou, e S. Lono
Kuʻu keiki o nā pali hāuliuli o ka ʻāina
Mai ka leo o ke kai wawalo i nā pali
Mai ka ua Pōpōkapa o ka ʻāina

This is a chant of mourning with much affection for you, S. Lono
My beloved child of the dark, verdant cliffs of the land
From the sound of the sea booming over the cliffs
From the Pōpōkapa rain of the land

From a kanikau, or lament, for S. Lono. Hawaiian source: Wala. English trans. by author.

16. *For another example of Pōpōkapa, see Lanipili (#11).*

Pōpōlehua. Rain associated with Hilo, Hawaiʻi. "Pōpō lehua" means "cluster of lehua flowers."

1. *See Polohīnano (#2).*

Pōpōua. Same as *Pōpōkapa*. Rain associated with Kawaikini, Kauaʻi. "Pōpō ua" means "rain bundle," as in the bundling up of something to protect it from the rain.

1. Kūnihi ka mauna i ka laʻi ē
ʻO Waiʻaleʻale lā i Wailua
Huki i luna ka Pōpōua o Kawaikini

Steep is the mountain in the calm
Waiʻaleʻale at Wailua
The Pōpōua rain of Kawaikini lifts

From a mele by Hiʻiakaikapoliopele outside of Malaehaʻakoa's house. Hawaiian source: Emerson, *Pele* 109. English trans. by author. Additional sources: Emerson, *Unwritten* 40; Hoʻoulumāhiehie, *Ka Moʻolelo* 178; Hoʻoulumāhiehie, *Epic* 167. *Note: The additional sources listed here contain other versions of this mele.*

Puakaiaulu. Also the name of a wind.

1. Rain name. Probably same as Kaiaulu.

Source: Pukui and Elbert, 1986, 346. *Note: Pukui and Elbert, 1986, 115 say that the Kaiaulu is "a pleasant, gentle trade wind breeze, famous in song, at Wai'anae, O'ahu."*

Puakukui. Rain associated with Moeawakea, Hawai'i. "Pua kukui" means "kukui flower."

1. Aloha ko'u hoa i ka ua Puakukui
Kui lehua o Moeawakea
Lei pua o Kalāhuipua

Love to my companion in the Puakukui rain
Stringing the lehua of Moeawakea
Flower lei of Kalāhuipua

From a mele by Hi'iakaikapoliopele warning Lohi'au to be on his guard against enemies. Hawaiian source: Emerson, *Pele* 213. English trans. by author.

Puanaea. Same as *Puanaiea*, *Puananaiea*, and *Puaneiea*. Rain associated with Pālolo, O'ahu.

1. He mau wāhine noho i ka ua Puanaea, 'o Ku'ialauahi me Huewa. *Ku'ialauahi and Huewa are women who dwell in the Puanaea rain.*

From a letter to the newspaper *Kuokoa* by Samuel Mānaiakalani Kamakau. Hawaiian source: Kamakau, "Nani Halo." English trans. by author.

Puanaiea. Same as *Puanaea*, *Puananaiea*, and *Puaneiea*. Rain associated with Pālolo, O'ahu. Also found at Waipuhia, O'ahu. "Puanaiea" means "feeble, sickly."

Rain of Makahuna, O'ahu

1. 'O 'oe kai Makahuna i Ku'ialauahi
E uhai ana i ka ua Puanaiea
I noho malihini aku i Palehala
Iō māua, hala i Huewa

That is you at Makahuna at Ku'ialauahi
Chasing the Puanaiea rain
Living in Palehala as a guest
Among us, passing Huewa

From a kanikau, or lament, for Abenera Kahekili. Hawaiian source: "He kanikau no Abenera Kahekili." English trans. by author. *Note: "Huewa" is the name of a pool in Pālolo.*

Rain of Palehala, Oʻahu

2. Huli aku nānā i ka ua Puanaiea
Ke hāliʻi maila lā i Palehala ē

Turn and see the Puanaiea rain
Covering Palehala

From a kanikau, or lament, for Kamehameha IV by one of his sisters. Hawaiian source: "He kanikau no ka Moi Iolani Kamehameha IV." English trans. by author.

Rain of Pālolo, Oʻahu

3. *See Kūkalahale (#11).*

Rain of Waipuhia, Oʻahu

4. Kuʻu kaikaina aloha i hala nei
I hele hoʻoao me ka ua Puanaiea
Me kuʻu hoa ua Waʻahila
Aia lā i luna o Waipuhia
Anu māʻeʻele i ke anu a ka makani

My beloved younger sibling who has passed away
Who passed the night with the Puanaiea rain
And my dear friend the Waʻahila rain
There above Waipuhia
A numbing cold in the chill of the wind

From a kanikau, or lament, for L. Keoua. Hawaiian source: Amamalua. English trans. by author.

Puananaiea. Same as *Puanaea*, *Puanaiea*, and *Puaneiea*. Rain associated with Alakaʻi, Kauaʻi.

1. ʻO ʻoe kā ia, e nā lehua i Alakaʻi
Ke pūhene ʻia maila e ka manu
He nui hoʻi na ka ua Puananaiea
He punahele na ka Lawelawemālie
I lāhui nō i ka uka o Kawaineki

It is you, O lehua at Alakaʻi
Teased by the birds
A darling of the Puananaiea rain
A favorite of the Lawelawemālie wind
Gathered together in the upland of Kawaineki

From a mele inoa, or name chant, by Keauka praising the child Kahelekūlani in the legend of Kamaakamahiʻai. Hawaiian source: Kaualilinoe, "Ka moolelo" 10/1/1870. English trans. by author. Additional source: Kaualilinoe, "Legend" 60–61.

Puaneiea. Same as *Puanaea*, *Puanaiea*, and *Puananaiea*. Rain associated with the *ʻEʻelekoa* rain.

1. *See ʻEʻelekoa (#4).*

Pūhenelehua. Rain associated with Makakolohe, Hawaiʻi, and with the *Kīhenelehua* rain. "Pūhene lehua" means "basket of lehua flowers."

1. *See Kīhenelehua (#2).*

Pūnāwaiea. Rain name. *Lit.*, spray (of) rain.

Source: Pukui and Elbert, 1986, 355.

Puni. Heavy rain of short duration; associated with Waiʻōhinu, Hawaiʻi.

1. *See Hāʻao (#7).*

pūnohu. Also known as pūnohu ʻula. Rising mist or smoke; low-lying rainbow; rain red in the sunshine. Also a pillar of cloud.

1. Kū ka pūnohu ʻula i ka moana
Hele ke ehu kai, uhi i ka ʻāina
ʻŌlapa ka uila, noho i Kahiki
ʻUʻina, nākolo
ʻUwā ka pihe
Lau kanaka ka hula
E Laka ē

The pūnohu ʻula mist rises over the ocean
The sea spray moves, covering the land
The lightning flashes, residing in Kahiki
Crackling, roaring
Shouting is the noise
A hula of four hundred people
O Laka

Chanted by hula dancers after having dressed in their costume and are entering the hālau. Hawaiian source: Emerson, *Unwritten* 55. English trans. by author.

2. I ke kāʻalo ʻana mai ma ʻaneʻi o Laeʻahi, aia hoʻi, ua kāhiko ʻia ʻo ia me nā hae o nā waihoʻoluʻu like ʻole, mai ke keʻa ihu a pinana i luna o nā kia a iho ma hope, me he lā he pūnohu ʻōnohi e neʻe ana i loko o ka ua kuakualau, ka ua mea he pipiʻo lua i ka lewa. I ia manawa nō i ʻanapu aʻe ai ka lono me he uila lā, "Ua holo ke Kuʻikahi Pānaʻi Like!"

As [the ship] passed this side of Laeʻahi, behold, it was adorned with flags of various colors, from the cross at the prow and climbing atop the mast and draping behind, like a low-lying pūnohu rainbow moving in a kuakualau shower, like a double rainbow in the sky. That was when the news flashed like lightning: "The Reciprocity Treaty was passed!"

From an article announcing the final passage by the U.S. Senate of the law to effectuate the treaty, news of which was brought to Hawaiʻi by the letter-carrying steamship *City of San Francisco*. Hawaiian source: "Huro!" English trans. by author.

3. *For more examples of pūnohu, see Kanilehua (#42), Kualau (#16), and Leikokoʻula (#9).*

Pūʻololehua. Rain associated with Mākaulele (or Mōkaulele), Hawaiʻi. "Pūʻolo lehua" means "bundle of lehua flowers."

1. Nani wale ia pua e hehia maila e ke anu
E loku ʻia maila e ka ua i ka nahele
Hele a luhi maʻule ia e noho nei
Eia lā i ka iwi hilo ke aloha e loku nei
Auē kuʻu kāne ē, kuʻu kāne mai ka ua Pūʻololehua
ʻO nā lehua o Mākaulele i kui paukū ʻia me ka hala
I lei heʻe nalu no kuʻu kāne i ke kai ʻo Huia

Beautiful is this blossom, trodden upon by the cold
Inundated by the rain in the forest
Sitting there, becoming weak and fatigued
Here in the very depths of the chest is love that pours forth
Oh, my dear husband, my man from the Pūʻololehua rain
The lehua of Mākaulele strung in sections with hala
A surfing lei for my husband in the sea of Huia

From a kanikau, or lament, for S. P. Koko. Hawaiian source: Honolulu, "He kanikau." English trans. by author.

Pupūhale. Rain famous at Hāmākua, Hawaiʻi. *Lit.*, (rain) remaining (near) house.

Source: Pukui and Elbert, 1986, 357.

pūʻukiʻuki. Mountain rain. "Pūʻukiʻuki" means "difficult."

1. Kuʻu hoa i ka ua pūʻukiʻuki anu o ka mauna
Ua hoʻouka ʻia mai au e ka leo
E pili ana ke ʻala me ke onaona
ʻĀ! ʻO ia onaona lā ua hiehie ē

My dear companion in the cold pūʻukiʻuki rain of the mountains
I am overcome by the voice

The fragrance joins the perfume
Ah! The scent is magnificent

From a mele inoa, or name chant, for Albert Ka Haku O Hawaiʻi by his father, Kamehameha IV. Hawaiian source: Iolani. English trans. by author.

2. Kuʻu hoa o ka ua pūʻukiʻuki o ka mauna. *My companion of the cold, unpleasant [pūʻukiʻuki] rain of the mountain.*

   From a chant. Source: Pukui and Elbert, 1986, 359.

3. *For another example of pūʻukiʻuki of the mountain, see Noenoe (#1).*

Puʻukoa. Rain associated with Kokomo, Maui. Also the name of a wind and of an ʻili in Paʻuwela, Maui. "Puʻu koa" means "hill of koa trees."

Rain of Kokomo, Maui

1. Ka ua Puʻukoa i Kokomo. *The Puʻukoa rain is at Kokomo.*

   From a mele concerning Maui and Molokaʻi winds by Kūapākaʻa. Hawaiian source: Kuapuu, "He wahi moolelo" 5/15/1861. English trans.: Nakuina, *Wind* 63. Additional sources: Kanepuu, "Kaahele"; Nakuina, *Moolelo* 68; Fornander, *Fornander* 5: 101. *Note: Since this is part of a wind chant, "Puʻukoa" may refer to a wind accompanied by rain.*

2. Kuʻu makua kāne mai ka malu hālau loa a ke kukui
   Mai ka moena hāunu ʻole o ka nahele ke noho ʻia
   A kāua e ʻalo ai i ka ua Puʻukoa o Kokomo
   Ua komo kō aloha, koni i ka puʻuwai
   Ke konikoni nei i kuʻu manaʻo

   My dear father from the shade of the long shelter of kukui trees
   From the mat of forest greenery that is to sit upon
   That we would experience together in the Puʻukoa rain of Kokomo
   Your love has entered, a fluttering in my heart
   Beating in my mind

   From a kanikau, or lament, for Edwin Miner. Hawaiian source: Miner. English trans. by author.

Rain associated with kuikui trees

3. Newa mālie i ka ua Puʻukoa ke oho o ke kuikui
   Maʻū ka panepoʻo ua Makawiliwili
   I nā kōʻai hoʻowali a ka ua Makahekili

   The leaves of the kuikui tree stagger slowly in the Puʻukoa rain
   The back of the head is wet from the Makawiliwili rain
   By the winding and mixing of the Makahekili rain

From a mele inoa, or name chant, for Heneri W. Auld. Hawaiian source: Hoopai.
English trans. by author.

Pu'ukolu. Rain associated with Haua'iliki, Kaua'i. Also the name of a wind. "Pu'u kolu" means "three hills."

1. 'Ike iā Wai'ale'ale, he anu Alaka'i
He nahele lipo na ka noe o Maunahina
Nani ka 'ō'ō a ka ua Pu'ukolu
Eia a'ela i Haua'iliki

Behold Wai'ale'ale, Alaka'i is chill
A forest made lush by the mists of Maunahina
Beautiful is the 'ō'ō bird and the Pu'ukolu rain
Here before us at Haua'iliki

From a mele kālai'āina, or political chant, for 'Emalani Kaleleonālani.
Source: Nogelmeier 200.

T

Tēhau. Same as *Kēhau*. Dew, mist; associated with Kaua'i and found on other islands.

1. *See Nāulu (#43).*

Tiu. Same as *Kiu*. Rain associated with Maluaka, Kaua'i. Also the name of a wind.

1. I lohia Wailua e ka maita'i
   I lohia ka wai hālau me ka tai
   I lohi lima 'ia ka nani o Maluaka
   Milimilika'a 'ia e ka Wai'ōpua
   I hamohamo lima 'ia wai na ke Tiu

   Wailua is sparked by beauty
   The expansive water and sea sparkle
   The beauty of Maluaka is ignited by the hand
   Caressed by the Wai'ōpua wind
   Waters soothed by the hands of the Tiu

   From a mele about Kaua'i, said to be composed by Kapa'akea. The chant is a mele inoa, or name chant, a mele 'āina, or chant for the land, and a mele hula, or hula chant. Hawaiian source: "Maikai Kauai." English trans. by author.

Tuahine. Same as *Kuahine*. Rain associated with Mānoa, Oʻahu.

1. Ke ʻai nei ʻo Kalani
I ka ʻōpae māhikihiki
Kiʻina ʻia e Fancy Lauliʻi
O ka ua Tuahine o Mānoa

The royal one is dining now
On the māhikihiki shrimp
Caught by Miss Fancy Dainty
From the Tuahine rains of Mānoa

From the song "He ʻai na ka lani." Source: Liliʻuokalani 119. Additional source: Holstein 24.

2. Kui ʻia kō lei
I koʻolua no ia uka
Ia uka ʻiuʻiu
Lanakoi a ka makemake
Hoʻopē ʻia
E ka ua Tuahine

Your lei will be strung
As a companion for these uplands
These lofty uplands
The desire is so keen
Drenched
By the Tuahine rain

From the song "Sunday Manoa" by Mrs. Eliza Holt. Hawaiian source: Holstein 139. English trans. by author.

3. ʻO ʻoe a ʻo wau i laila
I ka rain Tuahine o Mānoa
ʻO ka nihi a ka ua i Waʻahila
Kāhiko maila i Mānoa
Ua pono ʻoe, e ka ua tilihune
I ka hone a ka wai o Pualoke

You and I there
In the Tuahine rain of Mānoa
The soft creeping of the rain at Waʻahila
Adorning Mānoa
You are needed, O fine, delicate tilihune rain
In the sweet waters of Pualoke

From the song "Ka uka i Waahila" by Joe and Eliza. Hawaiian source: Holstein 154. English trans. by author.

4. He beauty i ka ua Tuahine o Mānoa
Pā aheahe ka makani
I ke ʻala o ka lauaʻe
Ke onaona o nā pua
Māhiehie i ka wao

A beauty is the Tuahine rain of Mānoa
Engulfed in the breeze
Bringing the sweet scent of the lauaʻe
And the fragrance of the flowers
Elegance in the forest

From the song "Ka beauty aʻo Mānoa" by Tony Conjugacion. Source: Wilcox et al. 90.

5. Mānoa is divided into two sections—the aliʻi or chief side by Waiʻoli Tea Room, and the kanaka or commoner side by Safeway in east Mānoa. The Tuahine rain moves faster than the Luahine rain and travels through the aliʻi side, stopping at the Cooke mansion. Next to the mansion is Kūkaʻōʻō Heiau, and under it is a cave where a brother and sister took refuge. They broke the ʻai kapu, or eating law, when they ate together, and thus the sister was turned into the Tuahine rain. Tuahine or Kuahine means "sister of a male." The brother was turned into the wind Kahaukāne, which pushes the Tuahine along.

From an email sent to the author by Kimo Alama Keaulana on April 4, 2013. *Note: The mansion was built in 1911 by Charles Montague Cooke Jr. and his wife, Lila Lefferts Cooke. The mansion, called Kūaliʻi, and Kūkaʻōʻō Heiau are listed on the National Register of Historic Places.*

U

Uaakalīpoa. Same as *Līpoa.*

1. Name of a fine, cold rain. *Lit.*, rain by the līpoa seaweed.

   Source: Pukui and Elbert, 1986, 361.

Uhiwai. Heavy fog; mist that is heavier than the noe, ʻohu, ʻehu, and ʻehuehu. Associated with Mānā, Hawaiʻi, and with Mānoa, Oʻahu, and found in other areas. "Uhi wai" means "water covering." It is both the name of a specific rain and a generally descriptive term; its various usages are determined by the context.

Mist of Mānā, Hawaiʻi

1. Mai kekahi hoa aloha mai o Waimea, Hawaiʻi, ʻo ke kākau ʻana mai i ka lā 10 nei, i loaʻa mai ai ka hōʻike ma hope iho no ke kūlana o ko Kamāliʻi Kaʻiulani maʻi. Penei mai nei kēia hōʻike:

   "I ka Pōʻaono, Feb. 4 ihola, ma luna o ka ʻauamo ʻāʻī, ua haʻalele akula ke kamāliʻi wahine ʻōpio i ka Uhiwai o Mānā no Waimea, no ka maʻi rumatika ʻehaʻeha loa, no ke ake o Kauka Walaka e loaʻa kahi mehana."

   It is from a friend of Waimea, Hawaiʻi, who wrote on the 10th that we receive the following news on the status of Princess Kaʻiulani's illness. This is what was reported:

"On Saturday, Feb. 4, carried on the shoulders, the young princess left the Uhiwai of Mānā for Waimea. Due to the extremely painful rheumatic disease, Doctor Walters wanted her to be in a warm place."

From a report on Kaʻiulani's illness. Hawaiian source: "Ke omaimai." English trans. by author. *Note: "Kauka Walaka" refers to Dr. St. David G. Walters.*

Mist of ʻĪao, Maui

2. Nae iki ʻĪao i ka uhiwai. *(Mount) ʻĪao is barely breathing in the heavy mist.*

   Source: Pukui and Elbert, 1986, 364. *Note: Pukui and Elbert say that the phrase refers to "one in dire distress."*

Mist of Mānoa, Oʻahu

3. Paʻa mai Mānoa i ka uhiwai
   Haʻaheo i ka uka lā o Kupanihi

   Mānoa is steadfast in the uhiwai
   Revered for the uplands of Kupanihi

   From the song "Hawaii i ka ehuehu" by Home Kauwila. Hawaiian source: *Buke Mele* 73. English trans. by author.

4. Kilakila Mānoa i ka uhiwai
   Haʻaheo i ka pua lā o ka lama

   Mānoa is majestic in the uhiwai
   Revered for its lama blossoms

   From the song "Aloalo ehuehu poka" by S. K. Kaloa. Hawaiian source: *Buke Mele* 16. English trans. by author.

5. *For another example of uhiwai of Mānoa, see Kuahine (#11).*

Fog of Kaʻala, Kauaʻi

6. *See Leikokoʻula (#9).*

ʻŪkiu. Same as *ʻŪkiukiu*. Rain associated with Makawao, Maui, and found on other parts of Maui. Also the name of a wind.

Rain of Kalena, Maui

1. Aia ka laʻi i Piʻiholo lā
   I ka ua ʻŪkiu o Kalena lā
   Ke nihi maila i nā pali lā
   I ka malu kukui o Kapela lā

   The calm is there at Piʻiholo
   In the ʻŪkiu rains of Kalena
   Sweeping along over the cliff tops
   And through the kukui shade of Kapela

From a mele inoa, or name chant, for Lele. Hawaiian source: Ku. English trans.: "Pā ka leo" 21.

Rain of Kula, Maui

2. *See Nāulu (#19).*

Rain of Liliko'i, Maui

3. Uhi ka 'ohu kūpouli, ne'e i ka piko kaulana o Haleakalā
He lā ho'okē 'ai kā nā ao Nāulu e ho'oulukū nei
'O kā ke kilokilo ia a ka 'Ūkiu e nihi hālo'ilo'i a'ela i loko o Liliko'i

The dazed 'ohu cloud engulfs, heading toward the famous summit of Haleakalā
A day of fasting for the agitated Nāulu shower clouds
This was foretold by the 'Ūkiu rain that, welling with tears, softly moves about in Liliko'i

From "Ka leo kaukau a nā moku o Kama," a voice of lament for Joseph Nāwahīokalani-'ōpu'u by people of the lands of Kama. Hawaiian source: Sheldon, *Ka Puke* xxi. English trans. by author. Additional source: Sheldon, *Biography* 015.

Rain of Makawao, Maui

4. Ua 'Ūkiu o Makawao—He ua li'ili'i me ka makani pū mai ke kuahiwi mai.

'Ūkiu rain of Makawao—It is a fine rain with wind that blows down from the mountains.

From an account of East Maui rains and their descriptions, as told to Thomas K. Maunupau by Josua Ahulii of Kaupō, Maui. Source: Maunupau, *Huakai* 27, 108.

5. Ma muli o ko Kauakahiapaoa akamai i ka hula a me kona u'i kanaka maika'i nō ho'i, ua lilo 'o ia i mea hiala'ai nui 'ia e nā po'e o ka 'ao'ao palupalu. A ma ka wā a ua po'e nei o ia pō'ai e ho'olale aku ai i ka ua 'Ūkiu o Makawao i mua o Kauakahiapaoa nei, ua hō'ole maila 'o ia.

With Kauakahiapaoa's skill in hula and his fine manly form, he was much admired by the gentler sex. When those of that circle flocked to him as the 'Ūkiu rain presses against Makawao, he would rebuff them.

From the legend of Hi'iakaikapoliopele. Hawaiian source: Ho'oulumāhiehie, *Ka Mo'olelo* 399–400. English trans.: Ho'oulumāhiehie, *Epic* 373.

6. Ku'u hoa o ka malu kuawa o Wailuku
Mai ke Kā'aha'aha la'i o Niua
'O ia kula a kākou e haele ai
Pulu i ka ua 'Ūkiu o Makawao ē

My dear companion of the shelter of Wailuku Valley
From the calm Kā'aha'aha wind of Niua
This plain where we would go
Soaked by the 'Ūkiu rain of Makawao

From a kanikau, or lament, for Kamehameha IV. Hawaiian source: B. Pauahi. English trans. by author.

7. Ka ua ʻŪkiu o Makawao. *The ʻŪkiu rain of Makawao.*

A traditional saying. Source: Pukui, *ʻŌlelo* 173.

8. Kuʻu kāne mai ka ua ʻŪkiu o Makawao
Ka ua Pehihala o Hāmākua
Kuʻu hoa hele o ke ala o Puʻuiki
Kanikau lā, he aloha nou, e Kamakaokalani

My beloved husband from the ʻŪkiu rain of Makawao
The Pehihala rain of Hāmākua
My dear traveling companion of the Puʻuiki trail
A lament, an expression of love for you, Kamakaokalani

Hawaiian source: Wiliama. English trans. by author.

Rain of Piʻiholo, Maui

9. He puʻepuʻe ke komo i ka nahele o Kokomo
E kuene ana i ke anu o Waiʻaleʻale
I pau pono Piʻiholo i ka ua a ka ʻŪkiu
Ua hele a ʻaeʻae me he koloa ʻau wai lā

Forcing through the entrance of the forest of Kokomo
Sustaining in the chill of Waiʻaleʻale
Piʻiholo is encompassed by the rain of the ʻŪkiu
Becoming smooth like a koloa duck swimming in water

From the song "Maikaʻi Kaʻuiki." Hawaiian source: Pukui, *Nā Mele Welo* 166. English trans. by author.

Rain mentioned with the word "hoʻolale" (to entice, incite, or suggest)

10. Pēlā ka hoʻoheihei ʻana a ua mau wāhine nei iā Lohiʻau a hiki maoli ai i kona wā i paulehia maoli ai iā lāua, a hoʻokahi wale nō mea nāna i kāohi mai i ka hoʻolale ʻana a ka ua ʻŪkiu, ʻo ia ka paʻa ʻia o ke kānāwai "Kaiʻokia" a ka uʻi o Puna i kau aku ai ma luna ona; akā, ʻo ka ʻono a me ka makemake, ke momoni wale ala nō ia mau mea i loko o kona puʻu.

Thus the two women flirted with Lohiʻau until he was fully enthralled by them, and only one thing offered restraint against their enticements, which coursed like the ʻŪkiu rain, and that was the imposition of the "Kaiʻokia" law, which the beauty of Puna had set upon him; the craving and the desire, however, were so tangible he could feel them in his throat.

From the legend of Hiʻiakaikapoliopele. Hawaiian source: Hoʻoulumāhiehie, *Ka Moʻolelo* 13. English trans.: Hoʻoulumāhiehie, *Epic* 12.

11. ʻAʻole anei ʻoe i hoʻomanaʻo i kahi wā o kou ola ʻana, i kou wā hoʻi e lohe ai i kekahi paeaea (olioli) a kekahi kāne a wahine maikaʻi o kāna hoʻāeae ʻana, ua ala mai nō nā hahaliʻa manaʻo kohu hoʻoheno ʻana na ka ua ʻŪkiu, ka mea i kaulana ma ke mele hoʻolale a ka ua ʻŪkiu?

    Don't you recall a time in your life when you heard a chant by some man or woman whose fine vocal prowess stirred delicious thoughts, like the tender ministrations of the ʻŪkiu rain, famed in song for its inciting qualities?

    From the legend of Hiʻiakaikapoliopele. Hawaiian source: Hoʻoulumāhiehie, *Ka Moʻolelo* 127–28. English trans.: Hoʻoulumāhiehie, *Epic* 120.

12. Hoʻolale a ka ua ʻŪkiu. *A suggestion of the ʻŪkiu rain.*

    A traditional saying. Source: Pukui, *ʻŌlelo* 117. *Note: Pukui says, "Go ahead and do what was suggested. The ʻŪkiu rain is cold enough to make one hurry and scurry."*

ʻŪkiukiu. Same as *ʻŪkiu*. Rain associated with Makawao, Maui. Also found on other parts of Maui and on Oʻahu, Kauaʻi, and Kaʻula. Also the name of a wind.

Rain of Kalena, Maui

1. ʻAʻole nō ʻoe e pakele
   I kahi wai huʻihuʻi o Kalena
   Ia wai huʻi a ʻumeʻume
   Me ka rain ʻŪkiukiu

   You shall not escape
   The tingling waters of Kalena
   These cool, alluring waters
   And the ʻŪkiukiu rain

   From the song "Wai hui o Kalena." Hawaiian source: Holstein 167. English trans. by author. *Note: In this rendering, "a ʻumeʻume" is used instead of "aumeume," which was printed in the Hawaiian source.*

Rain of Līhau, Maui

2. Laʻi ihola ka luna o Līhau
   I ka holo mau ʻia e ka ua ʻŪkiukiu
   He ua hoʻopiha i ke alo o nā kahawai

   The summit of Līhau is calm
   From the constant flow of the ʻŪkiukiu rain
   A rain that fills the banks of the streams

   From a mele ʻāina, or song of the land, for Maui. Hawaiian source: "Nani Lahaina." English trans. by author. Additional sources: Kanahele 39–40, 54, and 57; Kualu.

Rain of Liliko'i, Maui

3. Lamalama pua i ka wao kele
Ua ma'ema'e wale i ka lokua
E ka ua 'Ūkiukiu
O Liliko'i, o Liliko'i

Bright flowers growing by shaded nooks
By passing showers are fresher made
By the rain 'Ūkiukiu
Of Liliko'i, of Liliko'i

From the song "Liliko'i." Source: Lili'uokalani 234.

Rain of Makawao, Maui

4. Aloha 'o Makawao
I ka ua 'Ūkiukiu
He tiu na ka Nāulu
I ke tula o Kama'oma'o

Love for Makawao
In the 'Ūkiukiu rain
A sign of the Nāulu
On the plain of Kama'oma'o

From the song "'Auhea wale ana 'oe" by Nakulula. Hawaiian source: Pukui, *Nā Mele Welo* 188. English trans. by author. Additional sources: Kanahele 37–38; Reichel, *Kamahiwa*.

5. Me ka na'au i piha i ke aloha a me ke kaumaha mākou e komo pū aku nei i ke kūmākena pū 'ana no Hon. Iosepa Kaho'oluhi Nāwahīokalani'ōpu'u. I uē 'ia mai nei 'oe e ka ua 'Ūkiukiu o Makawao a me ka ua 'Ulalena o Pi'iholo āu i kipa malihini mai ai.

With hearts full of love and grief we join together in mourning with you all for Hon. Joseph Kaho'oluhi Nāwahīokalani'ōpu'u. You are wept for by the 'Ūkiukiu rain of Makawao and the 'Ulalena rain of Pi'iholo, lands which you have visited.

From a letter of sympathy written to Mrs. Emma 'A'ima Nāwahīokalani'ōpu'u and her children after the passing of Joseph Nāwahīokalani'ōpu'u. Hawaiian source: Sheldon, *Ka Puke* 213. English trans.: Sheldon, *Biography* 293.

6. 'Ohi e ka 'i'o o ka lā'au
No Makawao nō ia
'O ka ua 'Ūkiukiu
'Ohu'ohu nō
E aho nō e komo mai
I ka malu ma ke ao

The meat of the trees is gathered
It is from Makawao
The ʻŪkiukiu rain
Truly adorns and enhances
It is best to come
Into the shade during the day

From the song "Malu i ke ao." Hawaiian source: Holstein 146. English trans. by author. *Note: The first two lines refer to a Makawao custom of eating a type of tree fungus (de Silva).*

7. Kuʻu wahine mai ka ua ʻŪkiukiu o Makawao
Mai ka ua ʻUlalena o Piʻiholo

My beloved wife from the ʻŪkiukiu rain of Makawao
From the ʻUlalena rain of Piʻiholo

From a kanikau, or lament, for Luakauwahine. Hawaiian source: Pawai. English trans. by author.

8. *For another example of ʻŪkiukiu of Makawao, see ʻUlalena (#19).*

Rain of Māmala, Oʻahu

9. I ko Kaumualiʻi ʻike ʻana mai iā Kamehameha, e uluwela aku ana nā ʻau waʻa kaulua i ka ʻahu ʻula, me he ānuenue lā ka pipiʻo helohelo i ka ʻŪkiukiu.

When Kaumualiʻi saw Kamehameha, the double-hulled canoe fleet was crowded and colorful with feather capes, appearing like a red-arched rainbow in the ʻŪkiukiu rain.

From the story of Kamehameha I. Hawaiian source: Kamakau, "Ka moolelo o Kamehameha I" 8/10/1867. English trans. by author. Additional source: Kamakau, *Ke Kumu Aupuni* 176. *Note: In this rendering, "uluwela" and "helohelo" are used in place of "ula weli" and "hekoheko," which were printed in the Hawaiian source.*

Rain of Hikilei, Kauaʻi

10. *See Kapakapaʻahoa (#1).*

Rain of Kaʻula

11. Aʻiaʻi e lewa ka hua manu o Kaʻula
Ke hahai lā i ka manu walewale nui
I nā kulu ʻūpē a ka ua ʻŪkiukiu

Bright, the egg of Kaʻula floats
Following the bird that produces excessive phlegm
In the mucus drops of the ʻŪkiukiu rain

From the chant "Ike ia Limunui he aina ua." Hawaiian source: *Na Mele Aimoku* 282. English trans. by author.

Rain that drenches

12. 'O ka 'iwi leo nahenahe
    Inu wai o ka lehua
    I pulu 'elo i ka noe
    A ka ua 'Ūkiukiu

    The sweet-voiced 'iwi bird
    That drinks the lehua nectar
    Soaked by the mist
    Of the 'Ūkiukiu rain

    From the song "Nani pua o ke kiele." Hawaiian source: Holstein 84. English trans. by author.

13. 'Auhea wale ana 'oe
    E ka liko o ka 'ōhelo
    Ua pulu pē au
    I ka ua 'Ūkiukiu

    Where are you
    O bud of the 'ōhelo
    I am thoroughly drenched
    In the 'Ūkiukiu rain

    From the song "Ua lihi [*sic*] paha oi ala." Hawaiian source: Holstein 22. English trans. by author.

'Ula. Same as *Kaua'ula*. Rain found on Hawai'i, Maui, and O'ahu. Also the name of a wind. "'Ula" means "red."

Rain of Waimea, Hawai'i

1. *See Kīpu'upu'u (#32).*

Rain of Kanahā, Maui

2. Me he palahēhē maila ka uka o Kanahā
   Ke kīpalahe'a 'ula i ka liko o ka 'ōhi'a
   Ua lulumi wale ke au i ka ua 'Ula

   Like a running sore is the upland of Kanahā
   The reddish smear of the new 'ōhi'a leaf shoots
   The current in the 'Ula rain is overwhelming

   From a mele kālai'āina, or political chant, for 'Emalani Kaleleonālani. Source: Nogelmeier 190. *Note: Nogelmeier says that the 'Ula rain is a "famous wind and rain of Lāhainā, Maui, also known as the Kaua'ula."*

Rain of Lele (or Lāhainā), Maui

3. *See Nahua (#1).*

Rain of Wahiawā, O'ahu

4. 'O māua kai ka ua 'Ula o Wahiawā
He ho'olu'u moelua na ke Ki'oao
Ke ho'olu'u maila i uka o Kahui
I 'ale'ale a piha a hanini

We two in the 'Ula rain of Wahiawā
Striped dye of the Ki'oao rain
Coloring the uplands of Kahui
Which is filled and full and overflowing

From a chant originally composed for Lunalilo and inherited by Kalākaua. This portion of the mele was composed by Kalaiwahakapaa. Hawaiian source: *Na Mele Aimoku* 147. English trans. by author.

Another name for Koko rain

5. *See Koko (#3).*

'Ulalena. Same as *'Ulalenalena*. Probably related to *Lena*. Rain associated with Liliko'i and Pi'iholo, Maui. Also found on other parts of Maui and on Kaho'olawe, O'ahu, and Kaua'i. Also the name of a hill in Hāmākualoa, Maui. "'Ula lena" means "yellowish-red."

Rain of Maui

1. 'O Kaulua, pohā ka 'Ulalena, puka ka Huhui, 'o ia ka wā makani. *Kaulua, when the 'Ulalena gushes, the Pleiades appears; that is the windy period.*

From a Maui calendar describing the month of Kaulua. Hawaiian source: Manuokekula. English trans. by author. Additional source: Johnson 174.

2. Aloha ku'u hoa o ka ua 'Ulalena
Ke hāpai 'ia maila e ka Ho'olua, 'eā
'Elua kāua i ke anu a ke Ki'owao

Love for my companion of the 'Ulalena rain
Being lifted by the Ho'olua wind
The two of us in the cold of the Ki'owao rain

From a kanikau, or lament, for L. Kahuakai. Hawaiian source: Kaauwaepaa. English trans. by author.

Rain of Ha'ikū, Maui

3. E hoholo ana ka 'o'opu a Kai'aiki ē
He i'a 'ike ho'onowelo i ka pohu

He kupu ia mai ka poli o Waiheʻe
E holo ana i ka pō i Haʻikū
E pue ana i ka ua ʻUlalena
Ka hauwawā mai o Lilikoʻi
Ka ʻapaʻapa a ka wai i Kamaʻomaʻo

The ʻoʻopu fish of Kaiʻaiki run
A fish that knows where to find calm
A spawn from the bosom of Waiheʻe
Traveling at night to Haʻikū
Huddling in the ʻUlalena rain
The echo of Lilikoʻi
The lingering waters at Kamaʻomaʻo

From the chant "Pukaka na lehua." Hawaiian source: Kanahele 63. English trans. by author. *Note: In this rendering, "Kaiʻaiki" is used instead of "ka iʻa iki," and "Ka ʻapaʻapa" instead of "Kaʻapaʻapa," which were printed in the Hawaiian source.*

Rain of Haʻikū, Maui, and of Kaʻala, Oʻahu

4. A reddish-hued rain associated with Haʻikū, Maui, and Mt. Kaʻala, Oʻahu. Also a wind at Piʻiholo, Maui. Kapu ka luna o Kaʻala i ka ua ʻUlalena. *The uplands of Kaʻala Mountain are sacred with the red-yellow [ʻUlalena] rain.*

   From a chant. Source: Pukui and Elbert, 1986, 367. Additional source: Kalohelani.

Rain of Hāliʻimaile, Maui

5. I ke ao ʻana ua hele akula ʻo Kihaapiʻilani i ke noi lau ʻuala ma Hāmākuapoko, a ma Hāliʻimaile.... Hele maila ka lā ma ke kua, a ʻo ka ʻŪkiukiu ma ke alo, hana maila ka ua ʻUlalena i kāna hana, a ua hāliʻi ihola a ʻenaʻena i ka lau o ka ʻūlei.

   The next day, Kihaapiʻilani went to Hāmākuapoko and Hāliʻimaile to ask for ʻuala slips.... The sun came at his back, and the ʻŪkiukiu wind at his front. The ʻUlalena rain did its work and covered the leaves of the ʻūlei in a red glow.

   From the story of Kihaapiʻilani, who planted a large field of ʻuala slips because of a famine in Kula and Makawao. Hawaiian source: Kamakau, "Ka moolelo Hawaii" 12/8/1870. English trans. by author. Additional source: Kamakau, *Ruling* 24.

Rain of Hāmākualoa, Maui

6. He ʻeleʻele ko ke kukui noho malu
   He lena hoʻi ka lau o kekahi kukui
   ʻO ke kukui aku i wāhi i ka lā
   I ka ua ʻia e ka ua ʻUlalena

The kukui nuts in the shade are black
Some of the kukui leaves are yellow
Other kukui are cracked in the sun
Rained upon by the ʻUlalena rain

From a mele by Keaulumoku expressing his love for the Hāmākua districts of Maui, where he had lived with Kamehamehanuiʻailūʻau and Kahekili. Hawaiian source: Kamakau, "Ka moolelo o Kamehameha I" 2/23/1867. English trans. by author. Additional sources: Kamakau, *Ke Kumu Aupuni* 66; Kamakau, *Ruling* 115.

Rain associated with Noe of Kalena, Maui

7. *See Noe (#7).*

Rain of Līhau, Maui

8. Kāhiko ka nani i Līhau
I ka hale hīnano a nā manu
. . . .
Kāohi ke aloha peʻe ma loko
Pulu i ka ua ʻUlalena
I walea ia uka ʻiuʻiu
Lamalama nuʻa o ka palai

Beauty adorns Līhau
And the house of hīnano blossoms built by the birds
. . . .
Love withholds and hides within
Soaked by the ʻUlalena rain
That makes this lofty upland enjoyable
Aglow in thick clumps of palai ferns

From the song "Pili palua." Hawaiian source: Holstein 95. English trans. by author.

Rain of Lilikoʻi, Maui

9. Ua lono au i ka moe
E hoʻi ana au me ia lā
Me ka ua ʻUlalena
Kilihune o Lilikoʻi

I heed the dream
I will return with her
With the ʻUlalena rain
Light rain of Lilikoʻi

From the song "Nani pua o ke kiele." Hawaiian source: Holstein 84. English trans. by author.

10. Kuʻu hoa pili i ka ua ʻUlalena
He ua ʻUlalenalena Lilikoʻi

My intimate companion in the ʻUlalena rain
An ʻUlalenalena rain of Lilikoʻi

From a kanikau, or lament, for Batesepa Puhia, the wife of Davida Malo. Hawaiian source: Malo, "Eia." English trans. by author.

11. Ahu kupanaha a ka lā i ke kula o Kamaʻomaʻo
Nui ka hā Lilikoʻi i ka ua ʻUlalena

Marvelous is the sun on the plain of Kamaʻomaʻo
The water trough of Lilikoʻi swells in the ʻUlalena rain

From the song "Maikaʻi Kaʻuiki." Hawaiian source: Pukui, *Nā Mele Welo* 166. English trans. by author.

12. *For more examples of ʻUlalena of Lilikoʻi, see Līlīlehua (#11) and Nāulu (#19).*

Rain of Lilikoʻi and Piʻiholo, Maui

13. Kuleana Lilikoʻi ka ua ʻUlalena
He ua ʻUlalena anu no Piʻiholo
Holo aʻe ka ua, lena ke oho o ke kī
Lena kāhuli, ka lau o ke kukui
I hoʻoluʻu ʻia e ka ua Kiu o Makawao

The ʻUlalena rain belongs in Lilikoʻi
A cold ʻUlalena rain of Piʻiholo
The rain moves on, yellow are the leaves of the kī plant
Turning yellow are the leaves of the kukui tree
Dyed by the Kiu rain of Makawao

From a mele kālaiʻāina, or political chant, for ʻEmalani Kaleleonālani, credited to Nāhaolelua. Hawaiian source: Nogelmeier 211. English trans. by author.

14. I ka manawa i ʻike maopopo loa ʻia aʻe ai ua hāpai ʻo Hina, ua noho aku lāua ma ia ʻano a kokoke i nā lā nahunahu hānau keiki, ua ʻike ʻia nā ʻōuli o ka lani a me nā ao polohiwa e haʻakokōhi ana i loko o ka lewa nuʻu, a ke hoʻokūkū maila nā ʻōpua i ka ʻili o ke kai, a ke ʻōkūkū maila nā ʻale kūpikipikiʻō o ka moana, a ke neʻe paʻa aʻela ka ʻohu e uhi i nā awāwa a me ka piko o nā mauna, a ke ʻaneʻane maila ʻo Kūlanihākoʻi e hoʻohanini iho i kona wai lani, a ʻike ʻia akula ka lalapa o ka uila i ka lewa, a ke lohe ʻia akula nā leo pohāpohā nunulu o ka hekili e nākolo ana i ka honua, aia hoʻi nā ānuenue ke haʻaheo maila i loko o nā kilihune ua ʻawa, ka ua koko hoʻi ke haʻaheo maila ma nā awāwa i luna o nā lau lāʻau, "me he ua ʻUlalena lā no Piʻiholo e kīheʻaheʻa ana i kukui o Lilikoʻi."

When it was known for certain that Hina was pregnant, they lived the way they had before until the pangs of childbirth drew near; the heavenly portents were seen, as were the dark, threatening clouds in the atmosphere; the billowy clouds were piled on the surface of the sea; the turbulent waves of the ocean swelled and stood upright; the ʻohu clouds moved to mantle the valleys and the mountain peaks; Kūlanihākoʻi nearly poured out its heavenly water; lightning flashes were seen in the sky; the roaring, bursting sounds of thunder were heard rumbling the earth. Rainbows proudly displayed themselves in the drizzling ʻawa rain, with low-lying koko rainbows stretching proudly along the valleys, above the leaves of the trees, "like an ʻUlalena rain from Piʻiholo that stains red the kukui trees of Lilikoʻi."

From the story of the birth of Kahānaiakeakua, son of Kū and Hina, in the legend of Keaomelemele. Hawaiian source: Manu, *Keaomelemele* 8. English trans. by author.

Rain of Piʻiholo, Maui

15. ʻAuhea wale ana ʻoe
E ta ua ʻUlalena
Kāhito maila i uka
I ka nani o Piʻiholo

Where are you
O ʻUlalena rain
That adorns the uplands
In the beauty of Piʻiholo

From a mele inoa, or name chant, for Makaituaʻana by Nakulula. Hawaiian source: Pukui, *Nā Mele Welo* 188. English trans. by author. Additional sources: Kanahele 37–38; Reichel, *Kamahiwa*.

16. Alia ʻoe lā, e Haleakalā
E ālai nei iā Kaʻuiki
Ua maikaʻi ke alo o Piʻiholo
I ka noho a ka ua ʻUlalena
Haʻina ke aliʻi nona ka lei
ʻO Kapiʻolani i ka ʻiu o luna

Wait there, Haleakalā
Before you hide Kaʻuiki from view
Beautiful is the face of Piʻiholo
For there dwells the ʻUlalena rain
This is in praise of the chiefess whose lei chant this is
Kapiʻolani, so high above

Source: "Aia i Lihau."

17. Ka ua ʻUlalena o Piʻiholo. *The reddish-yellow [ʻUlalena] rain of Piʻiholo.*

A traditional saying. Source: Pukui, *ʻŌlelo* 173.

18. A he Kiu ka makani o ka ʻāina
Hoʻohae ana i ka Nāulu
Kāhiko i ka luna o Piʻiholo
Haʻaheo i ka ʻUlalena

A Kiu is the wind of the land
Teasing the Nāulu rain cloud
Adorning the heights of Piʻiholo
Proud in the ʻUlalena

From the song "Wai hui o Kalena." Hawaiian source: Holstein 167. English trans. by author.

19. Kuʻu makua kāne mai ka ua ʻUlalena hāliʻi mai i ke alo o Piʻiholo
Mai ka ua ʻŪkiukiu o Makawao

My dear father from the ʻUlalena rain that spreads over the face of Piʻiholo
From the ʻŪkiukiu rain of Makawao

From a kanikau, or lament, for Edwin Miner. Hawaiian source: Miner. English trans. by author.

20. *For another example of ʻUlalena of Piʻiholo, see ʻŪkiukiu (#5).*

Rain of Wailuku, Maui

21. Pau ʻole koʻu mahalo i ka laulā o Kamaʻomaʻo
Ka hālana maikaʻi a Keālia
Ka hemolele o ka ua ʻUlalena
Lena ka pua o ka māmane pala luhiehu i ka lā

My admiration is endless for the expanse of Kamaʻomaʻo
The fine rising of the waters of Keālia
The perfection of the ʻUlalena rain
Yellow are the blossoms of the māmane, soft and lovely in the sun

From a mele mākaʻikaʻi, or travel chant, for ʻEmalani Kaleleonālani by Kaleipaʻihala. Source: Nogelmeier 143.

Rain of Kahoʻolawe

22. Kū haili lani kapu o Kanaloa
ʻUlalena ka ua ke nihi nei
Kukū ka ʻale o ka makani
Holo lua, holo pili, holo kake

The sacred, heavenly spirit of Kanaloa rises
'Ulalena is the rain creeping over the land
The gust of wind strikes
Blowing in two directions, blowing close, blowing back and forth

From the chant "Kamohio Puna o Kaho'olawe." Hawaiian source: Kanahele 140. English trans. by author. *Note: In this rendering, "haili lani" is used instead of "ha'ili-lani," which was printed in the Hawaiian source. "Kanaloa" is another name for Kaho'o-lawe.*

Rain of Kahaukomo, O'ahu

23. I ko lāua ulele 'ana iho i ko lāua mau akamai, unuhi kai o Poke'o i ke keiki o Kaua'i—me he ānuenue lā ka hau o Kahaukomo, ka pipi'o i ka ua 'Ulalena, ka nōeo 'ula i ka ua Ki'owao i nā pali, ka holu pi'o i ke kāwelu o Lanihuli, ke kukū o ka lau o ka 'āhihi o Malailua, i ka welelau o ka makani 'o ka holo—'a'ohe wahi pa'a 'ole i ke keiki o Kaua'i.

    When they leaped at each other with all their skills, it was as if the long Poke'o prayer was recited perfectly by the boy of Kaua'i—the hau trees of Kahaukomo were like a rainbow arching in the 'Ulalena rain, the bright red brilliance in the Ki'owao rain on the cliffs, pliable and supple among the kāwelu grass of Lanihuli, standing firm like the leaves of the 'āhihi bush of Malailua, running at the tip of the wind—there was nothing unknown to the boy of Kaua'i.

    From a story about Nāmaka, a skilled lua fighter of Kaua'i during the time of Kalani-'ōpu'u. Hawaiian source: Kamakau, "Ka moolelo o Kamehameha I" 2/23/1867. English trans. by author. Additional sources: Kamakau, *Ke Kumu Aupuni* 64; Kamakau, *Ruling* 111.

Rain of Lanihuli, O'ahu

24. Ku'u wahine i ka ua 'Ulalena
    Kīhae pua kāmakahala
    Noho anaina o ke alanui
    Kānaenae pū'ā weuweu i Lanihuli

    My darling wife in the 'Ulalena rain
    That tears the kāmakahala blossoms
    Sitting in a group along the road
    Offering a humble clump of greenery at Lanihuli

    From the chant "Ku'u wahine i ka ua 'Ulalena." Hawaiian source: Kanahele 103. English trans. by author. Additional source: K. *Note: In this rendering, "kāmakahala" is used instead of "Ka makahala," and "pū'ā" instead of "pua," which were printed in the Hawaiian source.*

Rain of Maʻemaʻe, Oʻahu

25. Aloha ʻino, aloha wale
Ua ʻo Maʻemaʻemakahoukekula
I ke komoa e ka ua ʻUlalena
I ʻenaʻena nō i ka lā o Kaimukī
I nononoʻula i ke kai o Lēʻahi ē

What a misfortune, very sad
Maʻemaʻemakahoukekula
Penetrated by the ʻUlalena rain
Glowing red in the sun of Kaimukī
A red redness by the sea of Lēʻahi

From a kanikau, or lament, for Kahaukomo. Hawaiian source: Kahiamoe. English trans. by author.

Rain of Alakaʻi, Kauaʻi

26. Kuʻu hoa o ka ua ʻUlalena
O ka ua loku mai i ka nahele
Hāliʻi maila i ke pili
Pulu pē i ka Noe o Alakaʻi

My beloved companion of the ʻUlalena rain
Of the rain that pours down upon the forest
Spreading over the pili grass
Soaked with the Noe of Alakaʻi

From the song "Pua koolau lei o Kaiulani" by Kapoli. Hawaiian source: Holstein 58. English trans. by author.

Rain associated with the Leikokoʻula rain

27. *See Leikokoʻula (#10).*

ʻUlalenalena. Same as *ʻUlalena.* Probably related to *Lena.* Rain associated with Lilikoʻi, Maui. "ʻUla lenalena" means "yellowish-red."

1. *See ʻUlalena (#10).*

Uli. Rain associated with Haleleʻa, Kauaʻi. "Uli" means "dark."

1. He ua Uli ko Haleleʻa. *Haleleʻa has a dark [Uli] rain.*

From a makena, or lament, for ʻEmalani Kaleleonālani. Source: Nogelmeier 355.

Uluau. Rain associated with Hilo, Hawaiʻi. Also the name of a wind.

1. Ka Uluau comes with the Māluakiʻiwai wind.

   From a list of rain names and their descriptions. Source: Kelsey, "Rains." *Note: Pukui and Elbert, 1986, 234 say that the Māluakiʻiwai is a "sea breeze accompanied by showers, known at Hilo."*

2. ʻO Welehu ka malama
   ʻO ka malama ia o Hilo e hanini ai
   E haki a nalu ai ka ua i ka lani
   E popoʻi ai ka Uluau me ka Mālua

   Welehu is the month
   This is the month that Hilo overflows
   The rain breaks and forms waves in the sky
   The Uluau and the Mālua break in waves

   From a kanikau, or lament, for Kamehameha IV. Hawaiian source: Kualehelehe. English trans. by author.

Ulumano. Rain associated with Hilo, Hawaiʻi. Also the name of a wind.

Rain of Hilo, Hawaiʻi

1. Ka ua Ulumano of Hilo comes with the wind toward the land. It brings the ʻōhua fish, which the women catch.

   From a list of rain names and their descriptions. Source: Kelsey, "Rains."

Rain of Hilo Hanakahi, Hawaiʻi

2. *See Kanilehua (#32).*

Rain of Panaʻewa, Hawaiʻi

3. *See Kanilehua (#37).*

W

Wa‘ahia. Same as *Wa‘ahila*. Rain associated with Kona, O‘ahu. Also found on Hawai‘i.

Rain of Hilo, Hawai‘i

1. *See Pōpōkapa (#1).*

Rain of Kaimukī, O‘ahu

2. Kau ‘ōpua Lē‘ahi, oni me he ao lā
   Me he hālō nō kā ia la‘i lua ala i ka la‘i
   La‘i ua pā ka makani Hauālia
   Hamoa e ka Wa‘ahia, nani Kaimukī

   Lē‘ahi is placed upon the billows, appearing like a cloud
   Like a peering cloud in the calmest of calms
   In the calm, the Hauālia wind blows
   Caressed by the Wa‘ahia rain, Kaimukī is beautiful

   From a mele inoa, or name chant, for Keahi. Hawaiian source: Lohiau, "He inoa." English trans. by author.

Rain of Kona, O‘ahu

3. He pua ka lani, he pua laha ‘ole nei no nā moku
   He kamaha‘o ka lani na O‘ahu

I walea ka lani i Kona, i ka lulu
I ka pohu wale o ka ua Waʻahia
Ke hāliʻi maila i ke pili

The chief is a flower, a rare blossom of the islands
Magnificent is the chief of Oʻahu
The chief relaxes at Kona in the calm
In the soothing serenity of the Waʻahia rain
Covering the pili grass

From a kanikau, or lament, for Kahahana. Hawaiian source: Fornander, *Fornander* 6: 293. English trans. by author. *Note: Fornander,* Fornander *6: 301 says that a calm often attends a Waʻahia shower.*

Rain of Mānoa, Oʻahu

4. *See Waʻahila (#12).*

Waʻahila. Same as *Waʻahia*. Rain associated with Nuʻuanu, Oʻahu, and found on other parts of Oʻahu. Also the name of a wind and of a ridge between Mānoa and Pālolo.

Rain of Hālawa, Oʻahu

1. No laila, ʻo mākou o ka Ahahui Hooikaika Kristiano holoʻokoʻa o ka ua Waʻahila o Hālawa, ʻEwa, ma o ko mākou kōmike lā, ke komo pū aku nei e kaʻana pū i nā ʻīnea o kēia mau lā ʻehaʻeha me nā mākua i hoʻonele ʻia i ka lei ʻole, ka ʻohana a me nā pilikana me ke kau nui aku i maluhia mai ko kākou puʻuhonua a me ka ikaika mai.

   Therefore, we, on behalf of the entire Ahahui Hooikaika Kristiano of the Waʻahila rain of Hālawa, ʻEwa, through our committee, join in sharing the hardships of these tragic days with the parents, family, and relatives who have been deprived of their children, with hopes for peace and strength from our refuge.

   From a message of condolence from members of the Christian Endeavor Society. Hawaiian source: Moses et al. English trans. by author. *Note: "Puʻuhonua" or "refuge" probably refers to Jesus Christ.*

2. Nani Hālawa i ka ua Waʻahila
   Ke kīpū maila i luna o ʻAiea

   Hālawa is beautiful in the Waʻahila rain
   Remaining above ʻAiea

   From George M. K. Aekai o Kuloloia's response to a nane, or riddle, printed in the newspaper *Kuokoa Home Rula*. Hawaiian source: Aekai o Kuloloia. English trans. by author.

Rain of Halemano, Oʻahu

3. *See Pōpōkapa (#3).*

Rain of Kona, Oʻahu

4. ʻO ka wahine ʻalo ua Waʻahila o Kona
Nihi makani ʻalo ua Kūkalahale
Noho ānea kula wela lā o Pahua
Wahine holo ua Hāʻao Nuʻuanu ē, ia

The woman who resists the Waʻahila rain of Kona
Creeping softly like the wind, resisting the Kūkalahale rain
A bleak existence along the hot plains of Pahua
Woman traveling in Nuʻuanu's Hāʻao rain

From a kanikau, or lament, for Kaʻahumanu. Hawaiian source: Malo, "He kanikau." English trans. by author.

5. Eia au lā i ka lipo
I walea au i ke kui pua
Lililo ai māua i ka nahele
He kamaʻāina nō ka ua Waʻahila
ʻUhane peʻepeʻe ua Kūkalahale ʻoe

Here I am in the dense forest
Pleasantly stringing flowers
We become lost in the forest
The Waʻahila rain is our guide
You are the spirit that hides in the Kūkalahale rain

From a kanikau, or lament, for Abenera Kahekili. Hawaiian source: "He kanikau no Abenera Kahekili." English trans. by author.

Rain of the Koʻolau Mountains, Oʻahu

6. *See Kiʻowao (#17).*

Rain of Kou, Oʻahu

7. Ola ke awa o Kou i ka ua Waʻahila. *Life comes to the harbor of Kou because of the Waʻahila rain.*

A traditional saying. Source: Pukui, *ʻŌlelo* 272. *Note: Pukui says, "It is the rain of Nuʻuanu that gives water to Kou (now central Honolulu)."*

8. Ola ke kai o Kou i ka ua Waʻahila. *The land of Kou [Honolulu] lives by the Waʻahila rain.*

A traditional saying. Source: Pukui and Elbert, 1986, 375; brackets in the original.

Rain of Lanihuli, Oʻahu

9. Ai lā i luna ʻo Waipuhia
   Me aʻu āholehole i Lanihuli
   Ua holo a nui i ka ua Waʻahila

   There above is Waipuhia
   With my āholehole at Lanihuli
   Which swim in great numbers in the Waʻahila rain

   From a mele aloha, or song of love. Source: Fornander, *Fornander* 6: 544. *Note: Fornander says that Waʻahila is the "name of the rain on the tops of the mountains."*

Rain of Malailua, Oʻahu

10. Lei akula i ka hala o Kekele
    I nā hala moe ipo o Malailua
    Ua māewa wale i ke oho o ke kāwelu
    Nā lei kāmakahala o ka ua Waʻahila

    Wear the pandanus of Kekele
    And the sweetheart-pairing pandanus of Malailua
    In the kāwelu grass sway
    The kāmakahala wreaths of the Waʻahila rain

    From the song "Makalapua." Source: Wilcox et al. 164. Additional sources: Holstein 117; Elbert and Mahoe 76.

Rain of Mānoa, Oʻahu

11. Puʻipuʻi ka ua Waʻahila o Mānoa. *Stocky the Waʻahila rain of Mānoa.*

    From a song. Source: Pukui and Elbert, 1986, 350.

12. Kuʻu hoa o ka ua Waʻahila lā
    ʻIke wale i ka ʻāina
    ʻIke wale i ke one i ka ʻiliʻili
    A me ka lā laʻilaʻi
    A me ke kou o Molakia
    . . . .
    Auē kuʻu hoa, kuʻu hoa i ka makani he Aʻeloa o Honolulu
    Auē kuʻu hoa i ka ua Waʻahia o Mānoa

    My beloved companion of the Waʻahila rain
    See the land
    See the sand, the small pebbles
    And the calm, clear day
    And the kou trees of Molakia
    . . . .

Oh, my dear friend, my friend in the Aʻeloa wind of Honolulu
Oh, my beloved companion in the Waʻahia rain of Mānoa

From a farewell song for Nāhiʻenaʻena by Kini. Hawaiian source: Fornander, *Fornander* 6: 441–42. English trans. by author.

13. Auē kuʻu hānai
Kuʻu kaikūnane hoʻi
Mai ka ua Waʻahila o Mānoa
Ke hāliʻi maila lā i ke pili

Oh, my beloved hānai
My dear brother indeed
From the Waʻahila rain of Mānoa
Covering the pili grass

From a kanikau, or lament, for Kamehameha IV by one of his sisters. Hawaiian source: "He kanikau no ka Moi Iolani Kamehameha IV." English trans. by author.

14. The Waʻahila rain is found in the area of the University of Hawaiʻi at Mānoa, especially where the Kamakakūokalani Center for Hawaiian Studies building stands. Kauawaʻahila (the Waʻahila rain) and Kauakiʻowao (the Kiʻowao rain) were brother and sister, respectively.

From an email sent to the author by Kimo Alama Keaulana on April 4, 2013.

Rain of Nuʻuanu, Oʻahu

15. Waʻahila is said to have been the name of a favorite hula of Peleʻula, so called after a princess who, with her brother Kamanuwai, excelled in the performance of this dance. Her name has been perpetuated in an old saying that has come down to us: Ka ua Waʻahila o Nuʻuanu. This is a gentle rain that extends only as far down Nuʻuanu Valley as Wyllie or Judd Street.

From a note by Nathaniel B. Emerson regarding the legend of Hiʻiakaikapoliopele. Source: Emerson, *Pele* 170.

16. A ke waiho nei ʻo Luʻanuʻu i ka ua Waʻahila o Nuʻuanu. *Luʻanuʻu lies in the Waʻahila rains of Nuʻuanu.*

Hawaiian source: Kamakau, "Ka moolelo o Hawaii nei" 7/22/1865. English trans.: Kamakau, *Tales* 29.

17. Ua Waʻahila—Nuʻuanu; the rain that goes around the hills but does not reach the center of the valley.

From a list of rain names and their descriptions. Source: "Names . . . rain."

18. Ke [hoʻihoʻi] lā ka makani i ka wai a kau i luna
Kau i ka pali ʻo Kōnāhuanui
Nui ka ua Waʻahila papani i ke ala

Papani ʻokoʻa i ka wā o Nuʻuanu
Keʻa i ka hele pali o Malailua lā ē

The wind sends the rain back upward
Placing it atop the peak of Kōnāhuanui
The heavy Waʻahila rain blocks the trail
Completely blocking the gap of Nuʻuanu
Hindering travel along the cliff of Malailua

From an article titled "Haule aha ana" on why a prominent person in Hawaiʻi society converted to Catholicism. Hawaiian source: Ii, "Na hunahuna" 4/10/1869. English trans. by author.

19. Kāhiko i Nuʻuanu ka ua Waʻahila. *Adorned is Nuʻuanu by the Waʻahila rain.*

A traditional saying. Source: Pukui, *ʻŌlelo* 143. *Note: Pukui says, "The Waʻahila rain makes Nuʻuanu grow green and beautiful."*

20. Ani Nuʻuanu i ka ua Waʻahila
Walea ʻole iho nō Lanihuli
Huli ka makani
Naku i ke oho o ke kāwelu

Nuʻuanu waves in the Waʻahila rain
Lanihuli is not at ease
The wind turns
Trampling the blades of the kāwelu grass

From a mele for Kauilanuimakehaikalani Kamaialii. Hawaiian source: Kamakau, "He mau mele koihonua." English trans. by author.

21. Ke oho ʻāhihi lehua o Nuʻuanu
ʻO ke kī kāwelu a ka Waʻahila

Tresses like ʻāhihi lehua from Nuʻuanu
The flow of loose strands like the Waʻahila rains

From a mele hōʻihi, or honorific chant, praising the hair of ʻEmalani Kaleleonālani. Source: Nogelmeier 53.

22. Kuʻu hoa mai ka ua Waʻahila o Nuʻuanu
Mai ka ua Pōpōkapa o Kekele

My dear companion from the Waʻahila rain of Nuʻuanu
From the Pōpōkapa rain of Kekele

From a kanikau, or lament, for J. Henry. Hawaiian source: Kahinawe, "He kanikau." English trans. by author.

23. Nani ka hanu o Nuʻuanu i ka makani
I ka hoʻohanuhanu a ka ua Waʻahila

Splendid is the scent of Nuʻuanu upon the wind
The panting breath of the Waʻahila rain

From a mele by Kaumealani describing lovemaking. Hawaiian source: Hoʻoulumāhiehie, *Ka Moʻolelo* 151. English trans.: Hoʻoulumāhiehie, *Epic* 142.

24. Kuʻu kāne hoʻi
Kuʻu kāne ʻalo ua Waʻahila ē
I ka ua Pōpōkapa o Nuʻuanu
Ua anu au a māʻeʻele i kō aloha
I ka hele a ka hoʻopili, a ke kāne
Ma hea hoʻi ʻoe i hele iho nei
Auē kuʻu kāne ē

Oh, my beloved husband
My dear husband who endured the Waʻahila rains, oh
And the Pōpōkapa rain of Nuʻuanu
I am cold, numbed by your love
At the departure of the companion, of the husband
Where, oh where, have you gone
Alas! My beloved husband, oh

From an uē, or wailing call of love and grief, for Kamehameha IV by his wife, ʻEmalani Kaleleonālani. Source: Nogelmeier 330.

25. *For another example of Waʻahila of Nuʻuanu, see Pōpōkapa (#10).*

Rain of Waʻahila Ridge, Oʻahu

26. E *Ka Nupepa Kuokoa,* Honolulu, Oʻahu—Iaʻu ma Honolulu, ua lohe ihola au i kekahi moʻolelo kupua maikaʻi e pili ana iā Kaumana, he kupua kēia i lilo i pōhaku, ma ke kualapa ma waena o Pālolo a me Mānoa.

Ua kākau ʻia ka moʻolelo e aʻu ma ka ʻōlelo ʻEnelani, akā he mea pono nō e paʻa ma ka ʻōlelo Hawaiʻi.

ʻO koʻu hoa aloha ʻo Mr. Stokes kai hele aku e ʻike iā Mr. Solomon Kauai, ka mea nāna i hōʻike mai i kēia moʻolelo, aia kona wahi noho ma ka ʻaoʻao ma uka o ke alanui Waiʻalae, ma ka hui ʻana o ke alanui ʻo Kapahulu, e holo lā i Waikīkī mai kekahi wahi mai, e pili kokoke lā me ka pau ʻana o ke kualapa Kaumana. . . .

Inā paha no ka hiki iā ʻoe ke huli aku i ka moʻolelo ʻoiaʻiʻo, e lilo ana ia i mea hoʻohauʻoli mai i koʻu manaʻo, a ke hoʻopuka aku ʻoe ma loko o ka nūpepa, a laila e hoʻouna mai i kope naʻu. Penei iho ka moʻolelo e like me ia i loaʻa mai ai iaʻu. . . .

Ua ʻoi aku ke aloha o Kaumana i kāna keiki muli loa ma mua o nā mea ʻē aʻe a pau, no laila, hoʻomaka akula ʻo ia e ʻau no kēlā kapa o ka loko no ka hoʻākāka ʻana aku i kāna wahine i ka mea āna i hana ai, me ke kūmākena ʻana no kāna keiki.

Ma ka pō ʻana iho, ua pā maila kekahi makani ikaika, kai koʻo maila nō hoʻi ka moana, a i ke ao ʻana aʻe ma kekahi lā mai, aia hoʻi, ʻaʻole ka loko ma kona wahi i waiho ai, akā he one ke waiho mai ana, a lilo i alanui maikaʻi e hele ai.

ʻAʻole naʻe kēia i lilo i mea e hoʻohauʻoli ʻia aku ai ko Kaumana manaʻo, akā ua ʻoi loa aʻe kona minamina no kāna keiki, no laila hoʻomaka akula ʻo ia e pepehi i kona poʻe kānaka i hele mai ai mai Lāhainā mai, me ka pepehi pū ʻana i kāna wahine, a me kona mau mākua, a koe kāna mau kauā ʻelima, no lākou kēia mau inoa: Kauawaʻahila, Kauapalihala, Kauamakaiwi, Kauakuahine, a me Kaualīlīlehua, a lawe pū akula iā lākou e noho me ia ma ke kualapa ma waena o Pālolo a me Mānoa.

ʻO Kākuhihewa ke aliʻi o Oʻahu nei i ia manawa, e noho ana ʻo ia ma Ulukou, ma kahi e kū nei ka Hōkele Moana, a ma uka mai kāna loko iʻa, ma kahi o ka loko e ʻike ʻia nei i kēia manawa, he wahi hānai no nā manu kakā. Hāʻule maila he kuāua koʻikoʻi a hiki i ka nāhāhā ʻana o kapa o ka loko iʻa i ka wai, i ia wā i kamaʻilio aku ai nā kāula a me nā kāhuna i ke aliʻi, no Kaumana a me kāna poʻe kauā, ʻo lākou kēlā mau ua ma mua aʻe nei.

Ua ʻimi koke nā kāhuna i wahi e palekana ai ka loko iʻa a ke aliʻi, no laila lawe akula lākou he puaʻa hiwa, a waiho akula i mua o Kaumana, me ka noke ʻana i ka pule, a, maopopo iā Kaumana ua kokoke mai kona hopena, ʻo kona lilo aʻela nō ia i pōhaku, a ma ka ʻōlelo ʻia, ke waiho nei ia pōhaku a hiki i kēia lā.

ʻO Palihala, ʻo ia kahi kiʻekiʻe ma ka ʻaoʻao ma kai, mai ke awāwa mai ʻo Awāwaloa; Palikuahine, ʻo ia kēlā wahi e kiʻei ihola iā Waiakeakua ma Mānoa; Palilīlīlehua, aia ma kekahi ʻaoʻao, e huli lā i Pālolo; ʻo Makaiwi, ʻo ka ʻāina ia mai Palihala aku a i kahi i waiho ai ʻo Kaumana. Ua hōʻike ʻia mai iaʻu, aia ma ka mana ʻana o ke Alanui ʻElima e kokoke lā i ka uapo.

Ma ka hoʻākāka a Mr. Emekona, ma ka moʻolelo o Pele a me Hiʻiaka, ʻo ka ua Waʻahila, he ua kilihune ia mai [Nuʻuanu] mai, a hiki i kahi o Kauka, ma ke alanui Wyle. ʻO ka ua Līlīlehua, he ua ia mai Kaʻauhelemoa mai a hiki i Makaiwi. ʻO ka ua Kuahine, ʻo ka ua ia mai Kailua a hiki i ʻUalakaʻa.

Dear Ka Nupepa Kuokoa, *Honolulu, Oʻahu—While I was at Honolulu, I heard a good kupua (supernatural being) story about Kaumana, a kupua who turned to stone along the ridge between Pālolo and Mānoa.*

I wrote down the story in English, but it should be recorded in Hawaiian.

My friend Mr. Stokes is the one who went to see Mr. Solomon Kauai, the person who shared this story. He lives on the upland side of the Waiʻalae road where it meets with Kapahulu, which heads toward Waikīkī from a certain place very close to the end of the ridge of Kaumana. . . .

If you could search for the actual legend, it would greatly please me, and should you print it in your newspaper, then send me a copy. Here is the story as I received it. . . .

Kaumana's love for his youngest son surpassed his love for all others, so he began to swim to the other end of the pond to explain to his wife what he had done, and to lament his son.

That night a strong wind blew, and the ocean was rough. The next day, lo and behold, the pond was no longer where it used to be, but instead there was sand, which made a good road to travel.

However, this did not please Kaumana. It made him more regretful over his son, so he began to kill everyone who came with him from Lāhainā, including his wife and parents; all except his five servants, whose names were Kauawaʻahila, Kauapalihala, Kauamakaiwi, Kauakuahine, and Kaualīlīlehua. He took them with him to live on the ridge between Pālolo and Mānoa.

Kākuhihewa was the ruler of Oʻahu at this time. He was residing at Ulukou, where the Moana Hotel now stands, and his fishpond was inland of that, where we now see the duck-feeding pond. A heavy kuāua shower fell until the water broke the walls of the fishpond. And so the prophets and experts spoke with the aliʻi about Kaumana and his servants, the rains mentioned earlier.

The experts quickly searched for a way to save the aliʻi's fishpond, so they brought a black pig and placed it before Kaumana, continuing to pray. Kaumana knew his end was near, so he turned into stone, and it is said that this stone is still there to this day.

Palihala is the high point on the seaward side of the valley of Awāwaloa. Palikuahine is the area overlooking Waiakeakua in Mānoa. Palilīlīlehua is on the other side facing Pālolo. Makaiwi is the land between Palihala and Kaumana. It was shown to me, where Fifth Avenue splits, close to the bridge.

In the description by Mr. Emerson in the legend of Pele and Hiʻiaka, the ua Waʻahila is a gentle rain from Nuʻuanu to the area of Kauka (Judd) on Wyllie Street. The ua Līlīlehua is a rain from Kaʻauhelemoa to Makaiwi. The ua Kuahine is the rain from Kailua to ʻUalakaʻa.

From the legend of Kaumana. Hawaiian source: Kelsey, "He moolelo." English trans. by author. Additional source: Sterling and Summers 280. *Note: "Palikuahine" may be the same as "Paliluahine," a place described as "the foothills at the eastern corner of Mānoa Valley" and as the "small green hill" in Mānoa known for a moʻo named Luahine and her two sons, Kūmauna and Palihala, all of whom are stones (Sterling and Summers 290). Several old maps indicate that Paliluahine is in the area known as Kahaloa, below Kūmauna. "Kaumana," from the rendering above, may be the same as "Kūmauna," which is situated above the Luahine stone (Sterling and Summers 290) and above the Waʻahila and Kalaepōhaku Ridges, between Mānoa and Pālolo Valleys.*

Rain of Waikīkī, Oʻahu

27. Kuʻu kāne i ka ua noe
Noe hāliʻi a ka Waʻahila
Hoʻohila ka manaʻo, wehi i ka lau
Lau a ke aloha e piʻi ana i ka liko
Wā ihola, ke wā wale maila nō

My husband of the misty rains
Blanketing fall of the Waʻahila showers
Abashed, yet adorned by the outpour
An outpouring of love, rising to brightness
Boisterous, an uproar

From a mele by Hiʻiakaikapoliopele as she was leaving a house with noisy people playing the game of kilu in Waikīkī. Hawaiian source: Hoʻoulumāhiehie, *Ka Moʻolelo* 311. English trans.: Hoʻoulumāhiehie, *Epic* 290.

Rain of Waipuhia, Oʻahu

28. I Waipuhia ka wai, he waimaka no ka pua
No ke kāmakahala, no ka ʻāhihi lehua
Ua ʻula wale, ua lei wale nā wāhine a ʻohu
Mō ka ʻawe ka pua i ke alanui
Alanui ʻula i ka ua a ka Waʻahila

At Waipuhia is found the water, like teardrops of the flowers
Of the kāmakahala and lehua blossoms
Red are they and worn to bedeck the women
The lei bundles are opened up on the trail
Reddened is the trail in the Waʻahila rain

From a mele inoa, or name chant, for Liliʻuokalani. Source: Pukui, *Nā Mele Welo* 140–41.

29. Kaha pāweo ka Waʻahila i nā pali, ʻeā
Pali Waipuhia, paheʻe Kahaukomo, ʻeā

The Waʻahila rain sweeps away along the cliffs
Waipuhia is a cliff-face, and Kahaukomo is slippery

From a mele mākaʻikaʻi, or travel chant, for ʻEmalani Kaleleonālani by Kamaʻikuʻi. Source: Nogelmeier 152.

30. *For another example of Waʻahila of Waipuhia, see Puanaiea (#4).*

Other examples

31. Nani ʻiʻo nō ka lehua maka noe
    I ka ʻelo i ka ua Waʻahila
    Ona ʻia e nā manu
    I ka lupalupa i ke ʻala

    Truly beautiful are the lehua maka noe flowers
    In the drenching of the Waʻahila rain
    Alluring to birds
    For their lushness and fragrance

    From the song "Kilakila ihiihi Lanakila" by Keliimoewai and Kaleiluhia. Hawaiian source: Holstein 37. English trans. by author.

32. Pulu ʻole i ka ua noe ka maka o aʻu lehua
    Ua pēpē i ka ua Waʻahila ke oho o ka ʻawapuhi

    The bud of my lehua is not soaked by the misty noe rain
    The leaves of the ʻawapuhi are crushed by the Waʻahila rain

    From a mele inoa, or name chant, for Heneri W. Auld. Hawaiian source: Hoopai. English trans. by author.

33. Mai ʻalaʻala paha i ka ua o ka Waʻahila. *Almost received a scar on the neck, perhaps, from the Waʻahila rain.*

    A traditional saying. Source: Pukui, *ʻŌlelo* 221. *Note: Pukui says that the phrase means, "He just escaped trouble."*

34. Noe wale mai nō ka nahele
    He ua nihi pali
    Luhe ka lau o ka palai
    ʻElo i ka ua Waʻahila

    Now see the [noe] mist pass o'er the trees
    Now closely press near the cliffs
    Gracefully droops the leaf of the palai
    Laden with crystals of the Waʻahila

    From the song "Puīa ka nahele" by Liliʻuokalani. Source: Liliʻuokalani 73–74. Additional sources: Holstein 35; Elbert and Mahoe 86. *Note:* The Queen's Songbook *explains, "'Puīa ka nahele' is another of Liliʻuokalani's early, happy love songs in waltz time. The usual symbols for love appear here—fragrance, rain, mist, ferns, winds, the coolness they bring—and birds represent the lovers themselves."*

Waʻawaʻahia. Rain associated with Waipiʻo, Hawaiʻi. "Waʻawaʻahia" means "furrowed."

1. Ka ua Waʻawaʻahia o Waipiʻo. *The furrow-cutting [Waʻawaʻahia] rain of Waipiʻo.*

   A traditional saying. Source: Pukui, *ʻŌlelo* 173. Additional source: Pukui and Elbert, 1986, 375. *Note: Pukui says, "The rain of Waipiʻo, Hawaiʻi, sweeps along the gullies and gulches as it pours."*

Waiʻahulu. Also known as Kawaihulu. Rain associated with Waimea, Hawaiʻi. "Wai ʻahulu" means "discolored water." It is both the name of a specific rain and a generally descriptive term; its various usages are determined by the context.

1. *See Kīpuʻupuʻu (#32).*

Waikaloa. Also the name of a wind.

1. Cold rains at Honolulu.

   Source: Kent 379.

Wailehua. Same as *Alanilehua*. Rain associated with Panaʻewa and Puna, Hawaiʻi. "Wai lehua" means "lehua water" or "lehua nectar."

1. *See Alanilehua (#2).*

Waiolohia. Rain associated with Kauaʻi. Also the name of a wind. "Wai olohia" means "sounding water" or "water that moves back and forth."

Cold, misty rain of Kalanipuʻu, Kauaʻi

1. *See Paupua (#1).*

Rain of Puakei, Kauaʻi

2. Ua Waiolohia—Puakei, Kauaʻi.

   From a list of rain names and their descriptions. Source: "Names . . . rain."

BIBLIOGRAPHY

Note: Hawaiian-language newspaper articles can be accessed at papakilodatabase.com and nupepa.org. For this volume, research occurring between 2008 and 2013 utilized nupepa.org; research after 2013 utilized papakilodatabase.com.

Aekai o Kuloloia, George M. K. "He haina nane." *Kuokoa Home Rula,* buke 6, helu 51 (Dekemaba 18, 1908): 3.

Ahia, J. L., Kini Aki, and Mrs. William Kaina. "Hoalohaloha no Mrs. Hannah Kaunukapu Kaloi Kamelamela." *Ka Nupepa Kuokoa,* buke 63, helu 19 (Mei 8, 1924): 5.

Ahuihala, K. W. "He kanikau aloha no Paewahine." *Ka Hoku o ka Pakipika,* buke 1, helu 27 (Maraki 27, 1862): 4.

"Aia i Haili ko lei nani." Chant. MS SC Roberts, box 2.1, pp. 17–21c. Bishop Museum Archives, Honolulu.

"Aia i Hilo One ka pua o Kina." Chant. MS SC Roberts, box 5.3, pp. 27–28c. Bishop Museum Archives, Honolulu.

"Aia i Lihau ko lei nani." Chant. MS SC Roberts, box 2.1, pp. 14a–16b. Bishop Museum Archives, Honolulu.

Akahi. "O ka mapu kohai lau konale." Chant. MS SC Roberts, box 4.9, pp. 64b–65a. Bishop Museum Archives, Honolulu.

Akana, Collette Leimomi. *The O'ahu Exploits of Kamapua'a, the Hawaiian Pig-God: An Annotated Translation of a Hawaiian Epic from* Ka Leo o ka Lahui, *July 23, 1891–August 26, 1891.* Honolulu: Bishop Museum Press, 2004.

Alii Iolani. "He mau inoa no Ka Haku O Hawaii." *Ka Nupepa Kuokoa,* buke 1, helu 1 (Okatoba 1, 1861): 1.

"Aloalo poka i Daimana Hila." *Ka Makaainana,* buke 3, helu 12 (Mei 27, 1895): 1.
"Aloha au o kahi wai o Kepane." Chant. MS SC Roberts, box 2.4, pp. 170b–73a. Bishop Museum Archives, Honolulu.
"Aloha ka uka i ke onaona." Chant. MS SC Roberts, box 2.4, pp. 175b–79a. Bishop Museum Archives, Honolulu.
Amamalua, Pinaloke. "He mele kanikau no L. Keoua." *Ka Hoku o ka Pakipika,* buke 1, helu 4 (Okatoba 17, 1861): 3.
Ami. "He mele no Ka Haku Hawaii." *Ka Hae Hawaii,* buke 5, helu 17 (Iulai 25, 1860): 72.
Ana, Keoni. "Ke kai o Kaula kau aloha." Chant. HI.M.54, pp. 163–64. Bishop Museum Archives, Honolulu.
Andrews, Lorrin. *A Dictionary of the Hawaiian Language.* Rutland, VT: Charles E. Tuttle, 1974.
"'Ano'ai." MS GRP 329, box 1.45. Kimo Alama Keaulana Collection. Trans. Kimo Alama Keaulana. Bishop Museum Archives, Honolulu.
Apahu, M. "Hemele [*sic*] no Emalaina Kapualahaole." *Ka Hoku o ka Pakipika,* buke 1, helu 26 (Maraki 20, 1862): 1.
"Auhea wale oe e ka makani Kiu." Chant. MS SC Roberts, box 5.4, pp. 120b–22. Bishop Museum Archives, Honolulu.
"Auhea wale oe e ka ua noe." Chant. MS SC Roberts, box 5.3, pp. 67b–69. Bishop Museum Archives, Honolulu.
"Auhea wale oe e makani Kona." Chant. MS SC Roberts, box 5.4, pp. 111b–13a. Bishop Museum Archives, Honolulu.
"A Waialeale nui a ta hele kua." Chant. MS SC Roberts, box 5.3, p. 50. Bishop Museum Archives, Honolulu.
Barrère, Dorothy B., Mary Kawena Pukui, and Marion Kelly. *Hula: Historical Perspectives.* Pacific Anthropological Records 30. Department of Anthropology, Bernice P. Bishop Museum, Honolulu, January 1980.
Beckwith, Martha Warren, trans. and ed. *The Kumulipo: A Hawaiian Creation Chant.* Composed for Ka'īamamao by Keaulumoku. Chicago: University of Chicago Press, 1951. Facsimile reproduction with a new foreword by Katharine Luomala. Honolulu: University of Hawai'i Press, 1972.
Beni. "He kanikau no Ka Haku O Hawaii." *Ka Hoku o ka Pakipika,* buke 1, helu 51 (Sepatemaba 11, 1862): 4.
Board of Genealogy. Manuscript [Records of meetings 1882—Concerns heiau and winds]. GEN 12.2. Bishop Museum Archives, Honolulu.
Buke Mele Lahui. Honolulu: Hawaiian Historical Society, 2003.
"Calling the winds in a calm." Hawaiian Ethnological Notes (HEN) 1:1045. Edgar Henriques Collection. Bishop Museum Archives, Honolulu.
Charlot, John. *Moses Kuaea Nākuina: Hawaiian Novelist.* Lā'ie, O'ahu: Pacific Institute, Brigham Young University–Hawai'i, 2005.
Clark, John R. K. *Hawai'i Place Names: Shores, Beaches, and Surf Sites.* Honolulu: University of Hawai'i Press, 2002.

Craig, Mr. and Mrs. Alexander. "He puolo waimaka aloha, no kuu aunty heleloa Mrs. Kaiawe Helelani Punohu." *Ka Nupepa Kuokoa,* buke 64, helu 10 (Maraki 5, 1925): 5.

David, Kaua. "Ka make ana o H. Haumea." *Ka Hae Hawaii,* buke 6, helu 27 (Okatoba 2, 1861): 107.

Desha, Stephen Langhern. *He Moʻolelo Kaʻao no Kekūhaupiʻo: Ke Koa Kaulana o ke Au o Kamehameha ka Nui.* Vols. 1, 2. Hilo: Hale Kuamoʻo, Ka Haka ʻUla o Keʻelikōlani, 1996.

Desha, Stephen Langhern. *Kamehameha and His Warrior Kekūhaupiʻo.* Trans. Frances N. Frazier. Honolulu: Kamehameha Schools Press, 2000.

de Silva, Kīhei. "Lani Haʻahaʻa (Makawao)." Hālau Mōhala ʻIlima. Accessed January 14, 2016. http://www.halaumohalailima.com/HMI/Lani_Haahaa_(Makawao).html.

"Eia ana ka leo aloha." Chant. MS SC Roberts, box 4.6, pp. 62b–65a. Bishop Museum Archives, Honolulu.

"Eiae ka ua ke hele mai nei." Chant. MS SC Roberts, box 4.5, pp. 21, 24b–25a. Bishop Museum Archives, Honolulu.

"Eia ke weliweli nei." Chant. MS SC Roberts, box 4.6, pp. 58b–61a. Bishop Museum Archives, Honolulu.

Elbert, Samuel H., ed. *Selections from Fornander's Hawaiian Antiquities and Folk-lore.* Honolulu: University of Hawaiʻi Press, 1959.

Elbert, Samuel H., and Noelani Mahoe. *Nā Mele o Hawaiʻi Nei: 101 Hawaiian Songs.* Honolulu: University of Hawaiʻi Press, 1970.

Ellis, William. *Journal of William Ellis.* Honolulu: Advertiser Publishing, 1963.

Emalani. "He kanikau no Kamehameha IV!" *Ka Nupepa Kuokoa,* buke 3, helu 7 (Feberuari 13, 1864): 2.

Emerson, Nathaniel B. *Pele and Hiiaka: A Myth from Hawaii.* Honolulu: Honolulu Star-Bulletin, 1915.

Emerson, Nathaniel B. *Unwritten Literature of Hawaii: The Sacred Songs of the Hula.* Rutland, VT: Charles E. Tuttle, 1965.

Emma. "He kanikau no Ka Haku O Hawaii." *Ka Hoku o ka Pakipika,* buke 1, helu 51 (Sepatemaba 11, 1862): 1.

Emory, Kenneth P. *The Island of Lanai: A Survey of Native Culture.* Bernice P. Bishop Museum Bulletin 12. Honolulu: Bishop Museum Press, 1969.

Fornander, Abraham. *An Account of the Polynesian Race: Its Origin and Migrations.* Rutland, VT: Charles E. Tuttle, 1969.

Fornander, Abraham. *Fornander Collection of Hawaiian Antiquities and Folk-lore.* Ed. Thomas G. Thrum. Vols. 4–6. Honolulu: Bishop Museum Press, 1916–20.

Frazier, Frances N. *The True Story of Kaluaikoolau as Told by His Wife, Piilani.* Līhuʻe: Kauaʻi Historical Society, 2001.

"A ghost, O husband, it was taken unseen O wife." Hawaiian Ethnological Notes (HEN) News., *Ka Nupepa Kuokoa,* Okatoba 7–Dekemaba 23, 1921. Bishop Museum Archives, Honolulu. Translation of the Okatoba 28, 1921, installment in the "He akua e ke kane, he ike ole e ka wahine e!" series.

Haakulou. "He inoa hooilina no na moi o Hawaii, Kamehameha, mai ka wa kahiko mai." *Ka Nupepa Kuokoa,* buke 7, helu 44 (Okatoba 31, 1868): 4.

Hakau. "He kanikau no Ioseba Nawai." *Ka Hoku o ka Pakipika,* buke 1, helu 22 (Feberuari 20, 1862): 1.

Haleole, S. N. "Ka mooolelo o Kapakohano." *Ka Hoku o ka Pakipika,* buke 1, helu 7 (Novemaba 7, 1861): 1.

Haleole, S. N. *Ke Kaao o Laieikawai: Ka Hiwahiwa o Paliuli.* First published as "Ka moolelo o Laieikawai" in *Ka Nupepa Kuokoa,* 1862–63. Honolulu: Henry M. Whitney, 1863. http://ulukau.org/elib/cgi-bin/library?e=q-olaiekawai-000Sec--11haw-50-20-frameset-book-laieikawai-1-011escapewin&a=d&p2=book.

Haleole, S. N. *Lāʻieikawai.* Trans. Martha Warren Beckwith. Ed. Dennis Kawaharada, Richard Hamasaki, and Esther Mookini. Honolulu: Kalamakū Press, 2006.

Hanaia. "He mele no V.K.K. Kalohelani." *Ka Hoku o ka Pakipika,* buke 1, helu 16 (Ianuari 9, 1862): 4.

"Hana uso ole." *Ka Nupepa Kuokoa,* buke 4, helu 17 (Aperila 27, 1865): 2.

Handy, E. S. Craighill, and Elizabeth Green Handy. *Native Planters in Old Hawaii: Their Life, Lore, and Environment.* Honolulu: Bishop Museum Press, 1972.

Handy, E. S. Craighill, and Mary Kawena Pukui. *The Polynesian Family System in Ka-ʻu, Hawaiʻi.* Rutland, VT: Charles E. Tuttle, 1972.

Haui, C. H. "He mele no Ka Haku O Hawaii." *Ka Hoku o ka Pakipika,* buke 1, helu 15 (Ianuari 2, 1862): 4.

"He aihue ino loa." *Ka Nupepa Kuokoa,* buke 4, helu 46 (Novemaba 18, 1865): 2.

"He akua e ke kane, he ike ole e ka wahine e!" *Ka Nupepa Kuokoa,* buke 59, helu 43 (Okatoba 28, 1921): mahele elua, 1.

"He inoa no Kalani la." Chant. MS SC Roberts, box 2.4, pp. 118–22. Bishop Museum Archives, Honolulu.

"He inoa no Kaumualii." Chant. MS SC Roberts, box 4.8, pp. 130b–68a. Bishop Museum Archives, Honolulu.

"He inoa no Libe Maukaa." Chant. MS SC Roberts, box 4.9, pp. 35b–43a. Bishop Museum Archives, Honolulu.

"He inoa no Pauahi." Chant. HI.M.6, p. 54. Bishop Museum Archives, Honolulu.

"He kaao hoonaue puuwai no Puakaohelo, ke kaikamahine alii." *Ka Nupepa Kuokoa*:
buke 33, helu 11 (Maraki 17, 1894): 1
buke 33, helu 23 (Iune 9, 1894): 1.

"He kanaenae aloha no Aberahama K. Kalana mai na kupuna mai." *Ka Nupepa Kuokoa,* buke 62, helu 46 (Novemaba 15, 1923): 4.

"He kanikau no Abenera Kahekili kekahi alii o keia pae aina i make aku nei." *Ka Nonanona,* buke 3, pepa 6 (Augate 8, 1843): 30.

"He kanikau no Iosepa Nawai." *Ka Hoku o ka Pakipika,* buke 1, helu 17 (Ianuari 16, 1862): 3.

"He kanikau no ka moi i aloha nuiia, Kalani Iolani, Kualiholihoikekapu, Kamehameha IV!" *Ka Nupepa Kuokoa,* buke 3, helu 16 (Aperila 16, 1864): 4.

"He kanikau no ka Moi Iolani Kamehameha IV, ka moi i aloha nui loa ia e kona mau alii ame ka lahui o kona pae aina." *Ka Nupepa Kuokoa,* buke 3, helu 18 (Aperila 30, 1864): 4.

Heleahiki, Keiki. "Kanikau aloha no Kahananui." *Ke Au Okoa,* buke 2, helu 35 (Dekemaba 17, 1866): 3.

Helela, Kamehaitu. "He mele he inoa no ke onaona." Chant. MS SC Roberts, box 2.7, pp. 95c–99a, and box 5.3, pp. 54–55a. Bishop Museum Archives, Honolulu.

"He mele." *Ka Lei Momi,* buke 1, helu 2 (Iune 22, 1893): 3.

"He mele." *Ka Nupepa Kuokoa,* buke 1, helu 52 (Novemaba 22, 1862): 1.

"He mele he inoa no W. P. Leleiohoku." Chant. HI.M.3, p. 41. Bishop Museum Archives, Honolulu.

"He mele kapa no Liliu." Chant. HI.M.55, p. 106. Bishop Museum Archives, Honolulu.

"He mele no Keawenuiaumi." *Ka Hoku Loa,* buke 2, helu 1 (Iulai 1860): 4.

"He mele no Kekulei'ula." Chant. MS SC Roberts, box 4.9, pp. 49b–55a. Bishop Museum Archives, Honolulu.

"He molelo kaao no Kamapuaa." *Ka Leo o ka Lahui*:
buke 2, helu 221 (Iune 23, 1891): 4
buke 2, helu 222 (Iune 24, 1891): 4
buke 2, helu 223 (Iune 25, 1891): 4
buke 2, helu 228 (Iulai 2, 1891): 4
buke 2, helu 232 (Iulai 8, 1891): 4
buke 2, helu 238 (Iulai 16, 1891): 4.

"He moolelo kaao Hawaii no Kahalaopuna, ka nani oi kelakela a me ka ui i kona mau la." *Ka Leo o ka Lahui,* buke 1, helu 12 (Sepatemaba 3, 1889): 1.

"He moolelo kaao Hawaii no Laukaieie." *Ka Leo o ka Lahui,* buke 2, helu 849 (Ianuari 8, 1894): 3.

"He moolelo no Aahoaka ke koa a me kona hanau kupanaha ana." *Ka Nupepa Kuokoa*:
buke 16, helu 1 (Ianuari 6, 1877): 1
buke 16, helu 5 (Feberuari 3, 1877): 1
buke 16, helu 8 (Feberuari 24, 1877): 1
buke 16, helu 9 (Maraki 3, 1877): 1.

"He moolelo no Lonoikamakahiki." *Ke Au Okoa,* buke 1, helu 22 (Sepetemaba 18, 1865): 4.

"He oki no ke kuleiaula." Chant. MS SC Roberts, box 4.9, pp. 55b–57. Bishop Museum Archives, Honolulu.

"He uhane no Kealoha." *Ka Nupepa Kuokoa,* buke 1, helu 11 (Feberuari 8, 1862): 2.

"He wahi kaao no Mokulehua." *Ka Hoku o ka Pakipika,* buke 1, helu 10 (Novemaba 28, 1861): 1.

Hiilei, Mrs. Hana. "Kanikau no Kamakahoohie." *Ka Nupepa Kuokoa,* buke 3, helu 42 (Okatoba 15, 1864): 3.

Hoakimoa, D. "He uhane no Keliikuea [*sic*]." *Ka Nupepa Kuokoa,* buke 1, helu 46 (Okatoba 11, 1862): 4.

Hoapili, M. S. "He kanikau aloha no M. E. Manu, w. i make." *Ka Nupepa Kuokoa,* buke 1, helu 5 (Dekemaba 16, 1861): 3.

"Hoe Puna." Chant. MS SC Roberts, box 5.2, p. 35. Bishop Museum Archives, Honolulu.

Holaniku, S. W. "He inoa no Keliihelemauna." *Ka Nupepa Kuokoa,* buke 2, helu 34 (Augate 22, 1863): 1.

"Hole Waimea i ka ihe a ka makani." Chant. MS GRP 81, box 8.13. Mader Collection. Bishop Museum Archives, Honolulu.

Holi, P. R. "Na ionoa [*sic*] o na makani o Niihau." *Ke Au Okoa,* buke 1, helu 14 (Iulai 24, 1865): 2.

Holstein, Ed C., comp. *Ka Buke Mele o nā Hīmeni Hawai'i (The Lyric Book of Hawaiian Choral Songs).* Foreword by M. Puakea Nogelmeier. Honolulu: Bishop Museum Press, 2003.

"Honolulu, Dekemaba 24, 1864." *Ka Nupepa Kuokoa,* buke 3, helu 52 (Dekemaba 24, 1864): 2.

Honolulu, Mrs. "He kanikau aloha no S. P. Koko." *Ka Nupepa Kuokoa,* buke 7, helu 11 (Maraki 14, 1868): 1.

"Honolulu, Sepa. 6, 1862." *Ka Nupepa Kuokoa,* buke 1, helu 41 (Sepatemaba 6, 1862): 2.

Hoopai. "He inoa no Heneri W. Auld." *Ke Au Okoa,* buke 1, helu 10 (Iune 26, 1865): 1.

"Hooputa i kai i ka la Unulau." Chant. MS SC Roberts, box 5.3, pp. 28d–29a. Bishop Museum Archives, Honolulu.

Ho'oulumāhiehie. *The Epic Tale of Hi'iakaikapoliopele.* Trans. M. Puakea Nogelmeier. First published as "Ka moolelo o Hiiaka-i-ka-poli-o-pele" in *Ka Na'i Aupuni,* 1905–6. Honolulu: Awaiaulu Press, 2006.

Ho'oulumāhiehie. "He moolelo no Kamehameha I." *Ka Na'i Aupuni*:
buke 1, helu 27 (Dekemaba 27, 1905): 4
buke 1, helu 75 (Feberuari 22, 1906): 1
buke 1, helu 83 (Maraki 3, 1906): 1
buke 1, helu 90 (Maraki 12, 1906): 1
buke 2, helu 22 (Iune 27, 1906): 1
buke 2, helu 32 (Iulai 10, 1906): 1.

Ho'oulumāhiehie. "Ka moolelo o Hiiaka-i-ka-poli-o-pele." *Ka Na'i Aupuni,* buke 2, helu 123 (Okatoba 25, 1906): 4.

Ho'oulumāhiehie. *Ka Mo'olelo o Hi'iakaikapoliopele.* Edited by M. Puakea Nogelmeier. First published as "Ka moolelo o Hiiaka-i-ka-poli-o-pele" in *Ka Na'i Aupuni,* 1905–1906. Republished with modern orthography and an introduction by M. Puakea Nogelmeier. Honolulu: Awaiaulu Press, 2006.

Hopoiku, Lepano W. "He inoa no Kamehameha V." *Ke Au Okoa,* buke 1, helu 36 (Dekemaba 25, 1865): 1.

"Huro! no Hawaii me Amerika! He nu hou kaulana! Ua holo ke Kuikahi Panailike!" *Ka Nupepa Kuokoa,* buke 15, helu 35 (Augate 26, 1876): 2.

Ii, John Papa. "The death of Her Highness Princess Victoria Kamamalu Ka'ahumanu." Hawaiian Ethnological Notes (HEN) 1:2938–77. Bishop Museum Archives, Honolulu.

Ii, John Papa. "An expression of affection for Ka-'imi-haku (Lover-of-Chiefs)." Manuscript. MS SC Pukui, box 17.2. Bishop Museum Archives, Honolulu. Translation by Mary Kawena Pukui of "Kilikili hauna makani aloha la e" by John Papa Ii.

Ii, John Papa. *Fragments of Hawaiian History.* Translation by Mary Kawena Pukui of "Na hunahuna o ka moolelo Hawaii" by John Papa Ii. Ed. Dorothy B. Barrère. Honolulu: Bishop Museum Press, 1959.

Ii, John Papa. "Kanaenae aloha no ka mea i make." *Ka Nupepa Kuokoa,* buke 8, helu 3 (Ianuari 16, 1869): 1.

Ii, John Papa. "Ke ola a me ka make ana iho nei o Victoria K. Kaahumanu." *Ka Nupepa Kuokoa*:
buke 5, helu 23 (Iune 9, 1866): 1
buke 5, helu 24 (Iune 16, 1866): 1
buke 5, helu 26 (Iune 30, 1866): 1.

Ii, John Papa. "Kilikili hauna makani aloha la e." *Ka Nupepa Kuokoa,* buke 7, helu 49 (Dekemaba 5, 1868): 3.

Ii, John Papa. "Na hunahuna o ka moolelo Hawaii." *Ka Nupepa Kuokoa*:
buke 8, helu 15 (Aperila 10, 1869): 1
buke 8, helu 21 (Mei 22, 1869): 1
buke 8, helu 36 (Sebatemaba 4, 1869): 1
buke 8, helu 46 (Novemaba 13, 1869): 1
buke 9, helu 1 (Ianuari 1, 1870): 1
buke 9, helu 2 (Ianuari 8, 1870): 1
buke 9, helu 5 (Ianuari 29, 1870): 1
buke 9, helu 6 (Feberuari 5, 1870): 1
buke 9, helu 8 (Feberuari 19, 1870): 1
buke 9, helu 10 (Maraki 5, 1870): 1
buke 9, helu 12 (Maraki 19, 1870): 1.

Ii, John Papa. "Ola a make ana o ke kama alii." *Ka Nupepa Kuokoa,* buke 5, helu 33 (Augate 18, 1866): 4.

Ii, John Papa. "Victoria K. Kaahumanu." *Ka Nupepa Kuokoa,* buke 5, helu 28 (Iulai 14, 1866): 4.

"I ka Poalua iho nei." *Ka Lahui Hawaii,* buke 11, helu 26 (Iune 22, 1876): 3.

Ilihia, John K. "He moolelo kaao no Namakaokapaoo, ke ahikanana wiwo ole o Kula i Maui." *Ka Nupepa Kuokoa*:
buke 33, helu 20 (Mei 19, 1894): 4
buke 33, helu 21 (Mei 26, 1894): 4
buke 33, helu 41 (Okatoba 13, 1894): 1.

I Mua e nā Pōkiʻi: Go Forward Young Brothers. CD liner notes. Kaʻiwakīloumoku, Kamehameha Schools, 2003.

"Ino Koolau e, ino Koolau." Chant. MS GRP 81, box 8.13. Bishop Museum Archives, Honolulu.

Iolani. "He mau inoa no Ka Haku O Hawaii." *Ka Nupepa Kuokoa,* buke 1, helu 1 (Okatoba 1, 1861): 1.

"Ite i ta wai ula iliahi o Waimea." Chant. MS SC Roberts, box 2.7, pp. 120c–22a. Bishop Museum Archives, Honolulu.

James, H. "No ka mahiai." *Ka Hae Hawaii,* buke 2, helu 24 (Sepetemaba 9, 1857): 96.

Johnson, Rubellite Kinney. *Kukini ʻAhaʻilono: Carry on the News.* Honolulu: Topgallant Publishing, 1976.

Judd, Henry P. *Hawaiian Proverbs and Riddles.* Bernice P. Bishop Museum Bulletin 77. Honolulu: Bishop Museum Press, 1930. Reprint, Millwood, NY: Kraus Reprint Co., 1978.

Kaaihue, Marmionett M., ed. *Songs of Helen Desha Beamer.* Honolulu: Abigail K. Kawananakoa Foundation, 1991.

Kaaikaula, Kealalaina. "Welina puuwai kumakena." *Ka Nupepa Kuokoa,* buke 17, helu 47 (Novemaba 23, 1878): 1.

"Ka aina o Palesetina, Helu 5: Ke ano o ka noho ana a me na kau." *Ka Nupepa Kuokoa,* buke 5, helu 19 (Mei 12, 1866): 1.

Kaaipaka, R. "Ka make ana o Keawehunahala." *Ka Nupepa Kuokoa,* buke 15, helu 16 (Aperila 15, 1876): 1.

Kaakopua. "He kanikau no Hina." *Ka Hae Hawaii,* buke 6, helu 2 (Aperila 10, 1861): 8.

Kaalehu. "He kanikau aloha no Nuholani." *Ka Nupepa Kuokoa,* buke 1, helu 49 (Novemaba 1, 1862): 1.

Kaanaana. "He inoa no Lele." *Ka Hoku o ka Pakipika,* buke 1, helu 24 (Maraki 6, 1862): 1.

Kaanaana, Mrs. Lioe. "He wehi no Miss Ane Bell." *Ka Leo o ka Lahui,* buke 2, helu 1067 (Novemaba 14, 1894): 2.

"Ka anoai o ka aihue ana." *Ke Au Okoa,* buke 1, helu 15 (Iulai 31, 1865): 1.

"Kaao hooniua puuwai no Ka-miki." *Hawaii Holomua,* buke 1, helu 6 (Aperila 10, 1912): 14–16.

Kaapa, Joseph N. "Kanikau no Mrs. Maleka Ii." *Ka Nupepa Kuokoa,* buke 1, helu 6 (Ianuari 1, 1862): 4. English translation published in pamphlet format. Honolulu: Awaiaulu Press, 2007.

Kaapuwai, J. P. "Kanikau no Eda Kalua." *Ka Nupepa Kuokoa,* buke 59, helu 28 (Iulai 15, 1921): 2.

Kaaua. "He mele inoa no Kalakaua." Chant. HI.M.29, p. 187. Bishop Museum Archives, Honolulu.

Kaauwaepaa, J. H. "He kanikau aloha no L. Kahuakai." *Ka Hae Hawaii,* buke 1, helu 3 (Malaki 19, 1856): 11.

Kaauwai, William Hoapili. "Ka moolelo no ka makaikai ana a puni ka honua." *Ka Nupepa Kuokoa*:
buke 7, helu 38 (Sepatemaba 19, 1868): 1
buke 7, helu 49 (Dekemaba 5, 1868): 1.

Kaawa, P. W. "Ka hoomana kahiko." *Ka Nupepa Kuokoa*:
buke 4, helu 6 (Feberuari 9, 1865): 1
buke 4, helu 17 (Aperila 27, 1865): 1.

Kaena, Marie. "He inoa no Erisapeka." *Ka Hoku o ka Pakipika,* buke 1, helu 29 (Aperila 10, 1862): 4.

Kaeppler, Adrienne L. *Hula Pahu: Hawaiian Drum Dances.* Honolulu: Bishop Museum Press, 1993.

Kahaanapilo. "Kuu ipo hana maewaewa." *Ka Makaainana,* buke 1, helu 1 (Ianuari 1, 1894): 2.

Kahaekirikiano, T.W.P. "Ka Hoku Pakipika." *Ka Hoku o ka Pakipika,* buke 1, helu 4 (Okatoba 17, 1861): 3.

Kahalaia. "He mele no W. P. Leleiohoku II." Chant. MS SC Roberts, box 4.9, pp. 82b–87a. Bishop Museum Archives, Honolulu.

Kahalaia. "Ike i ka ua ulalena i ka maka aloha." Chant. MS SC Roberts, box 4.9, pp. 66b–67a. Bishop Museum Archives, Honolulu.

Kahale, Ezekiela. "He kanikau no Rode Koleiki." *Ka Nupepa Kuokoa,* buke 1, helu 49 (Novemaba 1, 1862): 4.

Kahananui, D. "Make i aloha nui ia." *Ko Hawaii Pae Aina,* buke 1, helu 47 (Novemaba 23, 1878): 4.

Kahananui, Dorothy M., and Alberta P. Anthony. *E Kamaʻilio Hawaiʻi Kakou: Let's Speak Hawaiian.* Honolulu: University of Hawaiʻi Press, 1970.

Kahāʻulelio, Daniel. *Ka ʻOihana Lawaiʻa: Hawaiian Fishing Traditions.* Trans. Mary Kawena Pukui. Ed. M. Puakea Nogelmeier. Honolulu: Bishop Museum Press, 2006.

Kaheiheimalie. "Ke kapu o Lani Manokalanipo." Chant. MS SC Roberts, box 4.6, pp. 1a–2a. Bishop Museum Archives, Honolulu.

Kahele, Mrs. Marie. "No Luahaku." *Ke Au Okoa,* buke 6, helu 39 (Ianuari 12, 1871): 4.

Kahiamoe, J. H. "He kino uhane ua hala, ua moe i ka moe kau moe hooilo." *Ko Hawaii Pae Aina,* buke 1, helu 25 (Iune 22, 1878): 1.

"Kahikinui hoa'ia'i." Chant. MS SC Roberts, box 4.6, pp. 65b–71a. Bishop Museum Archives, Honolulu.

Kahinalua, P. M. "He kanikau aloha no Ka Haku O Hawaii." *Ka Nupepa Kuokoa,* buke 1, helu 43 (Sepatemaba 20, 1862): 3.

Kahinawe, H. M. "Ka hope o ke ke [*sic*] kanikau o J. Henry." *Ka Nupepa Kuokoa,* buke 1, helu 22 (Aperila 26, 1862): 3.

Kahinawe, Mrs. H. M. "He kanikau aloha no J. Henry." *Ka Nupepa Kuokoa,* buke 1, helu 20 (Aperila 12, 1862): 4.

Kahiolo, G. W. "He moolelo no Kamapuaa." *Ka Hae Hawaii,* buke 6, helu 26 (Sepatemaba 25, 1861): 104.

Kahiolo, G. W. *He Moolelo no Kamapuaa.* Trans. Esther T. Mookini and Eric C. Neizmen, with David Tom. Honolulu: Hawaiian Studies Program, University of Hawaiʻi at Mānoa, 1978.

"Ka hoala." Chant. HI.M.21, p. 100. Bishop Museum Archives, Honolulu.

Kaholomoku, Aarona. "He inoa no Erisapeka." *Ka Hoku o ka Pakipika,* buke 1, helu 29 (Aperila 10, 1862): 4.

Kahoohulimoku, Mrs. "He inoa no Kawailahaole." *Ka Hoku o ka Pakipika,* buke 1, helu 21 (Feberuari 13, 1862): 1.

Kaiehu. "He inoa no Kaumakaokane." *Ka Nupepa Kuokoa,* buke 3, helu 22 (Mei 28, 1864): 1.

Kailiehu, L. S. "Na inoa o na makani o Hana." *Ke Au Okoa,* buke 1, helu 8 (Iune 12, 1865): 4.

Kailua Historical Society. *Kailua.* Kailua: Kailua Historical Society, 2009.

Kakelamaluikaleo, J.W.K. "He nane lau." *Ka Nupepa Kuokoa,* buke 58, helu 8 (Feberuari 20, 1920): 2.

Kalaaukumuole, L.Z.E. "Manao wehewehe." *Ka Nupepa Kuokoa,* buke 5, helu 13 (Maraki 31, 1866): 3.

Kalaaukumuole, L.Z.E. "Thought explanations." Hawaiian Ethnological Notes (HEN) Thrum 53. Bishop Museum Archives, Honolulu. Translation by Thomas G. Thrum of "Manao wehewehe" by L.Z.E. Kalaaukumuole.

Kalai, Mr. "He inoa no Lele." *Ka Hoku o ka Pakipika,* buke 1, helu 24 (Maraki 6, 1862): 1.
"Kalani Kaulilua." *Ka Leo o ka Lahui,* buke 2, helu 120 (Feberuari 2, 1891): 2.
Kalanikuihonoinamoku, B. V. "No na wahi a na'lii e makemake ai e noho i ka wa kahiko ma ka mopuni [*sic*] o Oahu nei." *Ke Au Okoa,* buke 1, helu 15 (Iulai 31, 1865): 4.
Kalanimoku. "Eia hou keia inoa o Kauikeaouli." *Ka Hoku o ka Pakipika,* buke 1, helu 21 (Feberuari 13, 1862): 4.
Kalehua, Mrs. H. "Make i walohia." *Ka Lahui Hawaii,* buke 3, helu 13 (Maraki 29, 1877): 3.
Kaleionehu. "O Haili, Haili ano ke'ehia." Chant. MS SC Roberts, box 4.6, pp. 18b–20a. Bishop Museum Archives, Honolulu.
Kaleo, Rebeka. "He uhane no Moenahele." *Ka Nupepa Kuokoa,* buke 1, helu 43 (Sepatemaba 20, 1862): 3.
Kalionui, J. W. "He kanikau no Esetera Wahahee." *Ka Nupepa Kuokoa,* buke 1, helu 27 (Mei 31, 1862): 3.
Kalohelani, V. K. "He mau inoa no Ka Haku O Hawaii." *Ka Nupepa Kuokoa,* buke 1, helu 1 (Okatoba 1, 1861): 1.
Kaloma, Mrs. "He kanikau aloha no Nalohaelua." *Ke Au Okoa,* buke 1, helu 22 (Sepetemaba 18, 1865): 3.
Kalua, D. "Kuu wahine aloha Eda Kalua ua hala." *Ka Nupepa Kuokoa,* buke 59, helu 28 (Iulai 15, 1921): 2.
Kamaiopili, D. K. "He inoa no Ka Haku O Hawaii." *Ka Hoku o ka Pakipika,* buke 1, helu 9 (Novemaba 21, 1861): 1.
Kamaka, Mrs. "He kanikau no Kaoo." *Ka Nupepa Kuokoa,* buke 4, helu 29 (Iulai 22, 1865): 4.
Kamaka, Mrs. "He mele no Keaopolohiwa." *Ka Hoku o ka Pakipika,* buke 1, helu 4 (Okatoba 17, 1861): 4.
Kamakau, Samuel Mānaiakalani. "Aloha ke kaha o Palaau." *Ka Nupepa Kuokoa,* buke 6, helu 4 (Ianuari 26, 1867): 3.
Kamakau, Samuel Mānaiakalani. "Beloved is the land of Palaau." Hawaiian Ethnological Notes (HEN) News., *Ka Nupepa Kuokoa,* Ianuari 26, 1867. Cataloged as "Untitled lament." Bishop Museum Archives, Honolulu. Translation of "Aloha ke kaha o Palaau" by Samuel Mānaiakalani Kamakau.
Kamakau, Samuel Mānaiakalani. "He mau mele koihonua a me na mele no Kauilanui-makehaikalani Kamaialii." *Ka Nupepa Kuokoa,* buke 7, helu 44 (Okatoba 31, 1868): 1.
Kamakau, Samuel Mānaiakalani. "He mele no Kualii, Kulanipipili, Kulanioaka, Kunui-akea &c." *Ka Nupepa Kuokoa,* buke 7, helu 20 (Mei 16, 1868): 4.
Kamakau, Samuel Mānaiakalani. "Ka moolelo Hawaii." *Ke Au Okoa*:
buke 5, helu 8 (Iune 10, 1869): 1
buke 5, helu 28 (Okatoba 28, 1869): 1
buke 5, helu 29 (Novemaba 4, 1869): 1
buke 5, helu 31 (Novemaba 18, 1869): 1
buke 6, helu 34 (Dekemaba 8, 1870): 1
buke 6, helu 36 (Dekemaba 22, 1870): 1
buke 6, helu 40 (Ianuari 19, 1871): 1
buke 6, helu 41 (Ianuari 26, 1871): 1.

Kamakau, Samuel Mānaiakalani. "Ka moolelo o Hawaii nei." *Ka Nupepa Kuokoa*:
buke 4, helu 25 (Iune 22, 1865): 1
buke 4, helu 29 (Iulai 22, 1865): 1
buke 4, helu 33 (Augate 19, 1865): 1–2
buke 4, helu 34 (Augate 26, 1865): 1–2.

Kamakau, Samuel Mānaiakalani. "Ka mo'olelo o Kahahana, ka hopena." *Ka Ho'oilina: Puke Pai 'Ōlelo Hawai'i (The Legacy: Journal of Hawaiian Language Sources)* 1, no. 2 (2002): 316–17.

Kamakau, Samuel Mānaiakalani. "Ka moolelo o Kamehameha I." *Ka Nupepa Kuokoa*:
buke 5, helu 50 (Dekemaba 15, 1866): 1
buke 5, helu 52 (Dekemaba 29, 1866): 1
buke 6, helu 1 (Ianuari 5, 1867): 1
buke 6, helu 8 (Feberuari 23, 1867): 1
buke 6, helu 19 (Mei 11, 1867): 1
buke 6, helu 23 (Iune 8, 1867): 1
buke 6, helu 32 (Augate 10, 1867): 1.

Kamakau, Samuel Mānaiakalani. "Ka moolelo o na Kamehameha." *Ka Nupepa Kuokoa,* buke 7, helu 39 (Sepatemaba 26, 1868): 1.

Kamakau, Samuel Mānaiakalani. *Ke Aupuni Mō'ī: Ka Mo'olelo Hawai'i no Kauikeaouli—Keiki Ho'oilina a Kamehameha—a me ke Aupuni āna i Noho Mō'ī ai.* Honolulu: Kamehameha Schools Press, 2001.

Kamakau, Samuel Mānaiakalani. *Ke Kumu Aupuni: Ka Mo'olelo Hawai'i no Kamehameha Ka Na'i Aupuni a me kāna Aupuni i Ho'okumu ai.* Honolulu: Ke Kumu Lama, 'Ahahui 'Ōlelo Hawai'i, 1996.

Kamakau, Samuel Mānaiakalani. "Mele hoolea I." *Ka Nupepa Kuokoa,* buke 7, helu 40 (Okatoba 3, 1868): 1.

Kamakau, Samuel Mānaiakalani. "Nani Halo i ka ua Waahila." *Ka Nupepa Kuokoa,* buke 6, helu 45 (Novemaba 9, 1867): 3.

Kamakau, Samuel Mānaiakalani. "No ke kaapuni makaikai i na wahi kaulana a me na kupua, a me na'lii kahiko mai Hawaii a Niihau." *Ka Nupepa Kuokoa,* buke 4, helu 24 (Iune 15, 1865): 1.

Kamakau, Samuel Mānaiakalani. *Ruling Chiefs of Hawaii.* Translation by Mary Kawena Pukui, Thomas G. Thrum, Lahilahi Webb, Emma Davidson Taylor, and John Wise of selected newspaper articles from the "Ka moolelo o Kamehameha I" and "Ka moolelo o na Kamehameha" series by Samuel Mānaiakalani Kamakau. Ed. Mary Kawena Pukui and Martha Warren Beckwith. Honolulu: Kamehameha Schools Press, 1961. Rev. ed. with an introduction by Lilikalā Kame'eleihiwa. Honolulu: Kamehameha Schools Press, 1992.

Kamakau, Samuel Mānaiakalani. *Tales and Traditions of the People of Old: Nā Mo'olelo a ka Po'e Kahiko.* Translation by Mary Kawena Pukui of selected newspaper articles from the "Ka moolelo o Hawaii nei" and "Ka moolelo o Kamehameha I" series by Samuel Mānaiakalani Kamakau. Ed. Dorothy B. Barrère. Honolulu: Bishop Museum Press, 1991.

Kamakau, Samuel Mānaiakalani. *The Works of the People of Old: Na Hana a ka Poʻe Kahiko.* Translation by Mary Kawena Pukui of selected newspaper articles from the "Ka moolelo Hawaii" series by Samuel Mānaiakalani Kamakau. Arranged and edited by Dorothy B. Barrère. Bernice P. Bishop Museum Special Publication 61. Honolulu: Bishop Museum Press, 1976.

Kamala, Mrs. "Kanikau aloha no H. Keowai." *Ka Nupepa Kuokoa,* buke 1, helu 37 (Augate 9, 1862): 4.

Kameʻeleihiwa, Lilikalā K. *A Legendary Tradition of Kamapuaʻa, the Hawaiian Pig-God.* Honolulu: Bishop Museum Press, 1996.

Kamehameha V. "He kanikau no ka Moi Iolani Kamehameha IV!" *Ka Nupepa Kuokoa,* buke 3, helu 1 (Ianuari 2, 1864): 1.

Kamoakeawe, Harata. "He kanikau no Ioseba Nawai." *Ka Hoku o ka Pakipika,* buke 1, helu 22 (Feberuari 20, 1862): 1.

"Ka moolelo kaao o Hiiaka-i-ka-poli-o-Pele." *Kuokoa Home Rula,* buke 6, helu 27 (Iune 12, 1908): 1.

"Ka moolelo o He'ma [*sic*], ke koaie ku pali o ka makani kaili aloha o Kipahulu, Maui."
Ka Nupepa Kuokoa:
buke 58, helu 36 (Sepatemaba 3, 1920): 2
buke 58, helu 40 (Okatoba 1, 1920): 7.

"Ka moolelo o Kepakailiula!" *Ka Nupepa Kuokoa,* buke 4, helu 9 (Maraki 2, 1865): 1.

"Ka moolelo walohia o Hainakolo ka ui pua ka-maka-hala o ka wailele o Hiilawe." *Ka Na'i Aupuni,* buke 5, helu 82 (Aperila 24, 1908): 1.

Ka Mooolelo Hawaii. Honolulu: Hawaiian Historical Society, 2005.

Kanahele, Elama, Emalia Licayan, and Virginia Nizo. *Aloha Niihau.* Trans. and ed. Elama Kanahele, Kimo Armitage, and Keao NeSmith. Waipahū: Island Heritage Publishing, 2007.

Kanahele, Pualani Kanakaʻole, head researcher/chanter. *Maui Chants.* Alu Like, n.d.

Kanaʻina, C. "Lei Kaukini i na lehua o Kaauwai." Chant. MS SC Roberts, box 4.6, pp. 22b–24a. Bishop Museum Archives, Honolulu.

"Kananaka." Song. MS GRP 329, box 3.75. Kimo Alama Keaulana Collection. Trans. Kimo Alama Keaulana. Bishop Museum Archives, Honolulu.

Kaneheleloa, Mrs. W. "Ke aloha ua hala." *Ko Hawaii Pae Aina,* buke 1, helu 47 (Novemaba 23, 1878): 4.

Kanepuu, Joseph H. "Eia ka makamaka maanei, no na hana pono nae." *Ka Nupepa Kuokoa,* buke 4, helu 8 (Feberuari 23, 1865): 1.

Kanepuu, Joseph H. "He moolelo no Pakaa." *Ke Au Okoa*:
buke 3, helu 30 (Novemaba 14, 1867): 4
buke 3, helu 31 (Novemaba 21, 1867): 4
buke 3, helu 32 (Novemaba 28, 1867): 4.

Kanepuu, Joseph H. "Kaahele ma Molokai." *Ke Au Okoa,* buke 3, helu 26 (Okatoba 17, 1867): 4.

Kanepuu, Joseph H. "Traveling about on Molokai." Hawaiian Ethnological Notes (HEN) 1:3013–24. Bishop Museum Archives, Honolulu. Translation of "Kaahele ma Molokai" by J. H. Kanepuu.

Kaniu. "He lei no W. C. Lunalilo." *Ko Hawaii Ponoi,* buke 1, helu 3 (Iulai 2, 1873): 1.

Kanoho, Mrs. K. "He mele no Keaopolohiwa." *Ka Hoku o ka Pakipika,* buke 1, helu 4 (Okatoba 17, 1861): 4.

Kanuha, G. Willie. "Mau mea i anoi ia." *Ka Nupepa Kuokoa,* buke 5, helu 11 (Maraki 17, 1866): 3.

Kanuha, G. Willie. "Some things greatly desired." Hawaiian Ethnological Notes (HEN) Thrum 52. Bishop Museum Archives, Honolulu. Translation by Thomas G. Thrum of "Mau mea i anoi ia" by G. Willie Kanuha.

Kapaalua, Helene. "He kanikau no Nehemia Hepa." *Ka Nupepa Kuokoa,* buke 1, helu 57 (Dekemaba 27, 1862): 4.

Kapakuhaili, H. K. "He lei no W. C. Lunalilo." *Ko Hawaii Ponoi,* buke 1, helu 3 (Iulai 2, 1873): 1.

Kapakuhaili, H. K. "Lei pua maikai ouli i ka ua a ka Nahua." Chant. MS SC Roberts, box 4.6, pp. 24b–26a. Bishop Museum Archives, Honolulu.

Kapihenui, M. J. "He mooolelo no Hiiakaikapoliopele." *Ka Hoku o ka Pakipika*:
buke 1, helu 16 (Ianuari 9, 1862): 1
buke 1, helu 17 (Ianuari 16, 1862): 1
buke 1, helu 21 (Feberuari 13, 1862): 4
buke 1, helu 28 (Aperila 3, 1862): 4.

Kapooloku. "He inoa no Ka Haku O Hawaii." *Ka Hoku o ka Pakipika,* buke 1, helu 9 (Novemaba 21, 1861): 1.

"Ka ua Kea." Song. MS GRP 329, box 3.36. Kimo Alama Keaulana Collection. Trans. Kimo Alama Keaulana. Bishop Museum Archives, Honolulu.

Kaualilinoe, J.W.K. "Ka moolelo o Kamaakamahiai, ka niuhi ai humuhumu o Kahului i Maui, ke puhi nau okaoka hoi o kona mau la koa, kahiluhilu hoi o kona wa ui." *Ka Nupepa Kuokoa*:
buke 9, helu 33 (Augate 13, 1870): 4
buke 9, helu 40 (Okatoba 1, 1870): 4
buke 9, helu 41 (Okatoba 8, 1870): 4
buke 9, helu 42 (Okatoba 15, 1870): 4
buke 9, helu 45 (Novemaba 5, 1870): 4
buke 9, helu 46 (Novemaba 12, 1870): 1
buke 9, helu 52 (Dekemaba 24, 1870): 1.

Kaualilinoe, J.W.K. "Legend of Kamaakamahiai." Hawaiian Ethnological Notes (HEN) 2:59–119. Bishop Museum Archives, Honolulu. Translation by Mary Kawena Pukui of "Ka moolelo o Kamaakamahiai" by J.W.K. Kaualilinoe.

Kauanoano, Henry. "Na hunahuna materia o ka paia nu'a i ka lehua." *Ka Makaainana,* buke 1, helu 16 (Aperila 16, 1894): 3.

Kauhi, Emma Kapūnohuʻulaokalani. *He Moʻolelo no Kapaʻahu: Story of Kapaʻahu.* Trans. Charles M. Langlas. Hilo: Pili Productions, 1996.

Kauinui, Mrs. "He kanikau aloha no Tonetakine Mapuhoe." *Ke Au Okoa,* buke 1, helu 12 (Iulai 10, 1865): 3.

Kauiomanoa, B. "Ka make ana o Keawehunahala." *Ka Nupepa Kuokoa,* buke 15, helu 16 (Aperila 15, 1876): 1.

"Ka uka holokia ahi manu Ola'a." Chant. MS SC Roberts, box 2.7, pp. 154b–56a, 157b. Bishop Museum Archives, Honolulu.

Kaulainamoku, Beni. "Hiku i ka nahele!" *Ko Hawaii Ponoi,* buke 1, helu 37 (Feberuari 25, 1874): 4.

Kauluhaimalama. "Eia hou keia inoa no Kauikeaouli." *Ka Hoku o ka Pakipika,* buke 1, helu 21 (Feberuari 13, 1862): 4.

Kauokoa, Kamehameha. "Au e hookano nei i ka moe." Chant. MS SC Roberts, box 4.6, pp. 8b–10a. Bishop Museum Archives, Honolulu.

Kawahamana. "He mele." *Ka Hoku o ka Pakipika,* buke 1, helu 18 (Ianuari 23, 1862): 3.

Kawahamana, Mere. "He inoa no Miki." *Ka Hoku o ka Pakipika,* buke 1, helu 23 (Feberuari 27, 1862): 1.

Kawahine, Mele. "He mele pualiinuwai no Panau." *Ka Nupepa Kuokoa,* buke 3, helu 18 (Aperila 30, 1864): 1.

Kawaikaumaiikamakaokaopua, Z.P.K. "Haina nane." *Ka Nupepa Kuokoa,* buke 61, helu 52 (Dekemaba 28, 1922): 7.

Kawailuna. "He inoa no Sarah Kukapu." *Ka Hoku o ka Pakipika,* buke 1, helu 15 (Ianuari 2, 1862): 1.

Kawao. "Pukaka na lehua o Mana." Chant. HI.M.58, p. 46. Bishop Museum Archives, Honolulu.

Kawelaakawai, P. "He hula ku-i ma Kawela, Molokai." *Ka Nupepa Kuokoa,* buke 7, helu 12 (Maraki 21, 1868): 4.

Kawelo, Hilda Hoohila. Interview by Clinton Kanahele. Recorded at Ka'alaea, O'ahu, Sept. 10, 1970. MP3 audio file Hilda Hoohila Kawelo 1, Clinton Kanahele Collection. Joseph F. Smith Library, Lā'ie, O'ahu. http://library.byuh.edu/library/archives/kanahele. Audio file of interview in Hawaiian; transcript and English translation available.

Kawelo, J. B. "He kanikau aloha keia no P. Kamohai." *Ka Nupepa Kuokoa,* buke 3, helu 20 (Mei 14, 1864): 4.

Kealakaa, A.B.W. "He mele no Waimea." *Ka Hoku o ka Pakipika,* buke 1, helu 8 (Novemaba 14, 1861): 1.

Kealakai, W. E. "Cultivation." Hawaiian Ethnological Notes (HEN) Thrum 125, pp. 12–16. Bishop Museum Archives, Honolulu. Translation by Thomas G. Thrum of "He wahi moolelo no ka mahiai ana o ka wa kahiko" by W. E. Kealakai. http://www.hawaiialive.org/resources/manuscript/227.pdf.

Kealakai, W. E. "He wahi moolelo no ka mahiai ana o ka wa kahiko." *Ka Hae Hawaii,* buke 6, helu 8 (Mei 22, 1861): 31.

Kealiikanakaole, W.J.B. "He kanikau no Miss L. Kahalepouli." *Ka Nupepa Kuokoa,* buke 1, helu 54 (Dekemaba 6, 1862): 3.

Kealoha, Mrs. Kanewahine. "Welina puuwai kumakena." *Ka Nupepa Kuokoa,* buke 17, helu 47 (Novemaba 23, 1878): 1.

Keawe, Mrs. Monika. "He hoomanao aloha!" *Ka Nupepa Kuokoa,* buke 56, helu 44 (Novemaba 1, 1918): 8.

Keaweamahi, Mr. D. "Huakai makaikai ia Kauai." *Ka Lahui Hawaii,* buke 2, helu 34 (Augate 17, 1876): 3.

Keelikolani, R. "He mau inoa no Ka Haku O Hawaii." *Ka Nupepa Kuokoa,* buke 1, helu 1 (Okatoba 1, 1861): 1.

Keelikolani, R. "Ua kalai ia ka nahele." Chant. MS SC Roberts, box 4.6, pp. 16b–18a. Bishop Museum Archives, Honolulu.

"Ke hea mai nei o Limaloa." Chant. MS SC Roberts, box 2.4, pp. 181b–83a. Bishop Museum Archives, Honolulu.

Kekauluohi, M. "Loohia i ka pono luna o Kawaikini." Chant. MS SC Roberts, box 4.6, pp. 20–21a. Bishop Museum Archives, Honolulu.

Kekaumehameha, M. "He inoa no Manoana [*sic*]." *Ka Nupepa Kuokoa,* buke 1, helu 29 (Iune 14, 1862): 4.

Kekuaiwa, P. H. "He inoa no Manoana [*sic*]." *Ka Nupepa Kuokoa,* buke 1, helu 29 (Iune 14, 1862): 4.

Kekuanaoa, M. "O ku'u hala lei i Hanalei." Chant. MS SC Roberts, box 4.6, pp. 20b–22a. Bishop Museum Archives, Honolulu.

Keliiahonui, A. "Lei Kuia i na hala a ka makani." Chant. MS SC Roberts, box 4.6, pp. 21b–23a. Bishop Museum Archives, Honolulu.

Keliihaleole, S. "No ka make ana o Kaawaloa." *Ka Nupepa Kuokoa,* buke 5, helu 34 (Augate 25, 1866): 4.

Kelsey, Theodore. "He moolelo kupua no Kaumana i lilo i pohaku." *Ka Nupepa Kuokoa,* buke 57, helu 27 (Iulai 4, 1919): 2.

Kelsey, Theodore. "Rains." Hawaiian Ethnological Notes (HEN) 1:807–8. Part of the T. Kelsey Collection. Bishop Museum Archives, Honolulu.

Kelsey, Theodore. "Winds." Hawaiian Ethnological Notes (HEN) 1:779–80. Part of the T. Kelsey Collection. Bishop Museum Archives, Honolulu.

Kent, Harold Winfield. *Treasury of Hawaiian Words in One Hundred and One Categories.* Honolulu: Masonic Public Library of Hawai'i, 1986.

Kenui, Barenaba. "He kanikau aloha no J. Paulo." *Ka Nupepa Kuokoa,* buke 4, helu 15 (Aperila 13, 1865): 3.

Keolanui, J.W.P. "He mau hale hina ma Honouliuli Ewa i ka makani." *Ka Nupepa Kuokoa,* buke 5, helu 7 (Feberuari 17, 1866): 3.

Keolanui, J.W.P. "He wahi rula kupono keia i na kaikamahine maoli, a hapa maoli hoi." *Ka Nupepa Kuokoa,* buke 4, helu 29 (Iulai 22, 1865): 2.

Keolanui, S. D. "Olelo paipai i na makaainana mai Maunalua a Moanalua." *Ka Hoku o ka Pakipika,* buke 1, helu 9 (Novemaba 21, 1861): 2.

Keolanui, S. D. "Olelo paipai i na mea a pau mai Hawaii a Niihau." *Ka Nupepa Kuokoa,* buke 1, helu 1 (Okatoba 1, 1861): 3.

"Ke omaimai o ke kama'lii." *Ka Makaainana,* buke 11, helu 7 (Feberuari 13, 1899): 5.

"Keopuolani." Chant. HI.M.63, pp. 131–32. Bishop Museum Archives, Honolulu.

Kepelino, Zepherin. *Kepelino's Traditions of Hawaii.* Trans. and ed. Martha Warren Beckwith. Bernice P. Bishop Museum Bulletin 95. Millwood, NY: Kraus Reprint Co., 1978.

Keukahi, D. W. "He mele aloha no *Ka Hoku o ka Pakipika.*" *Ka Hoku o ka Pakipika,* buke 1, helu 9 (Novemaba 21, 1861): 4.

"Kiekie Kau kua makani." Chant. MS GRP 81, box 8.13. Mader Collection. Bishop Museum Archives, Honolulu.

Kiliwehi. "O Aiwohi, kanaka o ka lipo." Chant. MS SC Roberts, box 4.6, pp. 8b–9a. Bishop Museum Archives, Honolulu.

Kinau. "Eia hou keia inoa o Kauikeaouli." *Ka Hoku o ka Pakipika,* buke 1, helu 21 (Feberuari 13, 1862): 4.

King, Pauline N. *A Local History of Kahoʻolawe Island: Tradition, Development, and World War.* Kahoʻolawe Island Conveyance Commission, Consultant Report No. 13, 1993.

Kirtley, Bacil F., and Esther T. Mookini. "Essays upon ancient Hawaiian religion and sorcery by nineteenth-century seminarists." *Hawaiian Journal of History* 13 (1979): 67–93.

Konaaihele. "He kanikau." *Ka Nupepa Kuokoa,* buke 1, helu 21 (Aperila 19, 1862): 4.

Ku, Mele. "He inoa no Lele." *Ka Hoku o ka Pakipika,* buke 1, helu 24 (Maraki 6, 1862): 1.

Kualehelehe, Wm. R. "He kanikau no ka Moi Iolani." *Ka Nupepa Kuokoa,* buke 3, helu 34 (Augate 20, 1864): 1.

Kualu, William. "Pau makemake ia Maleka." Chant. MS SC Roberts, box 3.1, pp. 37–50g. Bishop Museum Archives, Honolulu.

Kuapuu, S. K. "He moolelo kaao no Pakaa." *Home Rula Repubalika,* buke 1, helu 16 (Maraki 15, 1902): 5.

Kuapuu, S. K. "He wahi moolelo." *Ka Hae Hawaii*:
buke 6, helu 2 (Aperila 10, 1861): 8
buke 6, helu 3 (Aperila 17, 1861): 12
buke 6, helu 4 (Aperila 24, 1861): 16
buke 6, helu 6 (Mei 8, 1861): 24
buke 6, helu 7 (Mei 15, 1861): 25
buke 6, helu 9 (Mei 29, 1861): 36.

"Ku hau pio o ka la." Chant. MS SC Roberts, box 2.3, pp. 107–11b. Bishop Museum Archives, Honolulu.

"Kuluikului au ka po." Chant. MS SC Roberts, box 4.6, pp. 57b–60a. Bishop Museum Archives, Honolulu.

Kumalaua, S. "Make i aloha nui ia." *Ko Hawaii Pae Aina,* buke 1, helu 47 (Novemaba 23, 1878): 4.

"Kupaa ka manao me Liliu." *Ka Makaainana,* buke 3, helu 11 (Mei 20, 1895): 1.

"Kuu aloha la, kuu aloha." Chant. MS GRP 81, box 8.13. Mader Collection. Bishop Museum Archives, Honolulu.

L. "He inoa no Kekamalahaole." *Ka Nupepa Kuokoa,* buke 3, helu 23 (Iune 4, 1864): 1.

L. K. "Kuu wahine ua Lalena [*sic*]." Chant. HI.M.2, pp. 14–15. Bishop Museum Archives, Honolulu.

"Lanakila ka oiaio!" *Ke Alaula,* buke 1, helu 7 (Okatoba 1866): 27.

Laweliilii, G.W.K., Opio. "Na wahi pana o ka ua Kanilehua." *Ko Hawaii Pae Aina,* buke 3, helu 22 (Mei 29, 1880): 4.

Laweliilii, G.W.K., Opio. "Relating to the Kanilehua rain." Hawaiian Ethnological Notes (HEN) Thrum 107. Bishop Museum Archives, Honolulu. Translation by Thomas G. Thrum of "Na wahi pana o ka ua Kanilehua" by G.W.K. Laweliilii, Opio.

Lewaina, Mrs. J. "He inoa no Kaluaihaleola." *Ka Hoku o ka Pakipika,* buke 1, helu 14 (Dekemaba 26, 1861): 3.

Lihau, Mele. "He inoa no Manoana [*sic*]." *Ka Nupepa Kuokoa,* buke 1, helu 29 (Iune 14, 1862): 4.

Liholiho, A. "O Hilo aina ua Halaulani." Chant. HI.M.54, pp. 164–65. Bishop Museum Archives, Honolulu.

Lili'uokalani. *The Queen's Songbook.* Ed. Dorothy Kahananui Gillett and Barbara Barnard Smith. Honolulu: Hui Hānai, 1999.

Lohiau, S. "He inoa no Keahi." *Ka Hoku o ka Pakipika,* buke 1, helu 22 (Feberuari 20, 1862): 4.

Lohiau, S. "He mele no ka nupepa *Hae Hawaii.*" *Ka Nupepa Kuokoa,* buke 1, helu 7 (Ianuari 11, 1862): 3.

Lohiau, S. "He mele no Kawaikini." *Ka Hoku o ka Pakipika,* buke 1, helu 4 (Okatoba 17, 1861): 1.

Luahine, Iolani, chanter and hula dancer. Recorded at Hulihe'e Palace, Kona. 45 rpm, side B. "Kona kai opua i ka lai/Ua nani o Nuuanu." Waikiki Records, ca. 1960.

Luukia, Mrs. "He inoa no Keikeiloselani." *Ke Au Okoa,* buke 1, helu 14 (Iulai 24, 1865): 1.

Maele, Lahela. "He inoa no Manoana [*sic*]." *Ka Nupepa Kuokoa,* buke 1, helu 29 (Iune 14, 1862): 4.

"Ma'ema'e i ke kai ka pua o ka hala." Chant. MS SC Roberts, box 5.2, p. 66. Bishop Museum Archives, Honolulu.

"Maika'i ka oiwi o Kaala molale i ka malie." Chant. MS SC Roberts, box 4.5, pp. 44–49. Bishop Museum Archives, Honolulu.

"Maikai Kauai hemolele i ka malie." Chant. MS SC Roberts, box 2.7, pp. 122c–32d. Bishop Museum Archives, Honolulu.

"Maika'i Taula i ka mole o na moku." Chant. MS SC Roberts, box 2.7, pp. 99b–109a. Bishop Museum Archives, Honolulu.

Maiki: Chants and Mele of Hawai'i. CD. Maiki Aiu Lake and the Kahauanu Lake Trio and Singers. Hula Records, 1992.

Maila, Mary Kauhane. "Make i walohia." *Ka Lahui Hawaii,* buke 2, helu 28 (Iulai 6, 1876): 3.

Makaikai. "Ka la 4 o Iulai ma Ulupalakua." *Ka Lahui Hawaii,* buke 3, helu 30 (Iulai 26, 1877): 4.

Makuaole, William P. Z. "He moolelo no ka make ana o ka mea aloha H. Kamahiai." *Ka Nupepa Kuokoa,* buke 4, helu 22 (Iune 1, 1865): 4.

Malo, Davida. "Eia kona wahi mele kanikau." *Ka Nonanona,* buke 4, helu 24 (Maraki 18, 1845): 113–14.

Malo, Davida. *Hawaiian Antiquities: Moolelo Hawaii.* 1898. Translation by Nathaniel B. Emerson of *Ka Moolelo Hawaii* by Davida Malo. Bernice P. Bishop Museum Special Publication 2. Honolulu: Bishop Museum Press, 1951.

Malo, Davida. "He kanikau no Kaahumanu." *Ka Lama Hawaii,* makahiki 1, helu 17 (Augate 8, 1834): 3.

Malo, Davida. *Ka Moolelo Hawaii.* Manuscript. HI.L.18. Governor George Robert Carter Collection. Bishop Museum Archives, Honolulu.

Manu, Moses. "Ka moolelo o Kihapiilani, ka mea nana kipapa kanahele o Oopuloa a me ke ala pupu i Molokai." *Ka Nupepa Kuokoa,* buke 23, helu 10 (Maraki 8, 1884): 4.

Manu, Moses. *Keaomelemele.* Trans. Mary Kawena Pukui. Ed. M. Puakea Nogelmeier. Honolulu: Bishop Museum Press, 2002.

Manuokekula, G.W.K. "Ka hanau ana o na malama." *Ka Hoku o ka Pakipika,* buke 1, helu 7 (Novemaba 7, 1861): 4.

Maria, H. "He inoa no Erisapeka." *Ka Hoku o ka Pakipika,* buke 1, helu 29 (Aperila 10, 1862): 4.

"Maunakea." *Ka Makaainana,* buke 2, helu 12 (Sepatemaba 17, 1894): 3.

Maunupau, Thomas K. *Huakai Makaikai a Kaupo, Maui: A Visit to Kaupō, Maui.* Translation by Mary Kawena Pukui and Malcolm Naea Chun of "Huaka'i maka'ika'i no Kaupo, Maui" by Thomas K. Maunupau. Ed. Naomi Noelanioko'olau Clarke Losch. Honolulu: Bishop Museum Press, 1998.

Maunupau, Thomas K. "Huaka'i maka'ika'i no Kaupo, Maui." *Ka Nupepa Kuokoa,* buke 61, helu 24 (Iune 15, 1922): 4.

Mauuhilo. "Ke ala o ka mea pohihihi." *Ka Lei Momi,* buke 1, helu 19 (Dekemaba 11, 1893): 7.

McGregor, Davianna Pōmaika'i. *Nā Kua'āina: Living Hawaiian Culture.* Honolulu: University of Hawai'i Press, 2007.

"Me he pua no ka he aka i ka laau." Chant. MS SC Roberts, box 4.9, pp. 115b–17. Bishop Museum Archives, Honolulu.

"Mele kahiko." *Ka Nupepa Kuokoa,* buke 6, helu 5 (Feberuari 2, 1867): 1.

"Mikihilina Hainoa i ka awaawa." Chant. MS SC Roberts, box 4.5, pp. 34a, 35b–36. Bishop Museum Archives, Honolulu.

"Miki ke kai a ka ua Kipu'upu'u." Chant. MS SC Roberts, box 4.5, pp. 33, 34b–36. Bishop Museum Archives, Honolulu.

Miner, Mrs. Apia. "He kanikau no Edwin Miner o Makawao i make aku nei." *Ka Hoku o ka Pakipika,* buke 1, helu 1 (Sepetemaba 26, 1861): 3.

Moehonua, Mrs. Lucy, and Mrs. Hana Lilikalani. "He kanikau no ka Moi Alexander Kalanikualiholiho, Maka o Iouli, Kunuiakea o Kukailimoku, Kamehameha IV!" *Ka Nupepa Kuokoa,* buke 3, helu 4 (Ianuari 23, 1864): 4.

Mokumaia, J. K. "Holo kaapuni o Mr. ame Mrs. J.W.K. Kihe ia Oahu." *Ka Nupepa Kuokoa,* buke 59, helu 43 (Okatoba 28, 1921): 5.

"Mooolelo no Pamano." *Ka Hoku o ka Pakipika,* buke 1, helu 17 (Ianuari 16, 1862): 4.

Moses, Miss Anna, Miss Kalua Keka, Hookano Kaneaiakala, Master Charles Keka, Master Thomas Pahu, Miss Rose Kalimapehu, Miss Dora McKeagul, Mrs. Mele Kilo, and Mrs. H. K. Hoolapa. "Olelo hoalohaloha." *Ka Nupepa Kuokoa,* buke 45, helu 48 (Novemaba 30, 1906): 8.

Motta, Pi'olani. *Lena Machado: Songbird of Hawai'i.* With Kīhei de Silva. Honolulu: Kamehameha Schools, 2006.

"Na hiohiona o Kohala." *Ka Nupepa Kuokoa,* buke 7, helu 5 (Feberuari 1, 1868): 4.

Naihe. "Eia hou keia inoa o Kauikeaouli." *Ka Hoku o ka Pakipika,* buke 1, helu 21 (Feberuari 13, 1862): 4.

Nailiili, S. W. "Na inoa o na makani o Hilo." *Ke Au Okoa,* buke 1, helu 12 (Iulai 10, 1865): 1.

Nakuina, Moses K. *Moolelo Hawaii o Pakaa a me Ku-a-Pakaa: Na Kahu Iwikuamoo o Keawenuiaumi—ke Alii o Hawaii—a o na moopuna hoi a Laamaomao!* 1902. Facsimile reprint of the traditional legend collected from various sources, edited, and expanded by Moses K. Nakuina. Honolulu: Kalamakū Press, 1991.

Nakuina, Moses K. *The Wind Gourd of La'amaomao.* Translation by Esther T. Mookini and Sarah Nākoa of *Moolelo Hawaii o Pakaa a me Ku-a-Pakaa: Na Kahu Iwikuamoo o Keawenuiaumi—ke Alii o Hawaii—a o na moopuna hoi a Laamaomao!* by Moses K. Nakuina. Honolulu: Kalamakū Press, 1992.

Nā Leo Hawai'i Kahiko—Voices of Old Hawai'i: Historical Recordings of Chants and Songs. Liner notes for 33 rpm. Division of Audio-Recording Collections, Department of Anthropology, Bernice P. Bishop Museum, 1981.

Namahana. "Ka malu 'io lani, maka o ke 'lii." Chant. MS SC Roberts, box 4.6, pp. 1c–3a. Bishop Museum Archives, Honolulu.

Na Maka o ka Aina. "Mountain Deities" (includes excerpt of interview with Keawe Vredenburg). In *Mauna Kea: From Mountain to Sea.* Accessed 2013. http://www.mauna-a-wakea.info/maunakea/F1_mtndeities.html.

Nāmakaokeahi, Benjamin K. *The History of Kanalu: Mo'okū'auhau 'Elua.* Originally published in *Ka Nupepa Kuokoa,* 1900–1901. Translation by Malcolm Nāea Chun of "Mookuauhau elua o Hawaii" by Benjamin K. Nāmakaokeahi. Ed. Malcolm Nāea Chun. Honolulu: First People's Productions, 2004.

Nāmakaokeahi, Benjamin K. "Mookuauhau elua o Hawaii." *Ka Nupepa Kuokoa,* buke 39, helu 15 (Maraki 15, 1901): 3.

Namakeha, W. "He inoa no Kawahahala." *Ka Hoku o ka Pakipika,* buke 1, helu 24 (Maraki 6, 1862): 3.

"Na mea hou o Hawaii nei." *Ka Nupepa Kuokoa,* buke 1, helu 37 (Augate 9, 1862): 2.

Na Mele Aimoku, na Mele Kupuna a me na Mele Ponoi o ka Moi Kalakaua I: Dynastic Chants, Ancestral Chants, and Personal Chants of King Kalākaua I. 1886. Reprint, Honolulu: Hawaiian Historical Society, 2001.

"Names of different kinds of rain." Hawaiian Ethnological Notes (HEN) 1:1253–54. Edgar Henriques Collection. Bishop Museum Archives, Honolulu.

"Names of different winds." Hawaiian Ethnological Notes (HEN) 1:1341–51. Edgar Henriques Collection. Bishop Museum Archives, Honolulu.

Namuo, S. "He uhane no Keliikuea [*sic*]." *Ka Nupepa Kuokoa,* buke 1, helu 46 (Okatoba 11, 1862): 4.

"Nani hanohano ke kuahiwi o Kauiki." Chant. HI.M.5, pp. 113–15. Bishop Museum Archives, Honolulu.

"Nani Ka-u hanohano i ka makani." Chant. MS SC Roberts, box 2.4, pp. 183b–86. Bishop Museum Archives, Honolulu.

"Nani kuu maka e au wale nei e." Chant. MS SC Roberts, box 4.6, pp. 70b–73. Bishop Museum Archives, Honolulu.

"Nani Lahaina i ka ua Pa'upili." Chant. MS SC Roberts, box 4.1, pp. 2–4. Bishop Museum Archives, Honolulu.

"Naue ana e ike i ka mokupuni o Kaululaau." *Ka Nupepa Kuokoa,* buke 8, helu 2 (Ianuari 9, 1869): 4.

"Na wahi pana o Ewa." *Ka Loea Kalaiaina*:
buke 3, helu 21 (Iune 3, 1899): 2
buke 3, helu 22 (Iune 10, 1899): 4
buke 3, helu 26 (Iulai 8, 1899): 3
buke 3, helu 29 (Iulai 29, 1899): 3.

Niauehoe. "He kanikau no Kaahumanu-opio ka wahine a Kaumualii ka moi o Kauai." *Ka Nupepa Kuokoa,* buke 7, helu 12 (Maraki 21, 1868): 1.

Nishimura, Susumu, William Kaonohi Kauwenaole, and Maggie Nakihei Kauwenaole. Interview by Mary Kawena Pukui. Recorded on Lāna'i, May 5, 1963. HAW 142.11.1. Bishop Museum Archives, Honolulu. Audiocassette of interview in Hawaiian; transcript and English translation also available.

Nogelmeier, M. Puakea, ed. and comp. *He Lei no 'Emalani: Chants for Queen Emma Kaleleonālani.* Trans. Mary Kawena Pukui, Theodore Kelsey, and M. Puakea Nogelmeier. Honolulu: Queen Emma Foundation and Bishop Museum Press, 2001.

Noholoa. "He kanikau no Ioseba Nawai." *Ka Hoku o ka Pakipika,* buke 1, helu 22 (Feberuari 20, 1862): 1.

"No ko makou akua." *Ka Lama Hawaii,* makahiki 1, helu 13 (Mei 9, 1834): 4.

"Noted places of Ewa." Hawaiian Ethnological Notes (HEN) News., *Ka Loea Kalaiaina,* Iune 3, 1899–Ianuari 13, 1900. Bishop Museum Archives, Honolulu. Translation of the Iulai 8, 1899, installment in the "Na wahi pana o Ewa" series.

Ohiaha, John. "Haule ke'loha, ka pua a ka waimaka." *Ka Nupepa Kuokoa,* buke 9, helu 43 (Okatoba 22, 1870): 4.

"O ka leo o ka oo kani kuamauna." Chant. MS SC Roberts, box 4.6, pp. 55b–57a. Bishop Museum Archives, Honolulu.

"O ka wai halau i Wailua la." Chant. MS SC Roberts, box 5.4, pp. 87b–89. Bishop Museum Archives, Honolulu.

"O ke kane ka mea aloha." Chant. MS SC Roberts, box 5.1, pp. 8, 54–55. Bishop Museum Archives, Honolulu.

Olepau, W. "Kupaa ka manao me Liliu." *Ka Makaainana,* buke 3, helu 11 (Mei 20, 1895): 1.

"O Niolopua oe o ku'u kaikunane." Chant. MS SC Roberts, box 2.7, pp. 137b–41c. Bishop Museum Archives, Honolulu.

Opunui, D. N. "He inoa no Kamakai, kamalei a makou." *Ka Nupepa Kuokoa,* buke 5, helu 49 (Dekemaba 8, 1866): 4.

Paa, Wahineikeouli [Wahinekeouli Pa]. "Ka poli laua'e ka'u aloha." Chant. MS SC Roberts, box 5.3, pp. 65b–67a. Bishop Museum Archives, Honolulu.

Paahao, L.K.N. "Na mea hou o kauakea o Hana." *Ko Hawaii Pae Aina,* buke 7, helu 12 (Maraki 22, 1884): 3.

Paakai. "He inoa no Kalawaia." *Ka Nupepa Kuokoa,* buke 4, helu 6 (Feberuari 9, 1865): 4.

Paauilo. "He mele no Ka Haku o Hawaii." *Ka Nupepa Kuokoa,* buke 1, helu 4 (Dekemaba 2, 1861): 3.

Paglinawan, Richard Kekumuikawaiokeola, Mitchell Eli, Moses Elwood Kalauokalani, and Jerry Walker. *Lua: Art of the Hawaiian Warrior.* With Kristina Pilahoʻohauʻoli Kikuchi-Palenapa. Honolulu: Bishop Museum Press, 2006.

Pahee, R. K. "No ka ua." *Ka Nupepa Kuokoa,* buke 6, helu 20 (Mei 18, 1867): 2.

Pahee, R. K. "Of the rain." Hawaiian Ethnological Notes (HEN) News., *Ka Nupepa Kuokoa,* Mei 18, 1867. Bishop Museum Archives, Honolulu. Translation of "No ka ua" by R. K. Pahee.

"Pā ka leo: Hear the voice." Brochure for a presentation by Hoʻolaupaʻi: Hawaiian Newspaper Resource, May 5, 2007.

"Paku Keaau lulu Waiakea." Chant. MS SC Roberts, box 5.3, pp. 43–46a. Bishop Museum Archives, Honolulu.

Palaile, J. N. "He kanikau aloha no *Ka Hae Hawaii." Ka Hae Hawaii,* buke 6, helu 39 (Dekemaba 25, 1861): 155.

Pauahi. "Ea ka lani ku ka auhuhu hele moku." Chant. MS SC Roberts, box 4.9, pp. 59a–60. Bishop Museum Archives, Honolulu.

Pauahi, B. "He kanikau no ka Moi Iolani Kamehameha IV!" *Ka Nupepa Kuokoa,* buke 3, helu 4 (Ianuari 23, 1864): 1.

Pauahi, D. W. "He kanikau no Kapela." *Ka Nupepa Kuokoa,* buke 1, helu 45 (Okatoba 4, 1862): 2.

Paukaikena, Hana L. "He mele kanikau no David Kahalepouli Piikoi." *Ko Hawaii Pae Aina,* buke 1, helu 47 (Novemaba 23, 1878): 4.

Paulo, Mrs. Hakaleleponi. "Ka make ana o Paulo Ku." *Ka Nupepa Kuokoa,* buke 5, helu 47 (Novemaba 24, 1866): 3.

Pawai. "He kanikau aloha keia nou e Luakauwahine." *Ka Nupepa Kuokoa,* buke 1, helu 26 (Mei 24, 1862): 4.

Pilialoha, T. P. "Ka make ana o S. W. Kanalu." *Ka Nupepa Kuokoa,* buke 5, helu 31 (Aukake 4, 1866): 3.

Poepoe, Joseph M. "Ka-elemu-wai o Kalihi." Hawaiian Ethnological Notes (HEN) 2:155. Bishop Museum Archives, Honolulu. English translation of parts of "Na mea kaulana o ka lipolipo o Kalihiuka" by Joseph M. Poepoe.

Poepoe, Joseph M. "Na mea kaulana o ka lipolipo o Kalihiuka." Manuscript. HI.L.1.3.3. Bishop Museum Archives, Honolulu. In Hawaiian and English.

Puaaloa, A. "Na mea hou o Kipahulu, Maui." *Ka Nupepa Kuokoa,* buke 18, helu 16 (Aperila 19, 1879): 1.

Pualewa, W. N. "He moolelo e paipai aku ana ia oukou e ka poe e makemake mai ana e lawe i 'Ka Nupepa Kuokoa.'" *Ka Nupepa Kuokoa,* buke 1, helu 1 (Okatoba 1, 1861): 3.

Pukui, Mary Kawena, trans. *Nā Mele Welo: Songs of Our Heritage.* Honolulu: Bishop Museum Press, 1995.

Pukui, Mary Kawena, trans. *No nā Kamaliʻi.* CD. Bernice P. Bishop Museum, 1999.

Pukui, Mary Kawena, trans. *ʻŌlelo Noʻeau: Hawaiian Proverbs and Poetical Sayings.* Bernice P. Bishop Museum Special Publication 71. Honolulu: Bishop Museum Press, 1983.

Pukui, Mary Kawena, and Samuel H. Elbert. *Hawaiian Dictionary.* Honolulu: University of Hawaiʻi Press, 1971.

Pukui, Mary Kawena, and Samuel H. Elbert. *Hawaiian Dictionary.* Revised and enlarged edition. Honolulu: University of Hawai'i Press, 1986.

Pukui, Mary Kawena, Samuel H. Elbert, and Esther T. Mookini. *Place Names of Hawaii.* Revised and enlarged edition. Honolulu: University of Hawai'i Press, 1974.

Pukui, Mary Kawena, trans., and Alfons L. Korn, ed. *The Echo of Our Song: Chants and Poems of the Hawaiians.* Honolulu: University of Hawai'i Press, 1973.

R. "Ka Puali Inuwai ma Hilo i ka La Hape Nuia." *Ka Nupepa Kuokoa,* buke 6, helu 3 (Ianuari 19, 1867): 4.

Reichel, Keali'i. *Kamahiwa.* CD liner notes. Punahele Productions, 2005.

Reichel, Keali'i. *Ke'alaokamaile.* CD liner notes. Punahele Productions, 2003.

Richards, Lilia H., Charlotte A. Hanks, Lizzie Chung Hoon, Eliza N. Crowningburg, and Clarisa Jackson. "Na kamahele i ka uka o Puna." *Ka Lahui Hawaii,* buke 3, helu 13 (Maraki 29, 1877): 2.

Richardson, Ioane. "He kanikau no Doreka Richardson." *Ka Hae Hawaii,* buke 2, helu 11 (Iune 10, 1857): 43.

Saffery, Hattie. "Mahalo i ka nani o Kilauea." *Ka Nupepa Kuokoa,* buke 58, helu 7 (Feberuari 13, 1920): 2.

Samuela. "Halau Lahaina." Chant. HI.M.61, pp. 92–93. Bishop Museum Archives, Honolulu.

Savia, J. "He inoa no Wahinealoha." *Ka Hoku o ka Pakipika,* buke 1, helu 22 (Feberuari 20, 1862): 1.

Sheldon, J.G.M. *The Biography of Joseph K. Nāwahī.* Translation by M. Puakea Nogelmeier of *Ka Buke Moolelo o Hon. Joseph K. Nawahi* by J.G.M. Sheldon. Typescript. Honolulu: Hawaiian Historical Society, 1988.

Sheldon, J.G.M. *Ka Puke Mo'olelo o Hon. Iosepa K. Nāwahī.* Hilo: Hale Kuamo'o, Ka Haka 'Ula o Ke'elikōlani, 1996.

Silva, Noenoe K. *Aloha Betrayed: Native Hawaiian Resistance to American Colonialism.* Durham, NC: Duke University Press, 2004.

Soehren, Lloyd J., collector and annotator. *A Catalog of Hawaiian Place Names: Compiled from the Records of the Boundary Commission and the Board of Commissioners to Quiet Land Titles of the Kingdom of Hawaii* (online database). 2010. http://ulukau.org/cgi-bin/hpn.

Solis, Ron. "Kekahi mau Pule Mahi'ai." MA thesis, University of Hawai'i at Mānoa, 1999.

Spencer, Elizabeth. "Kuu pokii aloha, ua hala. Miss Mary Pe'a." *Ka Nupepa Kuokoa,* buke 63, helu 42 (Okatoba 16, 1924): 6.

Sterling, Elspeth P. *Sites of Maui.* Honolulu: Bishop Museum Press, 1998.

Sterling, Elspeth P., and Catherine C. Summers. *Sites of Oahu.* Honolulu: Bishop Museum Press, 1978.

"Story of Hema." Hawaiian Ethnological Notes (HEN) News., *Ka Nupepa Kuokoa,* Sepatemaba 3 and Okatoba 1, 1920. Bishop Museum Archives, Honolulu. Translation by Mary Kawena Pukui of "Ka moolelo o He'ma [*sic*], ke koaie ku pali o ka makani kaili aloha o Kipahulu, Maui."

Summers, Catherine C. *Molokai: A Site Survey.* Pacific Anthropological Records 14. Department of Anthropology, Bernice P. Bishop Museum, Honolulu, April 1971.

Tatar, Elizabeth. *Hula Pahu: Hawaiian Drum Dances.* Vol. 2, *The Pahu: Sounds of Power.* Honolulu: Bishop Museum Press, 1993.

Tava, Rerioterai, and Moses K. Keale Sr. *Niihau: The Traditions of an Hawaiian Island.* Honolulu: Mutual Publishing, 1989.

Thrum, Thomas G., comp. *Hawaiian Folk Tales.* Honolulu: Mutual Publishing, 1998.

"Ua loa Maunaloa, loa i ka la." Chant. MS SC Roberts, box 4.5, pp. 22, 24b–25a. Bishop Museum Archives, Honolulu.

Uaua, William Henry. "He moolelo kaao no Kaehuikimanoopuuloa." *Ke Au Okoa,* buke 6, helu 37 (Dekemaba 29, 1870): 4.

"Ula no weo [*sic*]." Chant. MS SC Roberts, box 5.4, pp. 119b–20a. Bishop Museum Archives, Honolulu.

Varez, Dietrich. *Hina the Goddess.* Hilo: Petroglyph Press, 2002.

Wahinealoha. "No ka 'Anoi.'" *Ka Nupepa Kuokoa,* buke 4, helu 48 (Dekemaba 2, 1865): 1.

Waialeale. "Ke hone ae nei." *Ka Makaainana,* buke 1, helu 41 (Dekemaba 17, 1887): 1.

Wailea, D. "He kanikau aloha no B. Mahune." *Ka Hoku o ka Pakipika,* buke 1, helu 16 (Ianuari 9, 1862): 4.

Wala, Mrs. "Haule ke'loha, ka pua a ka waimaka." *Ka Nupepa Kuokoa,* buke 9, helu 43 (Okatoba 22, 1870): 4.

Welina Mānoa. Funded by Hawai'inuiākea School of Hawaiian Knowledge, W. K. Kellogg Foundation, and The Judith Dion Pyle Dean's Chair in the Hawai'inuiākea School of Hawaiian Knowledge. "Mai Uka a i Kai Tour." http://welinamanoa.org/otherna-mea-%CA%BBe-a%CA%BBe/mai-uka-a-i-kai-tour. Accessed 2013.

Wilcox, Carol, comp.; Vicky Hollinger and Kimo Hussey, musical notation; and M. Puakea Nogelmeier, orth. and trans. *He Mele Aloha: A Hawaiian Songbook.* Honolulu: 'Oli'oli Productions, 2003.

Wiliama, Mrs. "He kanikau aloha no Kamakaokalani." *Ke Au Okoa,* buke 1, helu 6 (Mei 29, 1865): 3.

"Winds." Hawaiian Ethnological Notes (HEN) 1:826–28. Bishop Museum Archives, Honolulu.

Woodward, Mrs. Hellen Umiokalani. "He kanikau aloha no Daniel Manuhoa Woodward i hala." *Ka Nupepa Kuokoa,* buke 42, helu 29 (Iulai 15, 1904): 6.

INDEX

Note: This index includes Hawaiian place names and rain names. Every effort has been made to properly situate obscure place names.

Collette Leimomi Akana
has been teaching social studies at
Kamehameha Middle School since 1983.
She is kumu hula of Hālau Hula o Nā Momi Makamae
and has taught Hawaiian language at Windward
Community College and Kapiʻolani Community College.
She has a master's degree in Hawaiian history from
University of Hawaiʻi at Mānoa and is the author of
The Oʻahu Exploits of Kamapuaʻa, the Hawaiian Pig-God.
Her kulaiwi include Makapala, Hawaiʻi, and
Hāna and Moʻomuku, Maui. She raised her daughter
and *Hānau ka Ua* contributor, Kiele Gonzalez,
to share in her love of Hawaiian language and culture.
Kiele is a Hawaiian-language publishing specialist
at Kamehameha Publishing.